Yale University

Mrs. Hepsa Ely Silliman

Memorial Lectures

SILLIMAN MILESTONES IN SCIENCE

MAN ADAPTING

by René Dubos

with a new chapter by the author

Yale University Press, New Haven and London

Designed by John O. C. McCrillis,
set in Linotype Baskerville,
and printed in the United States of America.

Library of Congress Cataloging in Publication Data

Dubos, René Jules, 1901–
Man adapting.

(Yale University, Mrs. Hepsa Ely Silliman memorial lectures)
Bibliography: p.
Includes index.
1. Medicine—Philosophy. 2. Man—Influence of environment. 3. Adaptation (Biology)
4. Environmental health. I. Title. II. Series:
Hepsa Ely Silliman memorial lectures, Yale University.
R723.D77 1980 610 80-16492
ISBN 0-300-02580-7
ISBN 0-300-02581-5 (pbk.)

10 9 8 7 6 5 4 3 2

THE SILLIMAN FOUNDATION LECTURES

On the foundation established in memory of Mrs. Hepsa Ely Silliman, the President and Fellows of Yale University present an annual course of lectures designed to illustrate the presence and providence of God as manifested in the natural and moral world. It was the belief of the testator that any orderly presentation of the facts of nature or history contributed to this end more effectively than dogmatic or polemical theology, which should therefore be excluded from the scope of the lectures. The subjects are selected rather from the domains of natural science and history, giving special prominence to astronomy, chemistry, geology, and anatomy. The present work constitutes the thirty-ninth volume published on this foundation.

Silliman volumes in print

Jacob Bronowski, *The Origins of Knowledge and Imagination*
S. Chandrasekhar, *Ellipsoidal Figures of Equilibrium*
Theodosius Dobzhansky, *Mankind Evolving*
Ross Granville Harrison, *Organization and Development of the Embryo*
John von Neumann, *The Computer and the Brain*
Karl K. Turekian, *Late Cenozoic Glacial Ages*

Contents

Tables

Figures

Preface

> The design of a book is the pattern of a reality controlled and shaped by the mind of the writer. This is completely understood about poetry or fiction, but it is too seldom realized about books of fact. And yet the impulse which drives a man to poetry will send another man into the tide pools and force him to try to report what he finds there.
>
> JOHN STEINBECK AND EDWARD F. RICKETTS

MAN IS an animal linked to inanimate matter; but human life transcends its earthy origin. Knowledge of the physicochemical determinants of life is of course essential for the understanding of man's nature. In their present state, however, the exact sciences fail to account for the phenomena which are of most direct relevance to the human condition.

Biomedical problems must therefore be considered from two complementary points of view. From one angle, man appears as an assembly of organic materials, very similar in composition and properties to those found in all other living forms. Present-day biology has gone far toward unraveling the structures and mechanisms of the body machine. From another angle, man is seen as an organism responding to stimuli in a manner which sets him apart from the rest of creation. In most cases, his responses are determined less by the direct effects of the stimulus on his body fabric than by the symbolic interpretation he attaches to the stimulus. Human beings may suffer and even die under circumstances which seem highly favorable to physiological performance, whereas paradoxically others will prosper even though conditions appear almost incompatible with the maintenance of life.

The dominant theme of the present book is that the states of health or disease are the expressions of the success or failure experienced by the organism in its efforts to respond adaptively to environmental challenges. This theme obviously bears a close relation to the one developed by Professor T. Dobzhansky in his inspiring book *Mankind Evolving,* which discusses the interplay between genetic endowment and environmental factors. However, while Professor Dobzhansky's book emphasizes mankind as a whole, I have attempted to focus my attention on the individual human being, *l'homme moyen sensuel,* trying as best he can to meet the emergencies of the day and to prepare for the uncertainties of the future. He is *Man Adapting.*

The problems of human adaptation could be presented as a dialectic between permanency and change. To a surprising extent, modern man has retained unaltered the bodily constitution, physiological responses, and emotional drives which he has inherited from his Paleolithic ancestors. Yet he lives in a mechanized, air-conditioned, and regimented world radically different from the one in which he evolved. Thus, persistence of ancient traits does not mean that human history is a mere extension of the past; nor do changes in the ways of life imply a loss of the biological heritage.

The response that a particular person makes to a given situation is conditioned by his past; his evolutionary and experiential backgrounds sharply delimit the range of conditions within which his responses can be successful. But experience shows that human beings are not passive components in adaptive systems. Their responses commonly manifest themselves as acts of personal creation. Each individual person tries to achieve some self-selected end even while he is responding to stimuli and adapting to them.

The components of the body machine *react with* the environment, in the same way as do similar components in any other living or inanimate system. But living man *responds to* his environment. I have intentionally differentiated between *reaction* and *response* to emphasize that human adaptation can seldom be regarded as merely the result of man's body reacting with environmental forces. In fact, man's responses are not even necessarily aimed at coping with the environment. They often correspond rather to an expresssive behavior and involve the use of the environment for self-actualization. Health in the case of human beings means more than a state in which the organism has become physically suited to the surrounding physicochemical conditions through passive mechanisms; it demands that the personality be able to express itself creatively.

Human life is thus the outcome of the interplay between three separate classes of determinants, namely: the lasting and universal characteristics of man's nature, which are inscribed in his flesh and bone; the ephemeral conditions which man encounters at a given moment; and last but not least, man's ability to choose between alternatives and to decide upon a course of action. Ideally, the goal of medicine has always been to help man function successfully in his environment—whether he is hunting the mammoth, toiling for his daily bread, or attempting to reach the moon.

All aspects of human life being interrelated, I have found it necessary to consider in my discussion many topics concerning which I have neither firsthand knowledge nor even familiarity with the relevant literature. Wherever it has led me, I have followed the tortuous thread that links health and disease to environmental factors. I regret not having been able to extend my survey into social psychopathology, especially in view of the fact that this field provides so many striking illustrations of the effect that sociocultural determinants exert on medical problems. But I have tried to emphasize that all biological questions regarding man must be seen in the light of the fact that he is a social, thinking, sensitive, and ethical animal. These attributes are at least as important for understanding him, and for dealing with his medical problems, as are the chemical structures and properties of his body machine.

In my opinion, furthermore, it is unjustified, as well as a defeatist attitude, to pretend that the experiential complexities of living man cannot be studied scientifically, until his submicroscopic constituents and elementary reactions have been fully elucidated. Just as the performance of an electronic computer can be described without familiarity with the chemical composition of its wiring and housing, so can useful knowledge of the organism's responses be gained without full understanding of the make-up and operations of its submicroscopic parts.

Because of the peculiarities of the human condition, the problems posed by health and disease often involve a variety of facts which appear far removed from those studied by natural scientists. Indeed, medicine would soon lose much of its relevance to the welfare of man if it were to limit itself to problems that can be analyzed by the orthodox techniques of physicochemical sciences. Knowledge of man cannot come only from the reductionist analysis, however detailed, of the structures and fuels found in his body machine. Admittedly, the problems of organismic and environmental biology are still poorly defined, but they deserve nevertheless as much attention as that devoted at present to physicochemical biology. It is almost certain, in fact, that medicine will eventually flounder in a sea of irrelevancy unless it learns more of the relations of the body machine to the total environment, as well as to the past and the aspirations of human beings. The study of such problems will require the development of scientific methods far more sophisticated than those which constitute today the stock in trade of orthodox natural sciences. What is needed is nothing less than a new methodology to acquire objective knowledge concerning the highest manifestation of life—the humanness of man.

The very statement of the problems posed by organismic and environmental medicine accounts for the peculiar kind of schizophrenia which is an occupational hazard of work in the life sciences, especially those pertaining to man. Facts, large or small, are interesting for their own sake, and the investigator derives from their study the satisfaction of a day-to-day kind of scientific existentialism. But the discovery of new facts can also generate anxiety, because knowledge always creates the power to disturb the relationships between man and nature, or between man and man.

The widespread tendency to consider that scientific pursuits are activities apart from, and even dominant over, the other manifestations of human life is dangerous for the welfare of mankind and may even inhibit progress within science itself. In fact, the intimacy of the relationships between the scientific enterprise and the total social complex makes it probable that the very continuance of science depends on the ability of scientists to relate their professional interests to the main human currents and aspirations. As indicated in the last two chapters of the present book, I am convinced that medical scientists are approaching a crossroads in this regard. The selection of areas of emphasis can no longer be based only on criteria of personal interest; it will be increasingly influenced by social demands. Science is like an organism that can survive only by adapting itself to the changing conditions of the society in which it functions.

The Silliman Lectures provide one of the few occasions when it is socially acceptable for an experimenter to discuss in public the parascientific aspects of his professional activities. I wish to thank the President and Fellows of Yale University for having given me the opportunity to express in the Silliman Lectures, on the occasion of the hundredth anniversary of Yale University Medical School, my personal views concerning some problems which arise from the interplay between the natural sciences and the human aspects of medicine.

Before concluding I should perhaps briefly state the reasons that have progressively led me—a microbiologist not trained in medicine—to explore some of the biological and social implications of man's responses to his total environment. My concern with such problems emerged from an increasing awareness of the fact that the prevalence and severity of microbial diseases are conditioned more by the ways of life of the persons afflicted than by the virulence and other properties of the etiological agents. Hence the need to learn more of man and of his societies, in order to try to make sense of the pattern of his diseases.

Because of my professional specialization, I have naturally taken many examples from microbial activities to illustrate the problems posed by the interplay between man and his environment. The special emphasis on the microbiota of the digestive tract reflects my present investigative interests, but has also a deeper justification. Recent discoveries have revealed that the effects of this biota on animal and human life provide striking models of ecosystems, in which it is relatively easy to recognize and study the complex interrelationships between man and his physicochemical and biological environment.

In the fields outside my professional competence, I have cast my net so wide that I have been compelled to use many indirect sources of information. In most cases, furthermore, I have found it impractical to give credit where credit is due, because my text represents composite views derived from several authors. The large number of references to recent books and review articles should be regarded as an acknowledgment that much of my information is secondhand and, furthermore, uncritical. For these reasons, I have elected in many cases not to give in the text the origin of the facts and opinions I have discussed, since the documents I have consulted rarely constituted original descriptions of findings or hypotheses. The bibliography at the end of the book does not represent the product of a systematic search for the literature relevant to the topic, but merely reports the publications which chance brought my way.

I fully realize the dangers of attempting to discuss problems outside one's professional competence, and I expect to be criticized both for lack of knowledge and for errors of judgment. But I shall not mind the criticisms because the preparation and writing of these lectures have given me the profound pleasure which comes from self-education—however imperfect the outcome. With tongue in cheek, I shall answer my critics by slightly paraphrasing Michel de Montaigne:

> Bees cull their several sweets from this flower and that blossom, here and there where they find them, but themselves afterwards make the honey, which is all and purely their own, and no more thyme and marjoram: so the several fragments I have borrowed from others I have transformed and blended to compile a work that would be absolutely my own.

I. Man's Nature

1. Man's nature and man's history

a. Body machine and symbolic activities

Tribal and national conflicts, occurring all over the world from the Congo to South Asia and among the great powers of Western civilization, make it painfully clear that man's behavior is governed by forces which have their origin in his social and cultural past. There is much truth in Ortega y Gasset's statement: "Man has no nature; what he has is a history." Biologists shy away from these historical considerations when they attempt a scientific study of man. Many of them emphasize those aspects of man's nature which can be described as physicochemical systems and which unfold according to the information inscribed in the genetic code. Others attempt to trace the origins of the human species in the hope that knowledge of its evolutionary past will help in understanding its present condition. Still others are more interested in the physiological and psychological characteristics of each individual person, as conditioned by the environment. Different as they appear to be, all these aspects of the study of man are interrelated. Furthermore, all of them have a direct bearing on the problems of health and disease that will be discussed in the present volume.

Physicians and experimenters interested in the forces which determine the physical and mental state of man find it useful to differentiate between genetic, historical, and environmental factors, or between conscious and unconscious responses. In reality, however, man is an integrated entity and all these forces operate simultaneously in every event of his life. The body and the mind are the living records of countless influences which have shaped each individual person from the most distant past to the present instant. Man is indeed the product of his history, but this history is far more complex than Ortega's statement would suggest. The history of each human being includes his private experiences and personal decisions; it encompasses also the evolutionary as well as the social past. Man's physical and mental state, in health and in disease, is always conditioned by all the multiple determinants of his nature.

The expression "human nature," as commonly used in the English language, does not convey the complexity and interrelatedness of all the factors which make up "man's nature." In fact, the inadequacy of the former expression has been clearly recognized for more than half a century. In 1903, Elie Metchnikoff published under the title *Etudes sur la Nature Humaine* a book in which he attempted to provide explanations based on the biology of his time for the properties of the human flesh and for the manifestations of the human spirit. His translator, the zoologist P. Chalmers Mitchell, found it difficult to convey in English the full meaning of the expression *Nature Humaine* in Metchnikoff's title. As he wrote in the preface to his translation, "Human nature is not an exact equivalent of *la nature humaine,* for the latter phrase has a more complete significance, and definitely implies, not only the mental qualities of man, but his bodily framework, with its inherited and acquired anatomical structure and body functions." Mitchell thought that "the nature of man" was a more accurate rendering of Metchnikoff's title than "human nature." In 1930, the American biologist H. S. Jennings also recognized implicitly the limitation of the phrase "human nature" when he qualified it as *The Biological Basis of Human Nature* in a book which he devoted to "those aspects of experimental biology that are of most interest in considering the problem of human personality and society."

Throughout the present book I shall use the expression "man's nature" in order not to be restricted in my discussion of health and disease by the limited meaning of "human nature" implied in the English usage. Repeatedly, I shall emphasize the necessity to relate the instinctive, psychological, and moral attributes of man to all the physiological needs and urges which are woven in his flesh and bones and which he has retained from his evolutionary past.

In one of the short stories of her *Seven Gothic Tales* Isak Dinesen quotes an Arabic thinker as saying that man is "but a minutely set ingenious machine for turning, with infinite artfulness, the red wine of Shiraz into urine." The late American physiologist Homer Smith felt that this statement was on the whole a satisfactory way of stating that man is just a machine, infinitely more complicated than others but not different in principle. René Descartes is generally given credit for being the first to state clearly that animals are *only* machines, a concept which La Mettrie extended to human beings one century later. Molecular biology, electronic computers, and automatic self-regulating devices have indeed provided convincing analogies for the description of living organisms as machines. In fact, these analogies have been elaborated into theories of living functions, and even of human thought, which are compatible with the view that there exists an unbroken evolutionary chain from the elementary particles of physics to organic life, and from life to consciousness.

In practice, however, the actual problems posed by human life still remain beyond the range of even the most sophisticated physicochemical models and evolutionary doctrines. In Paul Valéry's words, "L'homme n'est pas si simple qu'il suffise de le rabaisser pour le comprendre." When the German philosopher and psychologist Benno Erdmann (1851–1921) was asked to express his views concerning modern trends in the sciences of man, he sadly replied, "In my youth, we used to ask ourselves anxiously: What *is* man? Today scientists seem to be satisfied with the answer that he *was* an ape."

Erdmann's disenchanted remark symbolizes with poignancy the dilemma faced by all those who try to apply scientific concepts and objective knowledge to the problems of human life. We have been immensely successful in describing the composition of the human body and its operations as a physicochemical machine. We have traced the evolutionary development of man from remote geological time, and have gone far toward defining his biological relations to other creatures. But stating that man has evolved from more primitive forms which themselves emerged from inanimate matter more than two billion years ago, and that a common physicochemical thread runs through all living forms, obviously fails to account for the traits that are peculiar to mankind. While man is clearly an animal, he possesses unique characteristics which obviously make him different from other animals. What matters most for understanding him is that "humanness" which sets him apart from the rest of creation. Human medicine is different from general biology and even from veterinary medicine precisely by reason of the fact that man differs qualitatively from even the most evolved animal.

Knowledge of ancient man is derived as much from the evidence which has survived of his mental activities as from the remains of his physical body. Skeletons in geological deposits acquire a special significance when they are found to be associated with artifacts indicating humanlike behavior. The finding of tools, and of offerings buried with the dead, is the final criterion on which to decide that the creature who occupied the site was a man and not a beast. The first vestiges that unquestionably show the presence of man on earth are the rough tools of the "pebble culture"; they symbolize a new event on earth, the entry of a conscious living organism capable of considering at the same time the goal to be attained and the development of an appropriate technique for reaching it.

The oldest picture of man discovered so far was painted some fifteen thousands years ago in the Trois Frères cave in the Dordogne, in France; it is commonly referred to as the Sorcerer. The most surprising thing about this prehistoric man, as we thus encounter him face to face acting the part of a Magdalenian "sorcerer" is that he resembles so much an Indian medicine man of today during a corn dance in the American Southwest. In fact, he does not differ much from a modern man who has costumed himself for some esoteric ceremony in a secret lodge anywhere in the Western world.

On a ledge 12 feet above the ground, at the end of a winding passage in the darkest recess of the Trois Frères cave, the Sorcerer seems to rule over the world of animals depicted on the walls below him. The animals are drawn in a marvelously naturalistic style, whereas he is shown as a weird creature, half beast, half man. The face is human but with a goatlike beard; the eyes are reminiscent of an owl; the antlers, of a stag; the ears, of a wolf. The body, with the claws of a lion and the tail of a horse, is in a lively dancing pose; the gaze is fixed on the onlooker.

Clearly, it was not for lack of skill that the Paleolithic artist produced such an unrealistic representation of man; the pictures of animals he drew on the walls of the same cave leave no doubt concerning the objectivity of his vision or the mastery of his technique. The portrait must therefore symbolize some form of magic. One may well imagine that the Paleolithic Sorcerer represents a personage who occupied a position similar to that of the shaman or medicine man among primitive people today—a member of the group who has considerable empirical knowledge of the fields, of the forests, and of the beasts, and who in addition has also the power to influence his fellow man. The Sorcerer in the Trois Frères cave represents man with his dual nature, or more exactly with the two complementary aspects of his nature—a physical body reacting to the forces of nature and an intellectual endowment capable of creating symbols which become forces on their own account. The ability to symbolize constitutes the one characteristic of man which most profoundly affects his own life and which makes possible his influence on the rest of creation. Unfortunately, as we shall see, this propensity is also the indirect cause of a large percentage of his diseases.

In many other Paleolithic sites from the shores of the Atlantic Ocean to the Ukraine there have been found statuettes of women that also reveal human traits strangely similar to those depicted in art forms being created today. The best known of these figurines is the one usually designated as the Venus of Willendorf, from the name of the locality near Krems in Austria where it was found. Like many other Paleolithic figurines, the Venus of Willendorf represents a naked woman with exuberant fleshiness. The sexual and maternal aspects are grossly exaggerated: the breasts are large and pendulous, the hips broad, the buttocks rotund. Consciously or unconsciously, the Paleolithic artist elected to emphasize the anatomical parts relating to femininity and neglected or barely suggested other parts, such as the hands and the feet, not related to the sense of female fulfillment.

The Venus of Willendorf obviously does not represent an actual woman, but is a symbol. She constitutes an interpretation of womanhood. She constitutes an interpretation of womanhood, and she probably had a symbolic significance in some form of fertility cult. Other Paleolithic figurines of women differ greatly from these fleshy symbols of sexual and reproductive life and represent instead a slim, almost ethereal, human type. Such is the Figurine à la Capuche, found in Brassempouy, France. Furthermore, there have survived from Magdalenian times representations of slender women engraved on stone or bone. The rock art of South Africa provides evidence that the later Stone Age people had a passion for dancing and for wearing masks representing animals, birds, and imaginary monsters. Sometimes games and musical instruments are also shown. Thus, one of the most ancient aspects of human life is the transcendence of simple biological urges. In obscure ways human life converts the physicochemical processes of purely biological existence into actions, representations, and aspirations which pose to the science of man problems not found to the same degree in the study of other living organisms.

It would be foolhardy, of course, to pretend that the nature and problems of modern man can be seen clearly in the hazy light diffusing from a few Paleolithic artifacts. But of one thing at least we can be certain. Today, as in the very distant past, the mark of man is the intense, complex, and so far unexplained interplay between his body, which is nothing but a complex form of matter, and the symbolic activities of his mind.[1] Because of this interdependence, anything that impinges upon him affects simultaneously both body and mind, and causes them to interact. Thus, to study the body machine and to neglect the symbolic activities which are inextricably enmeshed with it is to ignore the most characteristic aspect of man's nature.

In discussing prehistoric life, it is customary to emphasize man's struggle with the hostile forces of his environment, and the problems and ordeals which arose from the fact that he had to eke a precarious living out of meager resources. In reality, however, nonmaterial factors influenced his life as much as did the brute natural forces; the environment acted on him then as now indirectly at least as much as directly. Mechanical stresses, irritating materials, radiations, and temperature act directly on the human fabric just as they do on other nonhuman structures of similar composition, whether these be monkeys, oaks, amoebas, viruses, or inanimate substances. But in addition to their direct effects on the constituents of the body machine, environmental forces also have indirect effects on living things. In the case of man, the mind plays a dominant role in conditioning these indirect effects, both quantitatively and qualitatively.

Living forms respond in a manner which is determined not only by the nature of the stimulus itself but also by the indirect reac-

1. I realize that many philosophers, biologists, and physicians object to the use of words such as "mind," "meaning," or "mental." They find no justification in particular for using the word mind as if it were a noun, and urge that it be restricted to its grammatical function as a verb. According to these purists, it is correct to say "Mind the baby" or "Mind your own business," but not "He has a fine mind" or "His mind is deranged." Clearly, there is no evidence that "mind" refers to a location, a structure, or a substance in the body. Operationally, however, even the most materialistic philosopher has no doubt that his own "mind" is better than that of the village fool. For reasons of convenience, I shall therefore continue to use the word mind as a noun, even though I have no concept of the structures and processes thereby implied (see pp. 25–26).

tions that it mobilizes in them. This chain of indirect responses is of greatest importance in man because of his propensity to symbolize everything that happens to him, and then to react to the symbols as if they were actual environmental stimuli. As a consequence of this human trait, the response of a human being to any environmental factor is highly personal because it reflects all his own past experiences. The personal past is such a powerful force that it can distort and even metamorphose the most trivial as well as the most sublime experiences and thus convert them into ordinary physiological reactions. Marcel Proust immediately found all over the world an understanding public when he recounted how the taste of a cookie dipped in a cup of tea made him recapture in their full strength sensations and emotions long buried in the subconscious strata of his being. These sensations and emotions had been apparently forgotten, but in fact they had become part of his "nature." Indeed, the distortion of present experiences by the past reaches far deeper than the mechanisms set in motion by Pavlov's conditioning of his dogs, because symbolism involves the experience not only of the individual person but also of the group of which he is a part. Ulysses was speaking for all of us when he stated in Tennyson's poem, "I am a part of all that I have met."

The biological and social memory of the past is not the only factor that distorts the direct effects of environmental forces. Man's response to any stimulus is also profoundly affected by anticipations of the future—whether these be conscious or subconscious. Indeed, man's propensity to imagine what does not yet exist, including what will never come to pass, is the aspect of his nature which most clearly differentiates him from animals. The more human he is, the more intensely do his anticipations of the future affect the character of his responses to the forces of the present.

Thus, all the perceptions and interpretations of the mind become translated into organic processes. For this reason, the actual effects that the environment exerts on man commonly bear little if any resemblance to the direct effects that could have been expected from the physicochemical nature of the stimulus. The body machine reacts not only to the stimulus itself but also to all the symbols associated with the experiences of the past and the expectations of the future, symbols which are converted into effective stimuli by a particular event. Robert Burns expressed these human complexities in his poem "To a Mouse," which he wrote after startling the small helpless animal while plowing a field:

Thou art blest compar'd wi' me!
The present only toucheth thee:
But oh! I backward cast my e'e
On prospects drear!
An' forward tho' I canna see,
I guess an' fear!

b. Man the social animal

When *Homo sapiens* first becomes identifiable on the archeological scene, he already is found in fairly large social groups. We know mankind only as a crowd. Man is a social animal, dependent on other human beings for his physiological needs and his emotional satisfactions.

If instinct is defined as a specific adaptation to environment that does not have to be learned, then the human infant is singularly ill-equipped at birth and furthermore remains deficient in this respect for several years. Because his instincts are so inadequate, personal relationships are of paramount importance during his long period of development. Man, in fact, continues to require the support and encouragement of other men throughout his life. Moreover, the very process of civilization makes him even more a tribal creature. Thoreau preached the virtues of solitude, but he built his hut by Walden Pond within two miles of Concord Village, and he walked there several times a week to maintain contact with his mother and friends.

Throughout the living world, the association between different organisms brings out characters, properties, and problems which are different from those exhibited by an individual member of the group living alone. The emergence of new structures and new activities is most striking, of course, when organisms belonging to two different species form symbiotic partnerships (see Chapter IV.1). But biological innovations occur also whenever several members of one species associate to form a population, and this is particularly true of man. The individual organisms within a group interact in a variety of ways, and this interplay brings about the unfolding of potentialities that would remain unexpressed in the isolated state. For better or for worse, the interplay between individual members of the human species always changes the expression of their morphological and physiological endowments, and of course their behavior. Vague as they are, expressions such as "herd instinct" or "mob psychology" serve to emphasize that human beings have retained from their evolutionary past a propensity to function in groups rather than singly.

The sense of belonging and of togetherness engenders a feeling of security which usually increases the chances for biological success and also for happiness. On the other hand, deleterious effects are likely to ensue if the group becomes too large, especially if it grows too rapidly; crowd diseases and mob disorders constitute classical problems of medicine and of sociology (Chapter IV.2). For example, the expression "pathological togetherness" has recently been introduced to denote the lowered fertility and shortened life span of rats which elect to huddle like a farrow around a single food hopper when they are placed in a restricted enclosure.

Whatever its ultimate consequences, any association is bound to bring out characteristics qualitatively different from those manifested by the isolated members of the species. Growth, reproduction, general resistance, ability to learn, etc. are affected in a manner and to an extent not readily predictable or perhaps not predictable at all, even from the most detailed knowledge of the individual organisms involved. There are forms of behavior, of performance, as well as of disease which are determined by the size and structure of the group at least as much as they are by the characteristics of its individual components.

In human societies, furthermore, ethical considerations create attitudes and types of behavior that appear to be totally independent of biological needs and often conflict with biological urges. As pointed out by Freud, the law does not forbid that which man is not prone to do. "Thou shalt not kill" or "Thou shalt not covet thy neighbor's wife" are not prohibitions aimed at a few evil men, but at ordinary men with average propensities. The ability to perceive the difference between the "is" and the "ought" in personal and collective conduct constitutes one of the most essential requirements of human social life. As already mentioned, the human infant is completely dependent at the start, and he proceeds from this state not to independence but to the mutual interdependence of equals.

For a multiplicity of reasons, therefore, the social group is qualitatively different from the sum of its parts. This is not because of any mystical property, but merely because the characteristics of the group are determined by the manifestations of the interplay that takes place between its component parts under a particular set of circumstances. Man's nature includes not only his chemical and physiological properties but also all the attributes he derives from living as part of a complex social group.

c. Nature vs. nurture

Any discussion of man's characteristics is bound to evoke the ancient controversy of nature versus nurture. This controversy, however, constitutes a pseudo-problem because it is based on false premises. Whether the organism be microbe, corn plant, fruit fly, or man, all its characteristics are hereditary, and all are also determined by the environment. This apparent paradox applies to human health and disease as well as to all other manifestations of life. "Genes determine not 'characters' or 'traits' but reactions or responses—health and disease are manifested in the phenotype of the organism and the phenotype is, in principle at least, modifiable, and perhaps controllable, by the genotype as well as by the environment" (Dobzhansky, 1962b).

In the case of man, environmental control is at present of much greater importance than genetic control, but this does not mean that the human genetic endowment is immutable. It is true that the biological characteristics of man have not changed much since Paleolithic times, but the 20,000 pairs of genes that govern his constitution provide the opportunity for many important genetic shifts. In fact, man's genetic make-up is certainly still changing in detail, as environmental conditions favor the selection of one aspect or the other of his polymorphic nature. "The environment plays the role of a programming device and natural selection that of a regulatory mechanism" (Dobzhansky, 1963). In other words, it can be said in Toynbee's language that the environment presents challenges which give to the human species the opportunity to respond adaptively by altering its genetic composition.

The selection of the genes that control anatomic and physiologic characters, as well as mental traits, has always been influenced by the climate, the topography of the land, the kind of food most readily available, and other factors of the physicochemical environment. The need to compete with other animals must also have constituted a powerful force in the early phase of man's evolutionary development. All these factors certainly continue to affect today the genetic endowment of mankind, but they are decreasing in importance from the evolutionary point of view. For the past several millennia, in particular, the only forms of life that have been a real threat to man have been microorganisms, insects, and of course man himself!

Insects are a nuisance and the cause of great economic loss, but otherwise they do not affect human fate significantly, except as vectors of microbial disease. In contrast, microorganisms still create a selective pressure that can bring about genetic changes. Indeed, as claimed by Haldane (1932), infectious diseases may have been one of the most powerful factors in human evolution until the very recent past. The best studied of the genetic effects exerted by microbial life on man is the high incidence of the hemoglobin abnormality responsible for sickle-cell anemia in areas where malaria is prevalent. It is probable that the frequency of other genetic factors affecting red blood cells is also under the selective influence of the microbial environment. As we shall see later (Chapter VIII.2), epidemiological evidence strongly suggests that genetic variations are still occurring during the course of widespread epidemics.

The modern ways of life constitute other selective forces which influence the direction of evolutionary processes in our communities. Scientific medicine and especially public health practices are among the social factors that bid fair to exert an influence on human evolution in the future by allowing the survival of persons suffering from genetic defects. Thanks to insulin, for example, diabetics now have almost as long a life expectancy as normal persons and a comparable chance of begetting children. It can be expected, therefore, that the percentage of persons prone to suffer from diabetes will increase in communities which can afford to provide insulin for all who need treatment.

Human behavior is one of the social factors responsible for genetic changes. There was a time in certain parts of Southern Europe when young women suffering from goiter were regarded as especially attractive, with the result that sexual selection brought about an increase in incidence of the disease in these regions. Tastes, fashions, and habits certainly continue to act as selective factors today as in the past. Even emotional attitudes are probably either selected for or against, depending upon the dominant mores in each social group. Granted the complexity of the genetic determinants of traits such as those which make for aggressiveness or submissiveness, for originality or conformity, it can be taken for granted that a certain amount of genetic selection is brought about by social, economic, and political forces. The profound differences in culture and behavior between the Pueblo Indians in contrast to the warlike Indians of the Plains or Northwest could hardly fail to have expressed themselves in a selective advantage for the types best suited to the cultural pattern characteristic of each one of these Indian communities. Likewise, it is most probable that the shift from a competitive economy to a paternalistic welfare society will eventually affect in some undetermined manner the genetic endowment of the populations concerned.

Modern civilization increasingly provides a variety of niches for virtually every kind of physical and mental type. However, there is no evidence that the operation of natural selection has been entirely suspended or bypassed. It continues to operate in the new zone of adaptation into which man has moved, namely the new environments he is creating. Man is certainly losing some of the qualities, both physical and mental, that were needed for survival in the brutish environment in which his ancestors lived. He no longer has the physical strength and primitive resourcefulness of the caveman who had to fight savage beasts with bare fists or at best with clubs. But some of the characters he has lost are no longer of great use and might in fact handicap him in the environment of modern civilization; the caveman would probably not get along well in a congested city. Natural selection cannot maintain the state of adaptedness of human populations to an environment that no longer exists; but it continues to help them to become better adapted to the environment of today. As we shall see in Chapter X, however, adaptation is not without its own potential dangers.

While genetic changes still occur and will continue to occur in human populations, it is also true that the characteristics of man can be rapidly and profoundly modified by mechanisms that do not affect his genoplasm. The Mongols were harsh, uncouth nomads when they conquered China, as were the followers of Mohammed when they conquered Southern Spain; but the physical and mental stamina of the conquerors were rapidly lost as they adopted the Chinese or Andalusian ways of life. In both cases, racial behavior became modified so soon after the conquest that the changes could not have possibly been due to alterations of the genoplasm; they rather represented phenotypic responses to new environmental conditions.

Increasingly, man becomes adjusted to his environment through the social and technological methods characteristic of his culture;

these constitute adaptive mechanisms not available to the rest of the living world. The word culture is used here to signify what people do as a result of having been so taught. Adaptation by culture is vastly more rapid, and often more effective, than genetic adaptation. For example, a few decades usually suffice anywhere in the world to transform a rural population into city dwellers who have forgotten the ways of life on the farm; and the reverse can also be true, even though it occurs far less frequently. Thus, man has evolved a novel, parabiological method for adapting to his environment. He no longer relies exclusively on the forces of natural selection but instead increasingly uses sociocultural means. In final analysis, human evolution is now the resultant of the interaction between biological and sociocultural forces, and it involves a constant feedback between them. In this respect also man differs qualitatively from the rest of animal creation, including domestic animals.

Culture being a product of life so peculiar to mankind and operating independently of genetic and physiological mechanisms, it is natural to emphasize its importance in any discussion of the changes that occur in man and in his societies. But cultural forces are not the only factors that can produce profound and rapid phenotypic alterations in man's nature. As we shall see in subsequent chapters, almost every aspect of the biological and physicochemical environment can alter his body and his mind, even more rapidly and profoundly than do cultural influences. We shall now consider in somewhat greater detail the influences experienced during development in utero and during the early postnatal period, because a large part of man's nature is the outcome of the responses he makes to his total environment at the very beginning of life.

2. Prenatal and Early Postnatal Influences

Freud's writings have focused attention on the fact that early experiences condition man's whole living experience. In fact, recent studies have revealed that the range of the effects exerted by early stimuli is much wider than even Freud realized. Practically all events occurring in the very early period of development have a disproportionately greater influence on physiological characteristics and on behavior than have events of the same type or magnitude occurring later in life. The effects are greater; they are often irreversible; they extend to a large variety of characteristics, such as longevity, health, learning, and behavior.

Malformations of the fetus caused by chemicals that can cross the placenta at a particular stage of fetal development became a cause of acute public concern after the thalidomide tragedy; but in fact the generality of the phenomenon had been recognized and extensively studied long before this accident. Many pathological states resulting in mental deficiency, such as mongolism and microcephalism, have long been known to have their origin in early embryonic life; cretinism due to hypothyroidism and hydrocephalus caused by a failure of cerebral fluid circulation are also identified with mental retardation. Apart from gross anomalies, an adverse environment in the uterus can bring about subtle alterations in the developing nervous system that are reflected later in abnormal behavior. The fact that early influences also affect profoundly all characteristics of laboratory animals has greatly facilitated the experimental study of this aspect of human development.

a. Effects of early influences on animals

Toxic substances, nutritional deficiencies, infections, drugs, radiations, atmospheric changes, various hormones, etc. can indirectly produce physical anomalies such as hernias, deformation of the skeleton, cleft palate, and absence of limbs, tail, cranial bones, or eyes, etc. Changes in emotional states, as well as in learning and activity level, have also been produced in rats by exposing their mothers during pregnancy to intense sound, X rays, and conditions causing anxiety. There is no doubt, therefore, that both physical and psychological stress experienced by the mother can affect the physical characteristics and behavior of the offspring.

Female rats which had been conditioned to an electric shock in association with the sound of a buzzer were subjected to periods of anxiety by the use of the buzzer without the shock during pregnancy. In order to eliminate the possibility that these anxious mothers would "teach" their young to be anxious, half of the litters were exchanged immediately after birth with litters of control mothers for rearing ("cross-fostering"). Yet, the offspring of the stressed mothers responded differently from controls when placed in strange environments: they were more timid or "unforthcoming." Behavioral disturbances of the young, and lowered ability to learn a simple maze problem, can be produced also by administering epinephrine to the mother.

As we shall see later, crowding affects profoundly, and in a lasting manner, the qualitative characteristics and also the fate of populations (Chapters V.2 and XI.2). It is therefore of special interest that the offspring of mice intensely crowded during pregnancy exhibit abnormalities of behavior in comparison with those of uncrowded mothers; the differences persist for at least 100 days, and perhaps for the whole life span. This is not surprising in view of the fact that crowding brings about profound changes in hormone secretion, and that the fetus is probably susceptible to these hormones as it is to all other influences experienced during intrauterine life. In any case, it is clear that, since changes in behavior can be produced even when the offspring are reared by a foster, nonstressed animal, the fundamental disturbance occurs in utero, before the nervous system of the fetus is completely developed. Migration of hormones across the placental barrier probably plays a role in the effects that emotional stimuli experienced by the mother exert on the unborn.

Many experiments carried out with rats, dogs, and monkeys have shown that both restriction and increased level of environmental stimulation shortly *after birth* can have drastic and lasting effects on such traits as emotionalism, learning ability, degree of activity, social behavior, and perception.

The phenomenon of imprinting constitutes a striking example of the lasting effects of influences experienced during early life. This subject has been so often discussed that it needs no further elaboration here. It seems worthwhile, however, to illustrate the extent to which imprinting can affect behavior by quoting from an extraordinary account of the relation between wild animals and the Wai Wai people, a primitive Indian tribe that inhabits the forests north of the Amazon.

> How could one account for the Indians' success in taming, perhaps unparalleled elsewhere? . . . With baby creatures, attention, tenderness and care were undoubtedly the important factors. Creatures scarcely emerged from the womb or egg—parakeetlets the size of a finger-tip, baby humming-birds no bigger than peas—were taken back to the villages and reared to adulthood; newly-born mammals were suckled by the women, birds fed with pre-chewed cassava bread forced into their beaks, or even directly from the women's own mouths, in imitation of the mother birds; and if a creature

> were naturally shy or savage it was given to many people to handle so that it became accustomed to human beings.

The Wai Wai tame adult birds or other animals by a treatment designed to produce submission. The animals are

> put in darkness in a box or under a large bowl for a day or two without food or water, and at the end of that time, if still alive, are usually prepared to eat and make friends. Then their faces are rubbed into the owner's armpit or smeared with his red body paint so that they get to know his smell, and they are released. Nothing else seems necessary; thereafter, though perhaps wary of strangers, they are usually attached to their master and his family, come when called, allow themselves to be picked up, and return home even after long foraging trips into the forest [Guppy, 1958].

In contrast to this extraordinary picture of physiological intimacy between man and beast, it has been the general experience that animals raised from birth under various forms of isolation exhibit strikingly aberrant behavior when exposed later to normal sensory stimuli and environmental experiences. These findings give support to the statements made by ethnologists that there exist critical developmental stages during which certain behaviors and skills are acquired and after which they either completely fail to emerge or appear only in a distorted form. It has been suggested in this regard that certain biochemical templates become available only at special developmental stages and from then on shape the course of growth.

Most emphasis in the past has been placed on the deleterious effects of early influences. It is now realized, however, that the beneficial effects can be at least as important. Experiments in rats have shown that a certain amount of stress during early life commonly increases the chance of the animal to grow into a sturdy adult, to be emotionally stable, and to perform well in learning tasks. Thus, rats which have been gently stimulated grow more quickly than their litter mates which have been left undisturbed, their faster growth being due to more efficient utilization of food rather than to greater consumption. The learning ability of stimulated animals is also greater, as measured by conditional avoidance tests.

Handling during early life even increases general resistance to various forms of stress, for example, to prolonged deprivation of food and water, to cold exposure, and to immobilization caused by binding up the animals. The beneficial effects are most striking when the animals are stimulated by repeated handling during early life, for example, before the age of 20 days. In fact, the existence of such a sensitive critical period, which was first recognized in the case of imprinting in birds, has now been found to exist in most if not all animal species.

Stimulation during early life also affects subsequent behavior. On being placed in an unfamiliar open space, a handled animal explores confidently and shows little sign of exaggerated autonomic activation, whereas a nonhandled control tends to freeze as well as to defecate and urinate excessively. The handled animal is more prone to engage in exploratory activities even under normal conditions; its movements are wide-ranging and so oriented that a recently visited place is less likely to be entered than an unfamiliar one. Exploratory behavior of this type seems to occur independently of primary needs such as hunger, or thirst, or search for a mate, and could be qualified as pure curiosity.

Friendly, gentle stroking of the animals was thought at first to be essential for the achievement of these beneficial effects. In reality, however, the mere act of handling is per se the important stimulus. For example, resistance to stress can be increased simply by picking up very young rats, repeatedly putting them in a box, and replacing them in their cage. In other words, general stimulation rather than gentling is the significant factor in conditioning the animals for the future. Indeed, gentle electric shock or the mere lowering of body temperature can constitute a sufficient stimulus and improve behavior in later life.

The mechanisms of the effects exerted by early stimulation are still poorly understood. The view most widely held is that alteration in the balance of hormonal output plays a part in modifying behavior and in increasing resistance to stress, but the information on this score is rather limited. Handling per se does not increase the output of steroid hormones under normal conditions. Furthermore, this output rises both in handled and in control animals when the two groups are subjected to some painful stimulus; often the output is even somewhat greater in the nonhandled animals, reflecting perhaps their more acute response to any form of stress. The only difference so far detected that might be of significance is that the handled animals exhibit a much larger output of hormones immediately after stimulation, for example, during the first 15 minutes after a severe electric shock. It is possible, therefore, that prior stimulation endows the animals with the ability to respond more rapidly. An analogous speeding up of response is observed during development. Rats repeatedly handled during the first days of life respond effectively to stress by secreting corticotrophin at the age of 12 days, whereas nonhandled controls do not respond well until the age of 16 days.

Attempts have been made to relate the effects of early experiences to certain neural mechanisms. As is well known, two opposite types of response can be evoked by electrical stimulation of different areas of the brain; both of them are mediated through neural systems running from the anterior end of the brain to the brain stem. In rats and monkeys one type of response elicits a behavior, such as bar pressure, which aims at continuing the stimulation; in man its corresponding effect seems to be a pleasurable experience. The other type of response has opposite manifestations, namely refusal to continue the stimulus in experimental animals and unpleasant feelings in man. The two responses seem to have their final pathways in different areas of the hypothalamus and to be associated with differences in the activity of the autonomic nervous system. It would seem that the stimuli and experiences of very early life exert their effects through such channels and that their expression involves the hypothalamic-anterior pituitary-adrenal cortical axis. The activation of the neuro-endocrine mechanisms by early experiences seems to alter the threshold of response to stress and thus to condition the response to later life experiences.

As already mentioned, the beneficial effects of early and repeated handling should not be interpreted as being the equivalent of "giving love" or "mothering." Beneficial effects in animals can be obtained by stroking, vibration, tossing in the air, or even by mild electric shock. These manipulations have nothing to do with motherly love, yet the stimulation they produce renders the animal more able to cope with later stresses. If these findings are meaningful at all for human experience, they suggest that part at least of the beneficial effects derived by the newborn infant from his mother's fondling can be traced to early sensory stimulation.

In all animals, cutaneous stimulation is crucial for normal development. With monkeys, a particular type of cutaneous experience during early life has profound effects on growth, as well as on many aspects of subsequent behavior, including ability to mate and to take care of young. This has been demonstrated in a spectacular manner by the observations made on the colony of rhesus monkeys maintained since 1954 at the Primate Laboratory of the University of Wisconsin. By changing experimentally the tactile and other physical characteristics of "artificial mothers," it has been possible to modify at will not only the social responses of the infant monkey to other infants but perceptual and learning capacity as well. Furthermore, these influences continue to condition the animal's responses throughout its life; they affect critically its mating behavior and also the maternal behavior of the adult female when she succeeds in having infants of her own, often after much difficulty.

Altering the peer relationships at an early period in development also conditions the future emotional responses in rhesus monkeys. These animals develop bizarre affectional reactions when they are raised in total or partial isolation from peer groups. They are sexually incompetent, timid, lack defense reactions, and do not develop play techniques. These abnormalities can be reversed if the period of early isolation does not exceed 60 to 90 days, but they become essentially irreversible after six-months isolation. In some respects it appears that peer deprivation is even more important than mother deprivation. "Beyond a brief period of neonatal grace . . . every additional week or month of social deprivation increasingly imperils social development in the rhesus monkey" (Harlow and Zimmerman, 1958).

b. Effects of early influences on man

In animals, as we have seen, the fetus is profoundly influenced during intra-uterine life by the environmental stimuli that impinge on the mother. The gross anomalies so produced can probably also occur in man. It is naturally difficult to validate the varied folklore and beliefs concerning the effects exerted on the human child by the experiences of the mother during pregnancy. However, a number of recent observations are highly suggestive in this regard.

The production of congenital defects by drugs, poisons, infectious states, and nutritional deficiencies is now well documented. Evidence is also accumulating that medication during labor can affect the newborn. Furthermore, studies of children in Baltimore and Edinburgh indicate that a history of complications during pregnancy, especially preeclamptic toxemia, is correlated with poor behavior or performance during school life. Some evidence has been adduced that even variations in climate during pregnancy, if they are profound and unusual, can affect the intelligence of the child! Observations on identical twins also suggest that intra-uterine influences can affect not only anatomical structures and physiological functions but also subsequent behavior. For example, the case of the famous Dionne quintuplets has provided evidence that genetic identity can be compatible with phenotypic diversity at birth.

The Dionne quintuplets were a set of identical sisters, yet their attendants could tell them apart from the very beginning. Each one of the sisters soon developed a characteristic behavior pattern even though every effort had been made to provide them with identical postnatal environment. It is certain, of course, that unrecognized differences occurred in the early handling of the children and in their contacts with the physical environment after birth, but it is also very probable that some of the differences among them had become established before birth. Through the accidents of position in utero the conditions of development could not have been absolutely identical for the five fetuses during their seven critical months of intra-uterine life.

The results of a recent study of identical twins indicate that the events of intra-uterine life may play a role in the pathogenesis of schizophrenia. A search through the United States yielded five pairs in which one of the twins was schizophrenic, the other normal; it was established that the twins were identical and had been raised by their own parents, who were still living. A detailed study of the twins and their parents revealed profound differences that could account for the sick twin's illness. In each case, the twin who became schizophrenic weighed less at birth than the normal child; his parents feared for his life and showered him with special attention. The sick child was the more docile and dependent of the two, developed more slowly throughout childhood, and made poorer grades in school. These findings are not sufficient of course to explain why only one of the twins became schizophrenic, but they establish that the differences in prenatal conditions in the uterus can affect profoundly and lastingly many subsequent aspects of life.

It is tempting, of course, to account for the effects of experiences during intra-uterine life by the hypothesis that the endocrine system of the mother is a very significant factor in fetal development.

While there are no neural connections between mother and fetus, an intimate humoral interplay between them takes place through the placenta. This organ, furthermore, is not only a channel through which substances pass from the maternal blood stream to the fetus; it also secretes several hormones. The active substances which reach the fetus through the placenta are probably more numerous than is known, and their production is probably controlled by many factors that affect the mother during gestation. Thus, it would be surprising if the experiences of the mother, including her emotional disturbances, did not in some way affect fetal development.

Some of the effects of postnatal experiences are naturally easier to recognize than those of prenatal life. The inflammatory diseases of the brain membranes caused by viral infections and the demyelinating diseases are but a few among the many forms of damage incurred shortly after birth that can cause mental retardation. Other more subtle conditions of early postnatal life can also have serious although less obvious consequences, for example, deprivation of human contacts during early life. Several independent clinical studies have revealed that early institutional care, especially during the first six months of life, commonly exerts damaging effects on human infants. Their physical development is hampered and their resistance to disease reduced. The capacity for affective relationships and abstract thinking in later life also seems to be adversely affected by the experience of early institutionalization.

The many reports concerning feral children are often quoted to illustrate the effects of deprivation of human contacts during early life. Even without accepting in their details the stories that have been published about the lives of so-called wolf children, there are other reasons to believe that human beings are dependent on many different types of environmental and social stimuli to achieve and maintain a state considered normal. Thus, baby chimpanzees raised in complete darkness for a few months from birth never learn to use pattern vision; their eyes and brains are normal but they do not know how to see. Likewise, human beings who lacked pattern vision in childhood experience great difficulty in learning to distinguish visually a square from a circle after their vision has been restored, following removal of an opaque lens or transplantation of a clear cornea.

As already mentioned, young rhesus monkeys raised in contact with surrogate mothers grow well physically, but exhibit profound and lasting abnormalities of behavior. It can be taken for granted that in the case of human children also maternal deprivation is more than the lack of an emotional experience; it involves a crucial loss in the amount and variety of sensory stimulation essential for complete and orderly development.

Most forms of sensory deprivation, especially during the early phases of life, thus seem to prevent the organism from developing essential mechanisms and patterns of responses without which adaptability is severely handicapped throughout the rest of life. In other words, complete psychological development depends on a multiplicity of environmental stimuli. In their absence or if they are inadequate, intelligence does not develop normally and the personality becomes grossly atypical.

3. Sensory stimulation and personality structure

The need for stimulation does not stop in childhood. Many experiences, even though somewhat stressful, can improve performance, both physical and mental. It has even been asserted that mutilation in the young, as during primitive initiation rites or in face scarring, results in greater body growth, as well as in greater stamina. In any case, it is certain that man functions best when a fair percentage of his neurons are active, provided a sufficient number remain in reserve. Practical experience shows that the elimination of environmental stimuli leads not to Nirvana, but first to a vegetablelike existence and soon to a disintegration of personality. Life implies response and function.

In the course of normal life, organisms are constantly exposed to a large variety of stimuli, in part because their environment is continuously changing, and also because their own activities constantly modify their relationships to the external world. Since environmental stimuli have affected all aspects of evolutionary development, they are necessarily reflected in the pattern of physiological and psychological responses. The sort of "mutuality" that exists between organisms and their normal environment has become a requirement for physiological and psychological well-being.

Recent experiments have established in fact that a constant bombardment of stimuli is required for successful development and function. For example, animals raised in the germ-free state and maintained in a sterile environment throughout life exhibit anatomical, physiological, and nutritional abnormalities that would severely handicap them, and even prevent their survival, under the usual conditions of life. The adequate development and maintenance in a healthy state of certain anatomical structures and physiological activities require the morphogenetic stimulation created by the presence of the proper kind of microbiota in the digestive tract (see Chapter V.3). Strangely enough, a somewhat analogous conclusion is emerging from studies which have revealed that the development of the nervous system and even mental health depend upon a constant exposure to sensory stimuli and to new experiences.

The anatomical or chemical alterations through which the various organs respond to environmental changes are poorly understood, but they have been identified in a few cases. Thus, scattered observations indicate that both the development and the maintenance of nervous tissue are the outcome of metabolic phenomena that are conditioned by a proper intensity of stimulation. This dependence was brought out in a striking manner by markedly reducing visual stimulation or by withholding light entirely during the early postnatal life of chimpanzees. Atrophy of ganglion cells and of their optic nerve fibers occurred under these conditions, as revealed by the pallor of the disc, loss of essential constituents such as ribonucleic acid and nucleoprotein, and the eventual disappearance of light reflexes. The atrophic changes were readily reversed when stimulation was resumed by returning the animals to light after a short deprivation. However, some of the changes proved to be more lasting or even to be permanent when deprivation was more prolonged. (For reviews of this topic, see Fiske and Maddi, 1961.)

There is much empirical experience concerning the effects of restricted stimulation on adults. Thus, various types of mental aberration commonly occur among prisoners in solitary confinement or among otherwise normal men isolated during arctic explorations, as a result of shipwrecks, or even in the course of prolonged trips in small sailing vessels. These men are likely to experience abnormalities in feeling states, deterioration of ability to think, perceptual distortions, and vivid imagery often taking the form of hallucinations and delusions. Experimental work on this topic has been stimulated by the study of the effects of brainwashing on war or political prisoners, and also by observations of physiological and mental disturbances arising from the monotony of automated work and other dial-watching tasks. Many attempts have been made during recent years to reproduce experimentally some of the effects of isolation. The general technique has consisted in placing healthy human beings under conditions such that they are protected as completely as possible from sensory stimuli, or from patterned visual or auditory perceptions.

In observations that have become classical, vigorous male college students were placed in a room where they were fed on request and provided with a high degree of physical comfort, but were thoroughly sheltered from the kind of sensory information that man receives constantly in normal everyday life. The students rested on comfortable beds or floated in lukewarm water; they were not in darkness but wore goggles to prevent any form of patterned vision; they could not hear anything except a steady hum; they could not examine their environment tactually. These young men had planned to use their time in this peaceful atmosphere for reviewing their studies and for planning their research. Instead, they found themselves unable to carry on any orderly, connected thinking. Some had elaborate visual hallucinations and suffered sharp temper tantrums (reviewed in Hebb, 1949; Solomon et al., 1961).

Objective evidence of behavioral disturbances produced by sensory deprivation has been obtained from electroencephalographic studies. During 14 days of exposure to "unpatterned light and white noise," human subjects exhibited a progressive decrease in frequencies in the alpha range of brain waves, along with severe motivational losses and "an inability to get started doing anything." The electroencephalographic records were still abnormal one week after return to usual living conditions, and the motivational losses lasted even longer. These findings provide objective evidence that human beings kept isolated for any length of time are likely to suffer physiological and psychological disturbances that are severe and lasting.

Needless to say, the responses to sensory restrictions differ from subject to subject and also vary with the environmental circumstances. There is no doubt, however, that sensory deprivation is physiologically stressful, as revealed by the fact that it results in an increased secretion of epinephrine and norepinephrine. Indeed, the experience is always actually painful. Within a few hours, in all cases, the experimental subjects find it impossible to maintain logical thought or sustained attention on any topic. They commonly experience restlessness, irritability, disorientation, and anxiety often reaching a panic state. They lose the sense of structure of the outside world. Vivid imagery, illusions, and hallucinations are often striking, and beyond the subject's control. The term "regression" has been introduced to denote the breakdown in perceptual and cognitive differentiation resulting from sensory deprivation. In some respects at least the psychopathological states thus induced are similar to those caused by certain psychotomimetic drugs. Interestingly enough, mice kept completely isolated for a few days also develop a symptomatology similar to that observed in animals receiving lysergic acid diethylamide (reviewed in Boyer, 1963).

While the effects of sensory deprivation or of solitary confinement are so dramatic that they can be readily described and to some extent even quantitated, it is much more difficult to evaluate the response of human beings to the kind of relative solitude or removal from stimuli that is experienced under more ordinary conditions. One can hardly doubt, however, that disturbances somewhat similar to those caused by sensory deprivation occur in human beings confined for hours to a single, unchanging task under monotonous conditions. Thus, it is becoming apparent that monotony is objectionable not only because it is boring but even more because it can become an etiological agent for various types of mental disorders.

Ethical considerations greatly limit of course the experimental study of sensory restriction on human beings. Furthermore, experiments in animals concerning these effects are still so primitive and episodic that they do not lend themselves either to generalizations or to extrapolation of the results from one species to another and especially to man. Nevertheless, it is clear that whatever the precise mechanisms involved, sensory restriction can exert two forms of deleterious effects. In young organisms, it prevents the orderly development of essential mechanisms and patterns of responses without which adaptability is severely handicapped throughout the rest of life. In adults, it can bring about a disintegration of the personality, and this disintegration can be permanent if the sensory deprivation is prolonged. In other words, psychological development and health depend on a multiplicity of environmental stimuli.

In ordinary language, the word mind is used as if it referred to an attribute independent of sensory processes; the latter are thought of only as sources of information which the mind is free to use or to ignore (see footnote on page 6). Experiments on sensory deprivation indicate, however, that the very structure of the mind is determined by environmental stimuli. We know the mind only through its expressions in an accustomed social environment, particularly one which we consider to be normal. Like the body, the personality is being constantly molded and held in shape by environmental forces. It is this fact which gives literal truth to Emerson's remark that "men resemble their contemporaries even more than they do their progenitors."

Three main classes of phenomena which seem to emerge from the observations made so far may help to define the range of responses human beings make to environmental stimuli. The simplest situation relates to those stimuli which act directly on the organism without mediation from the nervous system. Light, sound, heat, and mechanical energy impinging on any young organism can condition development by altering the speed or direction of anatomical, physiological, and biochemical processes involved in the genesis of structures and functions. Thus, birds that are completely blind nevertheless respond to the day-night cycles because light penetrates their skulls. In any case, light can directly stimulate the anterior pituitary gland and the hypothalamus, thereby bringing about changes in hormone secretion and in activity levels.

More often, stimuli operate by triggering impulses into the nervous system through neurosensory mechanisms of two different kinds. It is likely that the effects of stimulation on the late fetus or very young animal occur mainly through arousal of a diffuse generalized activating system. This mechanism involves impulses from sensory receptors up to the midthalamus, from which point they are projected to all parts of the cortex. Later in life, when the animal can perceive the world in an organized and meaningful way, stimulation has effects mainly via the specific projection system; the pathway is then highly organized in the sense that a pattern of stimulation on the receptor is transmitted in a point-to-point manner to the thalamus and thence to a relatively well-defined cortical projection area.

On the basis of these considerations, it can be assumed that during the first few weeks of human life the mother keeps the infant aroused and alert through the generalized activating system. This alertness probably favors the building up of neural organization via the specific sensory pathways. Later on, the role of the mother is increasingly to aid the child in organizing his experience, by providing guidance for his behavior and helping him to focus his attention on certain aspects of the world (reviewed in Thompson and Schaefer, 1961).

4. Innate and conditioned responses to the environment

The response of mammals to an environmental stimulus or trauma can be described in terms of two components. On the one hand, each species exhibits a characteristic pattern of physiological and behavioral responses determined by the nature and intensity of the stimulus. On the other hand, all mammals, including man, tend to react to any stressful situation by a complex series of phenomena that are highly individual and almost independent of the nature of the inciting agent. The very vagueness of the stress concept presents advantages for a discussion of this problem.

According to the Oxford English Dictionary, the word stress was introduced into the language via Old French and Middle English as "distress" (later shortened to "stress" through slurring the first syllable). Then as now, it meant to impose strain on, to coerce or compel. A common pathological manifestation of most stresses is a series of associated physiological effects that have come to be known as "General Adaptation Syndrome": swollen drained adrenals, shrunken lymphatic tissue, gastrointestinal ulcers are among its most common characteristics. However, the word stress has been used so much and so carelessly that its meaning is becoming increasingly vague. Some physiologists use it to denote any procedure that causes an increase in adrenocortical activity, as the result of a reflex discharge of corticotropin from the pituitary.

In practice the response of an individual organism is not as stereotyped and predictable as would appear from these generalizations. The effect of a given situation or substance on a given living thing is conditioned by factors that have their origin not only in the racial but also in the individual history of the particular organism concerned. In other words, each organism has its own peculiar way of responding to the total environment.

The most fundamental discrepancies between the nature of environmental stimuli and their biological effects come from the fact that the conditions under which man lives now are very different from those which shaped his biological constitution during evolutionary development. Man can alter the external environment to fit his physical needs, foibles, and wishes, but, as we shall see in subsequent chapters, his innate responses are still those which were developed during his evolutionary past to adapt him to the conditions then prevailing but which no longer exist. Suffice it to state at this point that the persistence of these ancient responses can constitute a source of physiological conflicts under the usual conditions of modern life (Chapter II.2). Furthermore, man has also retained from his evolutionary past certain needs that no longer have a place in the world he has created, yet must be satisfied.

Many types of human behavior that are usually regarded as purely psychological are in reality so similar to traits found in animals that they are probably woven into the mammalian evolutionary fabric. In carnivorous animals, hunting is so essential to survival under natural conditions that this activity has become a drive in its own right. Cats need to hunt almost as much as they need to eat, even after they have become house pets. The hunting urge is a derived character, but its satisfaction has become an important factor of behavioral health. Its persistence can be recognized even in the most urbanized city dweller.

As to the importance of play, it needs only be mentioned that occupational therapy is now recognized as an essential component of good animal husbandry in zoological gardens. Animals may pine unless given opportunity to roam and play, and so is it probably for man. Even artistic expression may come to be regarded as a secondary biological need. Two chimpanzees, Alpha and Congo, painted organized patterns without the stimulation of any reward or social approval for their work; they threw tantrums when deprived of their pigments, paper, and brushes before they felt the picture was finished. When the chimpanzee was bribed with a food reward, and learned to draw for its supper, it lost interest in the organization of the picture. Any old scribble would do, and then the animal would immediately hold out its hand for the reward. Its previous careful attention to design was gone, to be replaced by a simian form of commercial art. For all we know, some form of artistic expression is essential for real mental health in man!

Ancient civilizations were aware of the profound effects that hidden physiological forces and psychological needs exert on human behavior, and they commonly symbolized these traits by a ferocious bull struggling against reason. In several empirical ways, they developed procedures to let the occult components of man's nature manifest themselves under somewhat controlled conditions. As shown by E. R. Dodds in *The Greeks and the Irrational,* the Dionysian celebrations, the Eleusinian mysteries, and many other myths and rituals served as release mechanisms for fundamental human urges that did not find adequate expression in the rational and classical aspects of Greek life. Even Socrates found it wise to participate in the Corybantic rites.[2] Many such ancient traditions still persist in the advanced countries of Western culture, even though in a distorted form. In the most modern city, as among the hills of Arcadia three thousand years ago, men and women perceive at springtime that nature is awakening and at work in their bodies, just as it is in the beasts and trees. Carnival is still celebrated when the sap starts running.

Another illustration of the fact that modern man retains essential traits of his evolutionary past is the persistence in him or hormonal and metabolic responses which were developed to meet threatening situations during his animal ancestry, but which no longer fit the needs of life in civilized societies. In a normal person as well as in animals, the physiologic and metabolic changes required for the successful performance of a physical effort begin before the actual effort is set in motion, indeed as soon as the need for them is anticipated by the body. This power to mobilize in an anticipatory manner and almost instantaneously the various bodily resources required for flight or for fight has been certainly of great advantage throughout evolution; it increased the chance of success in dealing with the harsh forces of the external environment, and in particular with predators and enemies of all sorts. But what was once an advantage is increasingly becoming a handicap under the conditions of modern human life.

The anticipatory mobilization of physiological processes comes into play even under the stimulus of emotional experiences unrelated to physical danger. These experiences have little in common with the threats encountered by primitive man when he had to

2. In Plato's dialogue *Phaedrus,* Socrates emphasizes the importance of the various forms of what he called the "divine madness" in human life. It appears from the text of the dialogue that the word madness is used by Socrates (or Plato) to denote those biological forces in man's nature not readily governed by reason.

struggle physically against natural forces and enemies. However, they mimic the effect of physical threats through the symbolic interpretation that man puts on them. Whatever the life situation, whether it corresponds to an actual physical danger or merely to an emotional crisis, the nature and intensity of the anticipatory changes the symbol elicits in the body have remained much the same in modern man as they were in his Paleolithic ancestor. Yet emotional states and even actual danger rarely call for the expenditure of vigorous physical effort under the usual conditions of modern life.

Since the circumstances which stimulate the endocrine system to mobilize the body resources rarely demand the actual expenditure of physical energy in the modern world, the usual result is that physiological preparation for action is not followed by action. The failure to act after the body has achieved a state of physiological and metabolic readiness constitutes a disturbing biological experience which is extremely common in civilized life, often frustrating, and likely to be deleterious. The competitive and tense situations that occur at the office or at the cocktail party are so trying, precisely because they cannot be released in physical exertion.

As is well known, the anticipatory state of readiness operates through the agency of endocrine and autonomic processes which affect a wide variety of visceral functions. Ever since Walter B. Cannon called attention to them, the physiological aspects of these anticipatory mechanisms have been studied in situations as varied as the preoperative period, the experience of flying, or psychosocial threats. Physiological processes involving adrenal activity have been specially emphasized. Thus, urinary excretion and blood levels of corticosteroids are usually increased under conditions which create sustained tension or which represent a threat even though it be only a symbolic threat. Excretion of epinephrine, norepinephrine, and aldosterone also seems to be affected (see Chapters IX.4 and X.1). In other words, emotional disturbances commonly increase the activity of both the cortex and medulla of the adrenal gland (reviewed in Hamburg, 1961, 1962).

Physiological measurements of the members of a boat crew before, during, and after a race have provided one of the classical demonstrations that unrelated physical and emotional types of stress can activate the same body mechanisms. Following their strenuous physical effort during a race, the oarsmen naturally exhibited a sharp fall in eosinophil levels as evidence of organic stress. More interestingly, however, the coach who had merely "sweated it out" on the sidelines had an eosinophil count even lower than that of any of the oarsmen. His anxiety for the outcome of the race had exerted the same stressful effect as had the actual and exhausting physical effort on the crewmen. The race also brought about an increase in excretion of corticosteroids, as had the practice runs beforehand. Here again, the coach and manager of the crew showed increased adrenocortical activity on the day of the race, even though they expended little physical energy. In their cases, psychological rather than physical factors had been responsible for adrenal stimulation.

Similar observations have been made more recently using blood pressure as an index. Arterial blood pressure has been found to rise to high levels in a number of different frustrating experiences that did not involve physical effort, for example, in the course of difficult arithmetic calculations. During a frustrating four-hour shamflight period, an airplane pilot's pressure rose to 220 mm. systolic and 150 mm. diastolic pressure; his normal levels at rest were 120 mm. systolic and 70 mm. diastolic. When physical activity is really needed, such anticipatory boost in the circulation constitutes an obvious advantage for survival. But the rise in arterial blood pressure is inappropriate, and almost certainly dangerous, when it occurs merely as a result of frustration.

Purely psychological stresses can elicit many other kinds of biochemical manifestations. On a metabolic ward, for example, marked variations in urinary nitrogen and electrolytes occurred among the patients following minor emotional disturbances of the kind associated with ordinary ward routine. In a student about to take an important academic examination, the fecal calcium greatly increased shortly before the time of the event and brought about a strongly negative calcium balance. Indeed, his calcium utilization during the period before the examination was so inefficient that balance could not have been maintained at any level of intake. Blood fibrinolytic activity, blood coagulability, blood levels of chylomicrons and lipid metabolism are other physiological characteristics that are altered by situations causing mental anguish or even mild anxiety (reviewed in Hamburg, 1961, 1962).

One of the biochemical effects of stressful situations that has been most extensively studied is the change in the lipid pattern of the blood. The stress caused in rats by placing them in a restraining cage decreased the ability of their livers to esterify cholesterol. In man, also, emotional as well as physical stress can increase the levels of serum cholesterol and of nonesterified fatty acids. Twenty students were tested for various biochemical activities before, during, and after a difficult examination. The urinary excretion of epinephrine increased markedly during the examination, as did the serum levels of glucose and of free fatty acids (see Chapter XII for the effect of anxiety on the fate of chylomicrons). These facts have naturally been regarded as suggestive evidence that stress plays a role in the etiology of cardiovascular disease through such biochemical effects. While this hypothesis is still unproved, nevertheless it deserves some elaboration here because it illustrates so well the indirect and devious routes through which the external environment can affect the welfare of man.

The increased secretion of adrenal hormones brought about by many forms of stress activates glycogenesis and lipolysis and thereby increases the amount of sugar and fat that can be used by the stressed person. In the evolutionary past, these processes constituted useful anticipatory responses because they made large amounts of energy readily available for physical effort. Now, the same responses take place, but since the fat mobilized is rarely used for actual physical exertion, it may get deposited in the internal lining of arteries of the individuals predisposed for this kind of biochemical pathology.

The view that stressful situations release fats which are not used metabolically is compatible with the claim that a correlation exists between behavior patterns and susceptibility to coronary disease. Men exhibiting excessive and competitive drive, with an acute sense of time urgency, were compared with another group of men exhibiting the opposite type of behavior. During working hours, the first group had a much greater increase in norepinephrine excretion, higher serum cholesterol levels, and faster clotting time; the incidence of clinical coronary disease was also much higher among them. Such individual differences might therefore account for the fact that atherosclerosis seems to be correlated with frequent and prolonged stressful situations, anl also with a sedentary way of life. Animal pathologists have also invoked social tensions and lack of exercise to account for the high incidence of vascular disease among animals (mammals and birds) maintained in zoological gardens.

Because of the increasing prevalence of cardiovascular diseases in Western societies, their psychophysiologic determinants are naturally attracting much attention. At a conference on the subject held in Oregon in 1963, the participants expressed general agreement that verbal and other symbolic stimuli could affect cardiovascular functions as much as do physical and chemical stimuli. However, the discussions failed to define the interplay between the autonomic, humoral, and higher nervous system components involved in the intermediate steps that convert symbolic stimuli into cardiovascular responses.

Many other situations could have been selected to illustrate that emotional states can affect biochemical functions and therefore the response of man to the environment. The mediation of the effects of emotion on the endocrine system occurs in the hypothalamus and, since there are many pathways between this organ and the higher centers, an anatomical basis exists for the obvious fact that emotional states affect endocrine activity. The more general point to be emphasized here, however, is that man's nature includes not only the biological attributes which can be studied by orthodox laboratory methods in laboratory animals but also the determinants of emotional states that can disturb endocrine homeostasis and thus affect all bodily responses.

Thus, it is evident that, while environmental factors act directly on man as they do on all other living things, some of their most important effects are so indirect that they bear little resemblance to what could have been expected from the nature of the inciting cause. These indirect effects are characteristic for each individual person because human beings differ not only in their genetic constitution but also in their past experience. Allergic reactions are examples of the importance of prior conditioning by environmental factors. Even more important, probably, is the fact that man responds not only to the stimuli that impinge on him but also to the symbols associated with these stimuli. Response patterns are so profoundly influenced by the prior conditioning of the body and of the mind that it is often more useful to study the phenomena of disease by regarding them as inadequate responses of the organism to a given situation than as direct effects of a noxious agent.

The effects of the environment on man will be discussed in the present volume by focusing attention both on the intrinsic properties of noxae or other environmental factors and on the meaning these have for each individual person. This meaning, as already emphasized, involves all the components of man's nature. A great deal is known concerning the genetic endowment of man, his physicochemical and physiological characteristics, the psychological, social, and religious determinants of his behavior. In contrast, there is hardly any knowledge of the manner in which man's response to the environment is conditioned by obsolete traits, needs, and urges he has retained from his evolutionary past, or by his unexpressed aspirations for the future. Yet these "subterranean" aspects of a given person's nature often play a decisive role in determining whether a given situation will result in a state of health or disease.

All the forces of the environment and all the aspects of man's nature are thus enmeshed in such an extremely complex manner that problems of etiology can never be as clear-cut as was thought a generation ago (Chapter XII.2). Medical sciences cannot afford to neglect the problems posed by the interrelatedness of the multiple aspects of human life. On the other hand, practical considerations dictate that the sciences of medicine and public health be focused on the environmental factors that can be manipulated most readily and effectively. As we review in the following chapters some of the environmental factors which are important for human life, we shall have to keep in mind that decisions must often be taken before all facts are at hand. Effectiveness of action must never be sacrificed at the altar of complete intellectual understanding.

II. Man in the Physical World

1. Of airs, waters, and places

Primitive peoples seem to accept as a reasonable explanation of their fate that human life is under the simultaneous influence of supernatural as well as natural forces. They are prone to trace their misfortunes and diseases to the evil doings of ghosts, witches, and other demonic powers; but they also know from experience that the body and the mind are directly affected by the elementary forces of nature that they can perceive with their senses. In fact, this dual attitude is not peculiar to primitive people. Even in the most sophisticated communities, belief in mysterious influences still coexists with the knowledge derived from rational philosophy and science. Lovers and poets are not the only ones to acknowledge that springtime awakens the senses, autumn engenders melancholy, and the stars inspire suprarational dreams.

Modern man has good reason for believing that his life is influenced by forces that are still mysterious. Whatever the technological development of the country in which he lives, he knows that his well-being, behavior, and fate are affected by many kinds of external factors that he cannot identify, let alone understand and control. All cultures have in their languages and traditions references to the biological effects of ill winds, the turn of the seasons, the phases of the moon, even to the influence of sunspots and stars. The naive folklore inherited from the distant past constitutes a kind of primitive biology, which, though grossly inaccurate, nevertheless has a basis of truth. In fact, the perennial faith in astrology is probably but the survival in a distorted form of an ancient and honorable knowledge derived from practical experience.

In its present form, belief in an influence of the celestial bodies on human fate is of course puerile and often pathetic; but it is not different from certain modern obsessions such as those involving microbes and antisepsis. Both attitudes are derived from genuine facts, and both have been distorted by a large body of nonsense. The astrological vocabulary is irrational because it reflects a mystical bent incompatible with objective knowledge, whereas the language commonly used to express fear of microbes appears rational because it has recent roots in scientific jargon. But historical origin does not make the former attitude completely devoid of factual substance, or the latter always valid. The belief in occult forces is often replaced among us by a kind of scientific religionism. Saying of a man that "he is moon-struck" often comes as near to the truth as the modern counterpart, "he has a virus."

The pre-Socratic philosophers made one of the first recorded attempts to account for health and disease in purely rational terms. For example, Alcmaeon of Croton taught a doctrine of natural causation of disease, which can be approximately rendered in the following words:

> Disease occurs sometimes from an internal cause such as excess of heat or cold, sometimes from an external cause such as excess or deficiency of food. It may occur in a certain part, such as blood, marrow, or brain; but these parts also are sometimes affected by external causes, such as certain waters, or a particular site, or fatigue, or constraint, or similar reasons. Health is the harmonious mixture of the qualities [translated by Miller, 1962].

From these visionary imaginings, it was but a step to the Hippocratic writings on "Airs, Waters and Places," which constitute the oldest known systematic account of the effects the environment exerts on health and on the temperament of people. The text of "Airs, Water and Places" was probably intended primarily to provide the Greek physician arriving in a new locality with a help to prognosis. The treatise gives specific instructions for predicting the kind of diseases most likely to be found under certain environmental conditions. To this end, it emphasizes the importance of the soil, the elevation, the water supply, the prevailing winds, exposure to the sun, and other physical characteristics.

Prognosis was all-important in the Hippocratic school. In Hippocrates' words, the physician who is an honor to his profession is the one

> who has a due regard to the seasons of the year, and the diseases which they produce; to the states of the wind peculiar to

> each country and the qualities of its waters; who marks carefully the localities of towns, and of the surrounding country, whether they are low or high, hot or cold, wet or dry; who, moreover, takes note of the diet and regimen of the inhabitants, and in a word, of all the causes that may produce disorder in the animal economy.

Because the Hippocratic teachings continued to dominate Western medicine until late in the nineteenth century, and because of their very direct relevance to the theme of the present book, it seems appropriate to quote here at some length a representative passage from "Airs, Waters and Places."

> Whoever wishes to pursue properly the science of medicine must proceed thus. First he ought to consider what effects each season of the year can produce; for the seasons are not at all alike, but differ widely both in themselves and at their changes. The next point is the hot winds and the cold, especially those that are universal, but also those that are peculiar to each particular region. He must also consider the properties of the waters; for as these differ in taste and in weight, so the property of each is far different from that of any other. Therefore, on arrival at a town with which he is unfamiliar, a physician should examine its position with respect to the winds and to the risings of the sun. For a northern, a southern, an eastern, and a western aspect has each its own individual property. He must consider with the greatest care both these things and how the natives are off for water, whether they use marshy, soft waters, or such as are hard and come from rocky heights, or brackish and harsh. The soil, too, whether bare and dry or wooded and watered, hollow and hot, or high and cold. The mode of life also of the inhabitants that is pleasing to them, whether they are heavy drinkers, taking lunch, and inactive, or athletic, industrious, eating much and drinking little.
>
> Using this evidence he must examine the several problems that arise. For if a physician knows these things well, by preference all of them, but at any rate most, he will not, on arrival at a town with which he is unfamiliar, be ignorant of the local diseases, or of the nature of those that commonly prevail; so that he will not be at a loss in the treatment of diseases, or

make blunders, as is likely to be the case if he does not have this knowledge before he considers his several problems. As time and the year passes he will be able to tell what epidemic diseases will attack the city either in summer or in winter, as well as those peculiar to the individual which are likely to occur through change in mode of life. For knowing the changes of the seasons, and the risings and settings of the stars, with the circumstances of each of these phenomena, he will know beforehand the nature of the year that is coming. Through these considerations and by learning the times beforehand, he will have full knowledge of each particular case, will succeed best in securing health, and will achieve the greatest triumphs in the practice of his art. If it be thought that all this belongs to meteorology, he will find out, on second thoughts, that the contribution of astronomy to medicine is not a very small one but a very great one indeed. For with the seasons men's diseases, like their digestive organs, suffer change.

In the second part of "Airs, Waters and Places," Hippocrates goes even further and asserts that the physical and temperamental characteristics of the various populations in Europe and Asia have been determined by the climate and topography of each particular region.

Inhabitants of mountainous, rocky, well-watered country at a high altitude, where the margin of seasonal climatic variation is wide, will tend to have large-built bodies constitutionally adapted for courage and endurance, and in such natures there will be a considerable element of ferocity and brutality. Inhabitants of sultry hollows covered with water-meadows, who are more commonly exposed to warm winds than to cold and who drink tepid water, will, in contrast, not be large-built or slim, but thickset, fleshy and dark-haired, with swarthy rather than fair complexions and with less phlegm than bile in their constitutions.

The belief that there exists a correlation between military qualities, political institutions, and environmental characteristics was apparently widespread among Hellenic thinkers in the latter part of the fifth century B.C. The theory is used in Chapter 16 of "Airs, Waters and Places" to explain the difference in military prowess between different communities of Asiatics, and also by Herodotus in Book V, Chapter 78, to explain the change that occurred in the Athenians after the expulsion of the Peisistratidae.

Until late in the nineteenth century, European physicians held to the Hippocratic view that human characteristics in health and in disease are conditioned by the external environment; but while they accepted the doctrine in principle, they recognized that its details were not entirely applicable everywhere. Greece, the country in which "Airs, Waters and Places" was written, differs from most other European regions especially with regard to the abruptness of its seasonal changes. As a consequence, the types of human responses described by Hippocrates often differed profoundly from those observed by the physicians of continental Europe.

Despite the fact that the practical teachings of Hippocrates had only limited applicability, his general doctrine long retained its authority because it was regarded as the source of medical wisdom. His treatise was used as a practical textbook of medicine and was reprinted for this purpose even as late as 1874. Thus, for two thousand years, the physicians under its influence regarded it a law of nature that, just as every country possesses its own plant and animal kingdoms, in the same way it possesses a characteristic disease kingdom.

"Airs, Waters and Places" ceased being reprinted as a practical guide to physicians after 1874 because the germ theory of disease and the science of nutrition began at that time to undermine its scientific authority. Many of the effects that location, weather, and seasons obviously exert on the incidence of particular diseases could now be traced to the action of tangible and rather simple agencies. Thus, swampy places are associated with malaria because they favor the breeding of mosquitoes; respiratory infections become more prevalent in winter because people congregate at that time in areas favorable to the survival and transmission of viruses and bacteria; hay fever is highly prevalent during the late summer because this is the period of ragweed pollination.

The discovery that certain vitamin deficiencies are prevalent at the end of the winter went still further in discrediting the vague concept of season and replacing it by more concrete etiological theories. Since small amounts of synthetic vitamins C and D were sufficient to correct some of the debilitating effects of a prolonged winter, it appeared justified to attribute the health-giving qualities of spring and summer entirely to reasons as simple as greater availability of vitamins derived from green vegetables and increased exposure to sunlight. As more and more seasonal effects became thus explainable in microbiological or chemical terms, scientific physicians progressively lost interest in climate, weather, and other environmental factors that did not fit the scheme and fashions of modern scientific medicine.

There remain, however, many examples of seasonal incidence of disease that are not so readily explained. Why, for example, does diphtheria appear in early winter and diminish in early spring, whereas scarlet fever begins later and lasts until early summer? Why does cholera show peaks of severity in humid weather, although its causative vibrio is widely distributed under other weather conditions? What is the reason for explosive autumn incidence of hog cholera and hog influenza in the farms of the American Middle West, although the respective viruses are widely distributed in the swine herds long before the outbreaks? Similar questions can be asked with regard not only to infectious processes but also to other types of disorders that display seasonal differences in their clinical manifestations, for example, diabetes and vascular accidents.

In 1937 the very matter-of-fact Danish microbiologist and epidemiologist Th. Madsen found it worthwhile to devote an important series of lectures delivered in the United States to the subject "The Influence of Seasons on Infection." Much in the spirit of the Hippocratic teachings, Madsen emphasized that many diseases exhibit cycles and seasonal peaks which are as yet unexplained, yet are of great practical importance for the physician and the public health officer.

As we shall see in Chapter II.3, a renewal of interest in bioclimatology is taking place at the present time. Moreover, recent studies are beginning to enlarge considerably the appreciation of the role played by drinking water in disease causation. It is claimed, for example, that in Devon (England) the prevalence of cancer is related to water supplies. In the United States, England, and Japan, the incidence of cardiovascular diseases is much higher in cities or areas where the drinking water is soft than where it is hard. In certain regions where the drinking water is so soft that its calcium and magnesium content is practically nil, the incidence of cerebrovascular and coronary heart disease is said to be very high, even though the amount of fats in the diet is negligible and the blood cholesterol level of the inhabitants is unusually low.

Thus, interest in the effect of cosmic and telluric factors on human health has never died out completely, but this field of thought is not popular among either practicing physicians or investigators. The clinical observations and experimental studies bearing on the problem are so sporadic that they provide at most stimulating ideas, but so far little convincing knowledge useful in practice. Typical of the field are the claims that human activity, both physical and mental, is greatest in areas where seasonal changes are sufficiently pronounced and where the proper variations in temperature, barometric pressure, sunlight, and wind velocity provide a stimulating but not overpowering environment. People living under such favorable climatic conditions, the claim goes, have been responsible for the greatest amount of technological and intellectual progress in the world. One of the most articulate exponents of this thesis was E. Huntington, professor of geography at Yale University. By an amusing coincidence, his study of this fascinating but very complex problem led him to conclude that the region around New Haven, Connecticut, where Yale University is located, provides the most ideal climate for the development of civilization!

Students of medicine and public health have naturally focused their attention on those factors of the physical environment which have unequivocal effects on man. The fact that temperature, humidity, barometric pressure, and the gaseous composition of the air are properties which can be readily measured has encouraged and facilitated the analysis of their physiological role. The medical aspects of these well-defined physical agencies have been extensively studied and are discussed at length in several modern monographs; but curiously enough, they receive little attention or are altogether ignored in standard textbooks of medicine and even of public health.

Yet, it is certain that the various techniques developed to control the physical forces of the external environment have had effects far deeper than the mere increase of human comfort. On the one hand, they have certainly contributed to the general improvement in health, at least as much as have medical advances and probably much more. On the other hand, some of them may have had deleterious effects and are probably playing an important role in the emergence of the so-called diseases of civilization. Today, as in Hippocrates' time, man's nature and fate are profoundly influenced by the various forces of his physical environment.

2. Biological rhythms

a. Effects of cosmic forces

The evolutionary development of all living organisms, including man, took place under the influence of cosmic forces that have not changed appreciably for very long periods of time. As a result, most physiological processes are still geared to these forces; they exhibit cycles that have daily, seasonal, and other periodicities clearly linked to the periodicities of the cosmos. As far as can be judged at the present time, the major biological periodicities derive from the daily rotation of the earth on its axis, its annual rotation around the sun, and the monthly rotation of the moon around the earth.

While daily, lunar, seasonal, and other periodicities are a universal feature of biological phenomena, their precise mechanisms are still a matter of controversy. The most common assumption is that each organism possesses a number of "internal clocks," which, like alarm clocks, indicate when a given biological activity should begin. Biological clocks, whatever their nature, would thus be the anatomical and physiological mechanisms of evolutionary adaptations to the periodicities of the physical world.

This hypothesis, however, is not universally accepted. In many cases, it is claimed, the organism's time sense is simply a reflection of changes in the environment. Even when placed under so-called constant conditions, living organisms are reacting in fact to a great number of fluctuating factors of the geophysical environment, from which it is difficult to shield them completely, such as cosmic rays, background radiation, and the earth's magnetic and gravitational fields. These fluctuating forces can act as determinants of biological rhythms.

New information concerning the mechanisms responsible for biological rhythms will probably come from studies carried out during prolonged space flights. On such flights the organisms will escape most of the earth's gravitational fields and also the effects of the earth's 24-hour rotation. These artificial life conditions should provide useful evidence in the continuing controversy over the mechanisms that account for the fact that diurnal physiological responses persist in darkness. The purpose of the following presentation, however, is not to discuss the controversies over the mechanisms of biological rhythms but to consider only some special aspects of these phenomena which may eventually throw light on the causation of disease.

Of all biological rhythms, the ones with a daily periodicity have been the most extensively studied. It must be emphasized at the outset, however, that this periodicity is not exactly a day's length. Daily rhythms usually differ significantly from 24 hours, and for this reason they are designated by the adjective "circadian" (from *circa dies*). The expression circadian is preferable to diurnal because it does not imply daytime as opposed to nocturnal activity, and refers only to an approximate 24-hour periodicity.

Circadian rhythms occur even in very simple forms of life, a fact which has made possible certain kinds of observation not possible in higher animals or man. For example, the intensity of light emission by dikaryotic cultures of basidiomycete fungi exhibits a minimum between 6 A.M. and 9 A.M. and a maximum 12 hours later; this pattern is constant whether the cultures are grown in total darkness, under constant illumination, or exposed to a normal day-night cycle.

The phototactic response of the alga *Euglena gracilis* likewise exhibits a circadian rhythm which persists even when this alga is kept in the laboratory under such conditions that there is no difference between night and day in the intensity of the light to which it is exposed. If the *Euglena* colony starts migrating at 6 A.M. one day, it will start migrating at the same time on subsequent days, within approximately 10 minutes, irrespective of the illumination; then 12 hours later, the *Euglena* cells, ignoring the light, will scatter. Furthermore, the rhythm persists for long periods of time even though the cells divide approximately every 16 hours. The migration proceeds, therefore, as if it were under the control of an inherited time mechanism. The alga seems to possess a kind of biological clock that ticks at a fairly regular rate, the operations of which are based on oscillations taking place within its cells. (For a general review of biological rhythms in plants, see Sweeney, 1963.)

The fact that circadian rhythms occur in unicellular organisms demonstrates that they do not require the complexities of nervous or multicellular organization. Several other findings suggest that periodicity phenomena are indeed very deep-seated in the cellular structure. For example, periodicity is largely independent of the stage in morphological development of the organism; thus, the oxygen consumption of embryonated eggs exhibits a sharply defined maximum on successive days, irrespective of the stage in the differentiation process. Periodicity is also largely independent of external temperature; the period of the phototactic rhythm in *Euglena* remains very close to 24 hours when the temperature ranges between 16° and 33° C.

The circadian rhythm in photosynthesis continues undisturbed over a number of cycles in *Acetabularia* even after the nucleus of this unicellular alga has been removed by severing the basal rhizoids. It appears, therefore, that the nucleus is not directly essential for the maintenance of rhythmicity and that the mechanism responsible for timekeeping must reside in the cytoplasm.

One of the most extensively studied of circadian rhythms in animals is that exhibited by the fiddler crab. When this Atlantic coast creature is removed from its native beach and maintained under constant conditions in some remote laboratory, it darkens its skin at a time which would be sunrise in its original habitat and it continues to run around at periods corresponding to those of the low tide in this habitat. In other words, the skin color of the fiddler crab is governed by the daily solar rhythm and its motion by the lunar rhythm of the tide.

A large variety of circadian physiological rhythms have been described for animals and man during the past two decades. (For recent reviews see Cloudsley-Thompson, 1961; Halberg, 1959, 1960a and b, and 1963.) These rhythmic activities include such different properties as blood levels of eosinophils, corticosterone, sugar, and iron; urine excretion of chloride and urea; adrenaline output; epidermal mitotic activity; fibrinolytic activity of the plasma; intraocular pressure; and deep body temperature and limb blood flow. Of special interest is the fact that the circadian rhythm persists even in regenerating tissues. When the liver was removed from mice, and the animals were sacrificed at intervals of 4 hours after the operation, a clear 24-hour rhythm was recognized by measuring the frequency of cell divisions and other characteristics of the regenerating liver tissue. A representative list of other recent studies of physiological circadian rhythms is presented in the bibliography for this chapter.

Biological rhythms have features differing profoundly from those of metabolically dependent physiological functions. On the one hand, as already mentioned, they are essentially temperature-independent. Furthermore, they are also insensitive to a variety of metabolic inhibitors. Actinomycin D is one of the few substances that seem to exert a selective inhibition of cellular rhythmicity.

In the marine dinoflagellate *Gonyaulax polyedra,* which has a maximum of bioluminescence late at night and of photosynthetic capacity during the day, rhythmicity is inhibited by this drug without complete interruption of metabolism and growth.

Studies in cockroaches have revealed that several independent circadian mechanisms can coexist in the same animal. Cockroaches secrete a hormone, according to a 24-hour cycle, which increases the activity of the animal. The production of this hormone can be stopped by chilling the secretory cells while keeping the rest of the body at normal temperature. This operation reveals the existence of a second rhythmic mechanism associated with the nervous system. Other observations to be discussed later strongly suggest that, as in the cockroach, there exist in the bodies of most living organisms several quasi-autonomous oscillatory systems. The facts presented later in this chapter indicate that such independent systems can cause difficult health problems during the adaptation of man to new environments.

Rhythms with periodicities longer than 24 hours have been studied chiefly in plants and animals. For example, monthly and yearly cycles seem to control the growth of potato plants maintained in a sealed chamber under constant temperature, air pressure, humidity, and light conditions. Similarly many plant seeds have an annual rhythm in their capacity to germinate; dried seeds exhibit an accurate annual rhythm whether stored at —8° or 130° F.

Although the essential characteristics of the internal environment in animals and man must remain within certain limits to be compatible with the maintenance of life, nevertheless, some of the most important biochemical activities of their tissues exhibit marked seasonal variations. A striking example of such seasonal periodicity is provided by the levels of ketonuria in rats kept without food for 48 hours. When measured from May to October in the United States, the amount of acetone bodies excreted by these animals as a result of fasting was three times greater than it was during the winter months (see Table 1, left column). Excretion of acetone bodies remained at a low level during the winter even when the rats were placed in a room maintained at summer temperature. It is clear, therefore, that factors other than heat were responsible for the higher level of fasting ketosis observed during the summer months.

These findings have been confirmed and extended in England,

TABLE 1. Seasonal Ketonuria in Fasting Rats *

Mg. acetone bodies per diem			
Seasonal Average per 100 gm. wt.		*Monthly average per rat*	
Spring-summer	6.2	April	32
		May	67
		June	51
		July	36
		September	12
		October	5
Fall-winter	1.9	November	7
		December	5
		January	12
		February	7
		March	7

* Condensed from G. and C. Cori, 1927 (seasonal averages), and from Burn and Ling, 1928 (monthly averages).

where it was found that the difference in fasting ketonuria between the spring-summer season and the fall-winter season was even greater than that observed in the United States. With the strain of rats used in England, the difference was of the order of tenfold (Table 1, right column). Furthermore, the amount of glycogen in the liver of rats after 24 hours' fasting also varied according to a seasonal pattern, being much higher in the winter than in the summer.

It is probable that the seasonal variations in metabolic activity have an evolutionary basis and derive from the fact that animals had to develop mechanisms which enabled them to withstand successfully long periods of starvation in order to survive during the winter. According to this view, energy requirements would be met by combustion of the fat stores during periods of scarcity, whereas this metabolic mechanism would be less likely to be called into play during the summer.

The high level of summer ketosis in rats is associated with a low capacity of the tissues to oxidize glucose and is probably due to a reduced functional activity of the pancreas. It appears, in other words, that some at least of the seasonal metabolic patterns have their basis in variations of hormonal activity. The seasonal changes in thyroid and adrenal activity are examples of this class of phenomena. In deermice, for example, radioiodine uptake by the thyroid gland was found to be high during July and December and low during March and October. (See Figure 1.)

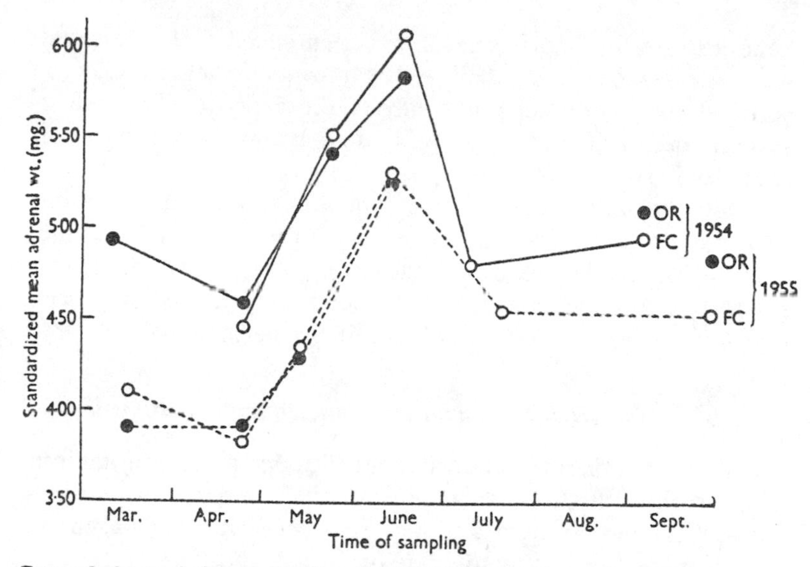

FIGURE 1. Seasonal changes in adrenal weight. Two populations of *Microtus agrestis* (solid and hollow circles respectively) were studied in successive years. Adrenal weights were highest in summer. (Reproduced from H. Chitty, 1961.)

Seasonal variations have many other effects on physiological processes, but only a few can be mentioned here. Thus, the irritability of the cardiac nerves, at least in the frog, changes with the time of the year. The blood pressure of man in temperate countries exhibits a peak in spring and a trough in the late summer. Urinary excretion of electrolytes, of nitrogenous compounds, and of 17 hydroxycorticosteroids exhibit summer-fall variations. Deep body temperature also seems to exhibit seasonal fluctuations. Hourly recordings of oral temperature were made in one subject living in a subantarctic environment from January to November on the same two days of the month. In addition to the expected diurnal variations, there occurred a rise in oral temperature in spring and summer, and a fall in autumn and winter.

Experimental evidence of a lunar periodicity in the common flatworm (planarian) has been derived from the finding that the period of the moon determines at a given time of the day the direction in which the animal will turn. Started south in the late morning, for example, the worm will move straight ahead on about the four days before the new moon and four or five days after the new moon, then will deviate predictably to right or to left at other times during the lunar cycle. It has been stated that this orderly rhythmic change can be abolished by a magnetic field as weak as 5 gauss! Lunar periodicity also affects the reproduction cycle of several species of *Rattus,* as well as of lemurs and lorisoids in captivity in New Haven.

While the influence of the moon on human physiology and behavior has been part of folklore since the most ancient times, it is still poorly documented by scientific observations. Only during recent years has statistical evidence been obtained suggesting that human reproduction exhibits a lunar periodicity.

b. Periodicities in responses to environmental stimuli

Animals and men maintained in a controlled and stable environment exhibit rhythmic variations in their response to stresses. Thus, resistance to any given toxic substance, infectious agent, or trauma is conditioned by the time of the day and the season of the year. In mice, for example, the response to substances as different as bacterial endotoxins or psychotherapeutic drugs exhibits marked circadian rhythms. Likewise, the hour of the day at which certain drugs are administered to man affects the results of the therapy, depending upon the time phase of the endogenous rhythm (reviewed in Tromp, 1964).

Of special relevance to human health is the fact that the marked seasonal variations in the incidence of certain pathological states may depend in part on the seasonal periodicity of physiological processes. The influence of seasons on the frequency and severity of infectious, circulatory, and metabolic disorders has long been recognized. However, there has been a tendency to dismiss the phenomenon as of little interest because, as mentioned earlier, many well-established facts seem at first sight to account adequately for seasonal variations in disease patterns.

There is no doubt, of course, that crowding, physical activity, availability of certain types of food, prevalence of parasites and their vectors, etc. are profoundly conditioned by the physical environment and play a role in disease. In addition to these obvious determinants, however, is the fact discussed above that the internal environment of man, as well as of animals, varies with the seasons. It would be surprising if these seasonal changes in metabolic and physiological activities did not affect resistance to various forms of stress. Even spring fever may be more than a mood inspired by higher temperature and longer hours of sunlight, and be rather the expression of a change in physiological state associated with the season. It has been claimed that with the advent of spring, there occurs a seasonal dilatation of the vascular bed, which demands an increase in blood volume and thereby results in hemodilution. The high tides of the oceans may have their counterpart in the high tides of our blood stream.

The response of rats to vasopressors exhibits a seasonal variation, the minimum effective does being lower in spring and summer than during the other seasons. Even the composition and activities of the intestinal flora seem to vary with the season. Marked seasonal differences have been observed also in the susceptibility of animals to various bacterial toxins and to infection with pneumococci or streptococci.

Needless to say, it is extremely difficult to establish convincingly the relation between internal clocks and pathological states in man. Nevertheless, it has been postulated on the basis of physiological and clinical evidence that three different types of clocks are of particular importance in medicine and psychiatry: peripheral mechanisms located, for instance, in the blood-forming tissues in the bone marrow, etc.; central mechanisms located in the brain, particularly in the hypothalamus and reticular formation; and homeostatic mechanisms involving target organs, endocrine glands, the pituitary, and the hypothalamus. Better knowledge of the clocks that govern the human body might help in the understanding of the so-called periodic diseases, involving periods of days rather than hours.

It seems worthwhile to single out for more detailed discussion some aspects of biological cycles concerned with the effects of light. The reason for this emphasis is that prolonged exposure to light constitutes a feature of modern life that has hardly been studied, yet might be of importance in certain pathological states.

In all animal species, gonadal activity is increased by light rays reaching the retina; this effect is mediated through the anterior lobe of the pituitary. As is well known, egg production by hens can be maintained throughout the year by nocturnal illumination of the hen house. Bird migration has in part at least a similar physiological basis, since it is under the influence of mechanisms controlled by light. Finch fanciers have long made use of the effect exerted by retinal stimulation on breeding; they expose the birds daily to artificial light, thus causing them to breed twice a year, in midwinter as well as in the spring.

Light increases sexual activities also in mammals. Ewes that have lambed in the northern hemisphere will again mate the same year if they are transferred to the southern hemisphere; the same is true of does. Fertile mating will also occur in ferrets during the winter if they are exposed daily to artificial illumination; in contrast, the onset of estrus activity is retarded or prevented in ferrets protected from light by a hood placed over their eyes. Similar observations have been made on rats, and as we shall see, the intensity and duration of exposure to light may have pathological consequences in animals and man.

As is the case for other biological cycles, interference with the natural cycles of light exposure can result in physiological disturbances. Severe pathological damage can be caused in plants by interfering with their 24-hour cycle through manipulation of light. In animals, the pathological potentialities of abnormal exposure to light have been observed in an experimental flock of chickens maintained throughout its existence under constant illumination. The birds proved to be so susceptible to disease that the flock became extinct in three years.

Mention must be made here of spectacular findings in the course of transplantation experiments with tissues taken from cockroaches. When the insects used in these experiments were out of phase with each other in their rhythms with regard to light, transplantation resulted in neoplastic growths which were transplantable. These did not occur when the two tissues were juxtaposed while the insects were in phase. Because of the potential significance of these findings, it seems worthwhile to present here some of the experimental details.

The hormone that appears to be involved in this effect is secreted by four specialized cells normally situated in a large nerve ganglion just under the esophagus. These cells can maintain their secretory activity when they are transplanted into the blood stream of another cockroach. A number of cockroaches were kept in light during the night and in darkness during the day until their rhythms had come to run 12 hours out of phase with those of animals living under normal conditions of light exposure. The secretory cells from the reversed day animals were then implanted into normal cockroaches. All the normal animals developed malignant tumors, most of them in the intestine. In the control experiment, in which both the donor and the receptor animals were in phase, no cancer was found. The obvious interest of these findings is that they point to the probability that pathological consequences can ensue from disturbances in rhythmicity.

Light plays a role in the hormonal activities of the human species just as it does in plants and animals. For example, Eskimo women are said neither to menstruate nor to conceive during the long polar night; and similar statements have been made concerning Patagonian women of the Antarctic.

Until the last century, man lived in the dark for long hours during the winter months, and this is still true in many primitive societies. Modern man, in contrast, is exposed to bright light for 16 hours a day throughout the year. In view of the fact that light rays can affect several hormonal activities, and that many, if not most, physiological functions are linked to circadian and seasonal cycles, it seems possible that this change in the ways of life will have long-range consequences for the human species.

c. Disturbances of biological rhythms

Like most young animals, the human infant is comparatively nonrhythmic during very early life, a fact of which parents are painfully aware. As the child becomes adapted to a normal rhythmic existence, the hormonal activities, the body temperature, the urine secretion, and other physiological processes increasingly take on the circadian character, and other rhythms also become established.

The progressive establishment and conditioning of the rhythms as a result of social experiences points to one of the greatest technical difficulties in the analysis of their determinants, namely the absolute necessity of excluding outside stimuli of social origin that can modify the effect of the cosmic and other environmental factors under study. For example, familiar noises, any trace of visible light, repeated daily routines, even though carried on outside the room in which the subject under experimentation is located, can act as stimuli and exert a marked influence on the system functions and on the various periodicities in a person thought to be placed under obsolutely isolated conditions. Despite these experimental difficulties, it has been possible to establish beyond any reasonable doubt that certain innate rhythms persist in human beings kept strictly isolated under conditions designed to exclude as completely as possible any manifestation of rhythm from the outside.

Biological rhythms are of great potential importance in medicine because they are not immutably set. For instance, inverting the light schedule of mice for two weeks produces shifts in daily rhythms with regard to blood eosinophils, mitoses in pinnal epidermis of liver, hepatic nucleic acid metabolism, and blood levels of corticosterone (reviewed in Halberg, 1959, 1960a and b, 1963). (See Figure 2.) In man also biological rhythms can be readily altered by changes in the ways of life and in the physical environment. The medical importance of these alterations resides in the fact that they cause profound physiological disturbances, as will be illustrated by a few examples dealing with phenomena of direct relevance to modern life.

The urinary excretion of 17 hydroxycorticosteroids is one of the physiological processes that normally exhibits a well-defined and fairly stable daily rhythm. When measurements of these adrenal hormones were made at frequent intervals during a 30-hour shift by air travel (propeller plane) from continental United States (Central Standard Time) to Japan and Korea, they revealed that the urinary excretion remained synchronized with Central Stand-

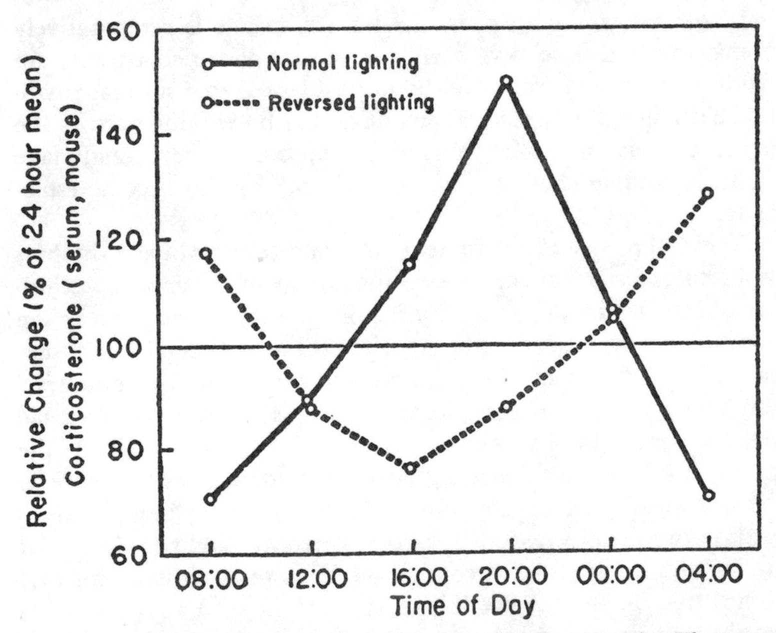

FIGURE 2. Reversal of circadian rhythm of serum corticosterone in mice. The curves present values for schedule of light from 6 P.M. to 6 A.M., alternating with 12 hours of darkness, compared with schedule of light from 6 A.M. to 6 P.M. (Reproduced from F. Halberg, 1958.)

ard Time even after arrival in Asia. Progressively, however, the timing of excretion changed; it eventually became synchronized with Asian time after nine days. The rhythm of execretion of sodium and potassium exhibited a similar pattern of change.

In another study, the diurnal temperature rhythm in man was studied during and after an airplane flight from Ontario to England. In this case, it took three to four days for the temperature rhythm to change from the Canadian timing and fall in step with European time. Needless to say, the physiological disturbances caused by rapid changes of latitude have become much more pronounced since the advent of jet airplane travel.

TABLE 2. Time Displacement of Potassium Urinary Excretion *

		Meq/3 hr. at indicated time [a]		
Place	*Date*	*0–3*	*9–12*	*12–24*
U.S.A.	8/1	3.1	15.8	5.7
Japan	8/5	21.8	6.5	11.4
Korea	10/14	4.5	12.7	4.6

* Data from E. B. Flink and R. P. Doe, 1959.
a. Time is recorded as local time.

Sleepiness and wakefulness are but two of the other characteristics that exhibit marked circadian fluctuations in man which can be altered by environmental changes. Biological rhythms often persist for a long period of time in night workers after the shift from day to night work; eventually they change and come to phase with the new conditions of life. However, the rhythms do not necessarily all come in phase at the same rate, and they consequently become dissociated one from the other. This dissociation constitutes further evidence of the existence in the body of several independent internal clocks.

A striking example of dissociation of rhythms was observed in a group of human subjects who spent seven summer weeks in Spitzbergen. The constant daylight and the uniform temperature made it easy to impose on them very abnormal routines of daily life and to observe the effect on different body functions. Of the several physiological rhythms that were investigated, those for sleepiness and body temperature seemed to respond the most rapidly to the change in environment, while the one controlling potassium excretion remained out of phase the longest time. Dissociation between the rhythms for body temperature and for pulse rate was observed in another group of persons also exposed to the long days of summer months in Norway (reviewed in Cloudsley Thompson, 1959).

The length of time required for adjustment of a given physiological function to a new environment naturally varies from person to person, but the dissociation of one rhythm from the other is always one of the most common and most striking features of the change. In one particular study, it took 6–8 days before the ketosteroid urinary output became adjusted to a 5-hour time shift, whereas the temperature rhythm came in phase in less than 4 days.

Little, if anything, is known of the mechanisms involved in these changes of rhythms, but there is some evidence that physical exertion, and perhaps high temperature, help in accelerating the adjustment to a new routine. In practice, however, it is very difficult to reset at will the clocks that govern physiological periodicities, as shown by the experience of those who travel great distances in modern aircraft. Moreover, some persons come into phase with new environmental conditions more readily and more rapidly than others; the slow adapters are likely to experience much stress when placed in an environment out of phase with their usual cycles. In fact, there may well be limits beyond which natural rhythms are not amenable to frequency-synchronization with new environmental periodicities.

From a practical point of view, it might be profitable for a person who plans to go to a new geographic location to attend a conference or to participate in a sporting event to take steps designed to minimize the physiological disturbances caused by the change. For example, adjustment might be facilitated by adopting a sleep-wakefulness cycle appropriate to the new location for several days before the trip; and it is possible that certain drugs might help in accelerating the physiological changes involved.

The facts outlined in the preceding pages make it clear that the responses of plants, animals, and man to cosmic forces present two different aspects. On the one hand, biological rhythms are the expression of some physiological characteristics inherent in the organism; one the other hand, they can be altered by modifying the environment. Rhythms are neither simply impressed from without nor simply persisting from within.

The interplay between the intrinsic biological nature of man and the environmental forces that impinge on him focuses attention on physiological problems which may have pathological implications. Modern man is living under conditions that differ more and more from those under which he evolved. He heats and cools his dwellings in an attempt to obliterate as completely as possible the environmental changes associated with the seasons; he carries his day into the night or vice versa; he moves by jet aircraft in a few hours from one latitude or longitude to another. These technological innovations certainly have profound effects on human physiology, and it would not be surprising if some of their effects played a part in the diseases characteristic of modern life.

Clearly, any discussion of the relation between disturbances of biological rhythms and human health is at present highly speculative. But the information available on this subject is sufficient to show that living things, including man, respond not only to heat, humidity, light, and other obvious climatic components that are readily perceived by the senses, but also to many other environmental factors not readily identified, and in part still unknown. Awareness of these complexities may not be helpful as yet in solving practical problems, but it teaches humility to the overconfident biologist.

It might be appropriate to close this discussion with Hamlet's words: "There are more things in heaven and earth, Horatio, than are dreamt of in your philosophy." However, the wise comments of one of my colleagues who read these pages should be quoted here:

> A lack of correlation with known environmental factors does not prove that unknown factors are operating. Primitive man explains natural phenomena with magic; when we are confronted with the inexplicable we too often fall back on mysticism, in somewhat more sophisticated language. This is a matter of personal taste, and on this point, I must confess to belong rather more with Horatio than Hamlet. Horatio no doubt was limited in his philosophy, but Hamlet believed in ghosts.

To which I can only answer that emphasizing the existence of unknowns or mysteries is not the same thing as believing in ghosts. To invoke ghosts constitutes a claim of explanation, whereas what I have tried to do in the preceding pages is to call attention to phenomena that are certainly of importance to human welfare, yet are grossly neglected by medical science.

3. Bioclimatology

a. Effects of some physical forces on living things

Acclimatization, acclimation, adaptation, and habituation are among the many terms commonly used to denote the various kinds of adjustments that man and animals make to changes of climate. The word acclimatization probably applies most accurately to nongenetic, rapid adjustments, and acclimation to somewhat more gradual processes, also nongenetic. In contrast, there is a tendency to limit the meaning of adaptation to the evolutionary processes with a Darwinian genetic mechanism. Habituation most commonly refers to processes involving the mind and resulting in a weakening of the normal responses or sensations; for example, the state of habituation can be abolished by chlorpromazine under the proper circumstances. In practice, however, acclimatization, acclimation, adaptation, and habituation are often used interchangeably because the processes these words are supposed to denote usually overlap and because the fundamental mechanisms involved are poorly understood.

The physiological responses that man makes to climatic stresses strongly suggest that his early evolution took place in a semitropical environment. In the absence of clothing, housing, and other means of protection, a temperature of 29.4° C (85° F), with moderate humidity and low air movement, seems most satisfactory for his comfort and least likely to cause physiological disturbances. Human beings make physiological adjustments much more readily to semitropical than to arctic conditions.

Exposure to heat mobilizes several independent physiological mechanisms. For rapid adjustment, enormous changes in blood flow through the capillaries can occur within a few minutes and regulate heat loss upward or downward as needed. When increased blood flow proves inadequate for rapid cooling, the sweating mechanisms comes into play and provides heat loss by evaporation; increase in elaboration of aldosterone affects the responsiveness of sweat glands to sodium retention. For more prolonged temperature control, as required by change of season or by adaptation to life in a warmer country, heat production can be regulated by lowering the metabolic rate. It has long been known that the size of the thyroid and of the adrenals is usually greater during the winter than during the summer in laboratory animals, and also that it can be altered almost at will by changing the environmental temperature.

Vigorous physical exertion in humid heat causes an increase in pulse rate, in skin and body temperature, and in sweat production. As acclimation proceeds sweat production may increase still further but its sodium content decreases. Further changes can occur in the cardiovascular system, in body temperature, in salt and water metabolism, and persist for two weeks or longer. As a result, the bodily functions can return to a normal level even under continued heat exposure; there is then less likelihood of heat cramp because of the lower salt concentration of the sweat.

In general, adjustment to life in a very cold climate depends chiefly if not exclusively on proper clothing and housing, as well as on cultural skills, rather than on physiological processes. The Eskimos among whom V. Stefansson lived maintained a tropical environment in their ice houses. This is also true of the Alakaluf, a dwindling tribe of Fuegian Indians, even though they possess an unusual capacity to resist cold temperatures. Recent studies indicate, however, that a small part at least of adjustment to cold has a truly physiological basis.

Man exposed to falling ambient temperatures meets the emergency first by an increase in body insulation through peripheral vasoconstriction, then by an increase in metabolic rate. Other physiological adjustments may also come into play. Thus, animals and men experimentally adjusted to cold exhibit an increase of heat production in the muscles without any electromyographic evidence of shivering (reviewed in Von Muralt, 1957, 1960). Furthermore, cooling of the hypothalamus in animals produces within a few seconds an increase of metabolism in the rest of the body. Finally, liver tissue from animals acclimatized to cold has an increased ability to oxidize fatty acids. Taken together, these facts suggest that there exists some form of adaptive response to cold which is not related to shivering and which is probably controlled by the temperature of the thalamus. Similar conclusions have been derived from studies of the Australian aborigines.

Any attempt at predicting the response to temperature and humidity must take the wind into account. Individual variation is another complicating factor. For example, it is said that the body temperature of vagotonics falls in warm dry air and rises in damp air whether warm or cool; whereas sympatheticotonics are said to experience the greatest body temperature rise in warm air and the greatest fall in cold air.

The various races and geographical groups of men differ markedly in their response to climatic conditions. Southerners commonly show greater ability than Northerners to work in heat, with less rise in metabolic rate, less expenditure of energy, and less profuse sweating; moreover, they tend to retain any acquired increased tolerance somewhat longer. It is a fact of common experience that Northerners who receive their army training in the deep South are at first more handicapped by the heat than are the soldiers of Southern origin. In contrast, Northerners are as a rule more resistant to cold injury than white Southerners, and still more than Negroes, as was revealed by a study of the incidence of frostbite among the American soldiers during the Korean War.

Striking as these differences are, it is probable nevertheless that they are not so much the expression of genetic differences as of acclimation brought about by life in a hot or cold climate. In fact, as already mentioned, genetic adaptation to cold is probably of small importance in man. By learning to build fires and to wear clothing, man became able long ago to survive in a cold environment without having to undergo profound evolutionary changes.

Contrary to common belief, it has not yet been convincingly shown that pressure per se or even changes of pressure can affect either the comfort or the health of man, except of course in the extremes of great depths or high altitudes. The problem remains open, however, in view of claims, to be reported later, that some of the physiological effects produced by the föhn wind are due to pressure oscillations. In insects, very slight reductions in pressure exert profound behavioral effects that appear to be independent of oxygen tension. Both in the laboratory and in the field the feeding habits of higher insects, as well as their rate of development and locomotor activity, are appreciably increased by slightly lowered or falling pressures. These conditions also seem to be associated with the sudden mass emergence of insects.

It is now generally accepted that animals not only perceive magnetic fields but can even distinguish their direction. Recent reports indicate furthermore that tomatoes placed between magnetic poles ripen faster than the controls and, in addition, the fruits nearest the magnetic south ripen faster than those nearest the magnetic north. If it is true, as has been suggested, that the cytochrome oxidase system can be influenced by a magnetic field, then the range of responses of living organisms to the physical environment is much broader than commonly assumed.

A few isolated observations made with microorganisms may be worth mentioning at this time to illustrate the wide variety of biological effects exerted by climatological factors. The metabolism of bacteria and yeasts seems to be attenuated during cyclonic periods and intensified during anticyclones. By recording automatically such activities as luminescence, motility, acid production, sporulation, it was found that the changes occurred within less than one hour of the weather change—in other words, so rapidly that they could not be correlated with the usual daily weather curves. In Switzerland, cultures of the spirochete *Borrelia duttoni* were found to behave abnormally during periods of föhn; they survived poorly in vitro under these weather conditions, yet multiplied rapidly in vivo when injected into mice.

The small ionized molecules of the air and the so-called "sferics" originating from natural electric discharges are other atmospheric factors that seem to have important biological effects; they are said to be deleterious in the case of positive space charges and beneficial in the case of the negative ones. Negative air ions enhance proliferation of tissue cells exposed to them in vitro. They also accelerate the tracheal ciliary activity of mammals both in vitro and in vivo, whereas positive air ions have an inhibitory effect on ciliary rate.

Very broad claims have been made for the benefits conferred on human beings by treatment with negative ion generators. These claims include: rise of blood pH and of carbon dioxide combining power, stimulation of the adrenals and thyroid, lowering of blood pressure, and promotion of growth and energy! Mountain air has been said to owe its stimulating effect to the fact that it contains more negative ions than are found at sea level. Confirmation of any of these claims would point to the need for a thorough investigation of a neglected aspect of biophysics, namely the biological effects of space charges of small ions.

b. Climate and disease

Needless to say, the regulatory mechanisms that bring about physiological adjustments of man to climatic changes are effective only within a limited range; any excessive demand on them will cause profound physiological disturbances. Even in acclimated persons, tropical fatigue and exhaustion will supervene if heat stress is long continued; and body temperature will fall to a dangerous level if cold exposure is too prolonged. A prisoner at Dachau survived with a rectal temperature of 25° C, but this is a most unusual case; men usually die of circulatory and respiratory failure under such extreme conditions.

Cold, heat, and humidity naturally influence health and disease in many different ways even under conditions far less extreme than those just mentioned. For example, patients at rest in the warm and humid atmosphere of a New Orleans summer had a cardiac output 57 per cent greater than when resting in a cool and air-conditioned ward. Respiratory and gastrointestinal diseases, rheumatic fever, skin infections, multiple sclerosis, diabetes, and vascular accidents are among the many conditions that have been claimed to be related to climatic factors, but the evidence is still *sub judice* in most cases.

Stormy weather, either warm or cold, is widely assumed to have pathogenic effects, but here again documentation is poor. Lung embolism, thrombophlebitis, and hemorrhage are said to occur predominantly in warm damp weather; whereas it is stated that migraine, colic, stroke, and epileptic seizure are more commonly associated with cool, damp weather. Twinges of gout and sciatica are popularly supposed to forecast rains. Most familiar is the conviction expressed in folklore that pains from scars and from arthritis sharpen during weather of the frontal type. Few are the persons indeed who do not believe that

> A coming storm our shooting corns presage,
> Our aches will throb, our hollow tooth will rage.

It is certain, on the other hand, that quite different types of weather can have similar effects on pathological conditions—witness the conflicting statements that myocardial infarction is most common when the weather is cold, hot, damp, or simply "bad." The truth is more likely to be that cardiac emergencies can be touched off by any continued climatic stress. High temperature is perhaps more dangerous for patients with cardiovascular disease because it augments blood volume through vasodilation, and thus puts additional demands on the heart muscle for increased output.

Although knowledge concerning the effects of bioclimatological factors on pathological processes is so uncertain, there are indications that objective tests can be developed in this field. Thus, quantitative observations show that warm fronts are associated with a decrease, and cold fronts with an increase, in capillary resistance; that physiological activities change when men move to high mountains; that nutritional requirements are affected by climatic stress (reviewed in Von Muralt, 1957, 1960).

In addition to their direct effects on health, climatic factors can also have indirect effects through their influence on susceptibility to infectious processes. Pasteur's celebrated experiments with anthrax in chickens, later confirmed on other animals, have demonstrated that lowering or raising the body temperature by a very few degrees can determine whether infection will become overwhelming or completely abort. In certain cases, the temperature effect is exerted directly on the infective agent itself, as in the case of those strains of mycobacteria which produce destructive ulcers only on the exposed parts of the body because of their low temperature optimum. Temperature can also have more subtle effects on virulence. For example, certain enterobacteria synthesize at 30° C a capsular substance that protects them against the bactericidal power of serum; but this protective material is not produced at fever temperatures.

Some remarkable temperature effects have been recently observed in viral infections. Thus, lowering the temperature by 2° C can bring about the selection in vivo and in tissue cultures of avirulent mutants of polio viruses. In rabbits infected with a certain strain of myxomatosis virus, simulated summer temperatures allowed 70 per cent of the animals to survive, whereas only 8 per cent recovered among those held at simulated winter temperatures.

It is probable that in many cases temperature influences the course of the infectious process indirectly by modifying the tissue response to the infective agent. This type of mechanism probably accounts for the finding that mice exposed to extreme cold temperatures developed a staphylococcal infection apparently of endogenous origin. One can assume that subtle physiological effects of this nature play an important role especially in the infectious processes caused by microorganisms that are ubiquitous, or normally present in the body. The greater frequency of upper respiratory infections in certain kinds of inclement weather probably has its origin in physiological changes caused in the host by the meteorologic environment.

Mention must be made of the pathological effects associated with special kinds of wind made up of descending subtropical air, for example the föhn of Switzerland and Southern Germany, the sirocco of Southern Europe, and perhaps the chinook of the Rocky Mountains. These winds become warm and dry on descending the leeward slopes of the mountains.

There are many reports that periods of föhn wind are associated with increases in the death rate, in circulatory as well as mental disorders, in suicides, and even in automobile accidents. These pathological effects seem to occur without any detectable change in any of the known geophysical surface elements; the patients may not even be aware of any bad weather in the usual sense. In fact, the complex of physiological reactions constituting the clinical entity known as "föhn" disease may subside when the wind actually supervenes, a finding which suggests that they are caused by physical factors which are still unidentified. In this regard, it may be of importance that föhn weather is preceded by a characteristic pressure oscillation with an amplitude of 5.3 mb. in periods of 3–15 minutes. Persons with a history of föhn disease are said to experience discomfort when exposed to oscillations of 0.1 to 0.2 mb. per 0.5 second to 2 minutes. It has been said also that the föhn brings down large amounts of ozone from the upper atmosphere, a peculiarity that might account for some of its physiological effects.

The claims that have long been made concerning the relation between the föhn and certain pathological conditions are typical of the state of knowledge in bioclimatology. Many facts have been observed, some have been fairly well documented, few are clearly defined, and in hardly any case is there adequate understanding of the mechanisms involved. The problem is real, it is certainly important, but investigation of it is at best episodic. Only one generalization seems fully justified at present.

Evidence derived from epidemiological, clinical, and experimental observation strongly suggests that sudden atmospheric changes, rather than any particular climatic factor per se, are the events most likely to be associated with exacerbation of pathological states. As Hippocrates wrote 2,500 years ago: "It is changes that are chiefly responsible for diseases, especially the greatest changes, the violent alterations both in the seasons and in other things. But seasons which come on gradually are the safest, as are gradual changes of regimen and temperature."

III. Man's Food

1. Nutritional habits and requirements

Gorillas in the wild are primarily herbivorous, but in captivity they readily take to meat and may eventually prefer it to their natural fare. It has been claimed that this change in dietary regimen is associated with a marked change in their intestinal microbiota. As we shall see, similar but better documented observations have been made with other animals (Chapter V.1). Man, like the gorilla, is also highly adaptable with regard to nutrition. His dentition and the anatomy of his intestinal tract indicate that he was omnivorous during a large part of his evolutionary development. Originally he probably was chiefly herbivorous, but he seems to have been primarily carnivorous during Paleolithic times, before the emergence of Neolithic agriculture.

Human groups with predominantly herbivorous or carnivorous nutrition still exist throughout the world today. In fact, it is certain that customs and taboos are as important as physiological needs in determining the dietary habits of people. In East Africa, for example, the diet of the Masai tribe consists chiefly of raw milk and blood, whereas the Akikuyu, who live near them in Tanganyika, consume almost exclusively cereals, roots, and fruits. The contrast is just as great between the diet of the meat- and fat-eating Eskimos among whom V. Stefansson lived and that of orthodox Hindus and other strict vegetarians. According to Stefansson, the Eskimos he observed resented having to eat berries and vegetables; they did so only in times of dire necessity, before eating their dogs, the stage that immediately preceded cannibalism. The nutritional prejudice against beef among Hindus and against pork among Arabs has its counterpart in the unwillingness of English-speaking nations to eat dog or horsemeat. Oysters were once the diet of the poor in England but became fashionable among the wealthy after they had become less abundant.

Primitive people who have lived for a long time in a certain region, and who have had only limited contacts with the rest of the world, may experience periods of food shortage and even of famine, but they rarely suffer from specific nutritional deficiencies. Through empirical experience, most of them have learned to take advantage of a great variety of the natural food resources in their environment, and thus manage to achieve a fairly complete and balanced diet even under conditions of scarcity. When livestock or wild game is available, they consume not only the muscles but also the internal organs, which are rich in certain growth factors, especially vitamin B_{12}. Furthermore, they eat many kinds of wild plants and small animals. Certain insects are regarded as great delicacies by people living under primitive conditions.

Tubers of all sorts, grubs, ant eggs, bird nests, mushrooms, berries, and teas brewed from the buds of evergreen and other plants have supplied diets adequate in proteins and vitamins to countless generations of people who had acquired empirically a nutritional wisdom similar to that of wild animals. The diet of primitive people, quantitatively limited and monotonous as it may be, thus includes many items not recorded in textbooks of nutrition. Its very complexity minimizes the likelihood that it will result in specific deficiencies.

In many areas, empirical but useful culinary practices have been developed to improve the nutritive value of certain products. Despite the scarcity of meat and milk in most of China, each local community had worked out a diet of adequate nutritional value, with the proper amino acid composition, through the skillful mixture of plant products readily available in that particular region. The extensive use of sprouts—the germinated beans, peas, and lentils of China and India—also has nutritional merit, since essential amino acids, fats, and vitamins become more available in the sprouts as a result of germination. Extensive nutritional experiments have revealed, for example, that the nutritive value of soybeans is significantly greater after germination than before, as indicated by nitrogen and sulfur retention and by their effect on the growth of experimental animals. This improvement in nutritive value occurs despite the fact that the amino acid composition and trypsin inhibitor activity do not change appreciably when the beans are converted into sprouts during germination. Mexican Indians cook corn with chalk and thus supplement their calcium intake, which would otherwise be insufficient. Pulqué, the fermented juice of the agave plant, is not only a pleasant alcoholic beverage for the Mexican peon; it is also a source of the various vitamins contained in the suspended particulate matter that accounts for its milky appearance.

The Otomi Indians provide one of the most striking examples of nutritional adjustment to extremely difficult situations. Although the Mesquital Valley of Mexico, in which they live, is a semidesert with many months of drought yearly, they exhibit little, if any, clinical evidence of malnourishment. Their consumption of meat, dairy produce, fruit, and conventional vegetables is extremely low. They have few of the foods usually considered essential for a balanced diet; but they eat purslane, cactus fruit, and pigweed; they use sow thistle for green stuffs and drink pulqué along with their tortillas. When tested a decade ago, their diet was found to supply a better nutritional coverage than that eaten by many city dwellers in the United States examined at the same time.

Most primitive people repeatedly and rapidly change from one type of diet to another, depending upon the circumstances. Their supply of food may be severely limited in quantity and extremely monotonous during the late part of the dormant season, but varied and abundant after harvest time. The Andaman Islanders will live for a time on pork, next on fish, then be restricted to honey and wild fruit. Central Africans are likely to have a quantity of fresh vegetables available at one season, pass to a diet of mushrooms and fruit at another, then be limited to porridge, dried peas, and beans. Few are the primitive people who limit themselves to a vegetarian diet out of choice. When game becomes available, those who had been on a plant diet out of necessity are likely to consume enormous amounts of meat at a sitting.

Anthropologists who have lived among primitive people report that sudden changes in diet often result in severe gastrointestinal disturbances, which the natives usually accept as a matter of course. Such obvious effects, however, are probably less important than are others of a more subtle nature associated with the physiological adaptations required by profound changes in nutritional regimens under primitive living conditions (Chapter III.4). Needless to say, sudden nutritional changes rarely occur now in the Western world since a large and varied supply of food is generally available to all throughout the year as a result of advanced farming practices and transportation facilities.

Despite the wide range in human eating habits, the nutritional requirements of man are essentially the same everywhere. Of all the medical sciences, none has reached a higher level of development than that dealing with the composition of foodstuffs and the nutritional requirements for rapid growth. The essential nutrients, some fifty in number, have been identified chemically and most of those of an organic nature have been synthesized. Nutritional requirements have been measured for all ages of life; the pathological changes likely to result from quantitative or qualitative changes in the nutritional regimen have been recognized and can be predicted to a large extent. Animals of many species have been raised on semisynthetic regimens. Indeed, rats and mice have been brought to full development on mixtures of completely soluble and chemically defined diets in which synthetic amino acids constitute the sole source of nitrogen. Such completely synthetic nutrition has been achieved even with germ-free animals. As a result of the precise knowledge of nutritional requirements that has been thus acquired, the deficiency diseases which used to be so common have been all but eliminated wherever social and economic factors make it possible to apply the teachings of scientific nutrition to daily life.

Much emphasis has been placed during recent years on the need for and utilization of proteins and amino acids. For optimum utilization by adult man, the eight essential amino acids must be present in the diet in a proper ratio. According to the formula of the Food and Agriculture Organization (FAO), the following relative proportions in mg. per gm. of nitrogen seem to be adequate: isoleucine, 270; leucine, 306; lysine, 270; methionine and cystine, 270; phenylalanine and tyrosine, 180; threonine, 180; tryptophane, 90; and valine, 360. The requirements for tryptophane and leucine seem to be much higher in infants.

The protein constituents of milk, eggs, fish, and meat have an amino acid pattern that corresponds closely to the FAO formula. As these proteins are readily digested, they are highly suitable for growth. Proteins from algae, green leaves, and some microbial species also seem to be fairly satisfactory from the point of view of amino acid composition, but unfortunately some are not readily digested. In contrast, the proteins from seeds, tubers, and other storage parts of plants are usually deficient in one or several amino acids.

Ever since the classical studies of nutritional requirements early in this century, it has been widely accepted that the minimal intake of high biological quality protein required to maintain adults in nitrogen balance is approximately 25 gm. daily, provided of course that the caloric needs are adequately covered. An intake of 1 gm. per kg. of body weight from the time of maturity throughout adult life is usually regarded in the United States as near the optimum.

TABLE 3. Estimated Minimum Daily Protein Intake *

Age (*in years*)	*Protein intake* (*gm./kg. body weight*) *To prevent protein deficiency*	*To provide full labile protein stores*
0	(1.6) [a]	(4.8) [a]
2	1.2	3.6
4	0.9	2.7
8	0.7	2.1
16	0.7	2.1
18	0.45	1.3
21	0.35	1.0

* Adapted from E. T. Mertz, 1959.
a. Based on human milk; other values based on whole egg protein or equivalent.

Except during periods of pregnancy, lactation, and recovery from disease, there is no apparent advantage in going above this level of protein consumption. In fact, the American nutritionist R. H. Chittenden (1856–1943), who remained in excellent health to the time of his death at 88 years of age, is said to have limited his protein intake to 40 gm. daily throughout his adult and late life.

Needless to say, the infant's requirements for growth are very much higher than those of the adult. According to many authors, they are of the order of 2.2–2.5 gm. of protein per kg. of body weight during the first two months of life. However, opinions differ widely on this score. Since the average intake of human milk during breast feeding provides only 1.6 gm. of protein per kg., it has been suggested that this amount must be the optimum, because it corresponds to the intake naturally arrived at in the course of evolution. In any case, protein requirements fall progressively to 1.2 gm. per kg. at the end of the second year of life, and to 0.7 gm. at 8 years. They seem to remain at this level until the late teen years, although the requirements may increase somewhat at the time of puberty.

Despite the fact that dietary habits differ profoundly from one person and one social group to another, there seems to exist throughout mankind a fundamental similarity in physiological need for the different classes of nutrients. This is illustrated by the identical results obtained in two independent self-selection studies carried out with human beings in two entirely different age groups. Newly weaned infants were allowed to select their foods ad lib. from a wide range of plant and animal products, for a 12–month period. On the average, they consumed 17, 36, and 47 per cent of their caloric intake respectively as proteins, fats, and carbohydrates, and all of them apparently did well on this ad lib. regimen. The other self-selection study was carried out with soldiers, yet gave almost the same result, namely 17, 40, and 43 per cent for the three classes of foodstuffs.

Most of the biochemical problems of nutrition having been solved, it would seem at first sight rather easy to apply in practice the existing nutritional knowledge. In fact, however, faulty nutrition constitutes today the largest single cause of disease in the world. This state of affairs has two very different causes. The most common cause of malnutrition in underdeveloped countries is a shortage of good quality protein. There is no relief in sight for this shortage in the near future because a way has not yet been found to incorporate the know-how of scientific agriculture into the economies of destitute populations. In contrast, the form of malnutrition most prevalent in prosperous countries has its origin in the very abundance of the dietary regimens. Overnutrition brings about maximum growth rate, but seems to be unfavorable to longevity and in general to the maintenance of health.

2. Malnutrition

a. Amino acid imbalance

As already mentioned, populations who have long lived undisturbed in a given area empirically make efficient use of its natural resources and usually manage to achieve satisfactory nutritional adjustments (Chapter III.1). These adjustments, however, are extremely precarious; they are rapidly upset by almost any change in the ways of life. Thus when primitive people are displaced from their ancestral grounds, or when their habits are suddenly changed by contact with Western civilization, the nutritional wisdom developed through centuries of trial and error is no longer helpful because the environment and conditions of existence become so different. Most commonly the first effect of change to a more "civilized" way of life is to make people dependent on a few crops that provide unbalanced diets, even though they supply larger caloric yields.

A recent study of malnutrition among Southern African Zulus revealed the havoc wrought by the substitution of corn for millet in their diet. Millet had been the indigenous cereal of these people before the arrival of the white man, but it was rapidly displaced by corn when the natives were shown that this crop yielded more abundant food for less work. Other changes occurred at the same time. In the past, the large areas of unsettled land had provided the Zulus with game, milk, berries, and other sources of food that eventually disappeared from their fare as white people settled in South Africa. The modifications of diet were sufficiently slow that the Zulus themselves were unaware of them. They could not believe that the replacement of millet by corn and other changes in food habits which they regarded as of minor importance could have had anything to do with their disease problems and with their failure to equal the strength and health of their ancestors.

Malnutrition has naturally become more acute and more widespread as a result of the rapid population increase in underdeveloped countries, and it often involves caloric shortages as well as unbalanced diets. In many places, the malnutrition problem has been further aggravated by the substitution of cash crops, such as cotton or sugar cane, for the balanced food crops that used to be the basis of primitive agriculture. It has been estimated that 10 to 15 per cent of the world population suffers from actual hunger, and approximately one-third from some kind of nutritional deficiency. Malnutrition is furthermore increased by the prevalence of internal parasites. According to a recent statement, the population of China alone harbors some 130,000 tons of intestinal worms!

Throughout the underdeveloped areas of the world, major reliance must be placed on one or more of the five major staples—namely, rice, wheat, corn, cassava, and millet. Malnutrition is to be expected wherever the population depends almost exclusively on one of these products for its principal source of food, along with a few other starchy materials, such as bananas, sweet potatoes, and especially manioc. On the whole, adults do reasonably well on these predominantly carbohydrate diets, but not the infants, preschool children, and pregnant or nursing women. These vulnerable groups show various signs of malnutrition because they have difficulty in obtaining from starchy foods enough of the required nutrients and especially the essential amino acids, for which their needs are greater than those of adults. The cereals mentioned above are low in protein content and, furthermore, the proteins they contain are of low qualitative value because they are unbalanced with regard to amino acid composition. This is true not only of cereals but also of most other reserve plant products, such as the various kinds of tubers or beans. Many experiments in animals have revealed that growth is depressed when the diet is unbalanced from the point of view of amino acid composition. The growth depression is most marked in young growing animals.

As dairy products and meat are in short supply almost everywhere except in prosperous countries, it is not surprising that the most common and most important form of malnutrition in the world today is protein malnutrition. Low protein intake and amino acid imbalance are naturally often complicated by other nutritional deficiencies, as well as by insufficient caloric intake. For this reason, many different names have emerged to denote the physiological disorders caused by malnutrition. In the Americas, they are commonly referred to as "multiple deficiency syndrome in children" or "sugar baby"; in Asia as "nutritional distrophy"; in Europe as "nutritional edema"; in Africa as "m'buaki" or "kwashiorkor." The last name is probably the one most commonly used at the present time. But the increasing awareness that the determinants of malnutrition are always multiple has led many nutritionists to favor the expression "protein-calorie malnutrition."

In a more or less acute form, kwashiorkor or protein-calorie malnutrition prevails in large parts of Africa, Asia, and Latin America, especially among children. Its extreme form has been described as follows in the 1963 World Health Organization publication, *Malnutrition and Disease:*

> The children affected are nearly always not only small for their age, with hair and skin of a pale color, but also exhibit feet and legs swollen from an accumulation of excess fluid; their appetites are capricious, and they are easily liable to digestive upsets. Those more severely ill may have hair of any color to greyish-white, very pale skins, and swellings of the legs, thighs, hands, and face. In the most advanced state the hair is so loosely embedded that it can be pulled out in tufts without causing pain; the eyes may be closed with swelling, which occurs in nearly every part of the body, and the

> skin may break down as though it had been burnt. At this stage the child appears to be desperately unhappy or sunk in apathy. He will not stand or walk and will not willingly move in his bed except to pull the covers over his head. He resists any interference, even feeding.

The fact that the basic signs of kwashiorkor disappear or are much improved when the patients are given milk, purified casein, or even the proper mixture of amino acids indicates that the disease is primarily caused by protein malnutrition. It has not been ruled out, however, that a deficiency in nitrogen, rather than in essential amino acids, can also account for the disease. In any case, pathological lesions similar to those found in kwashiorkor have been produced experimentally in rats by force-feeding them certain plant foodstuffs and thus creating in them an amino acid imbalance. In the human species, kwashiorkor often develops when the mother tries to compel the recently weaned infant or young child to eat large amounts of food containing proteins of poor biological value.

The lack of milk, meat, poultry, fish, eggs, or of mixtures of plant proteins properly balanced soon leads to retardation of growth, followed typically by edema, diarrhea, changes in skin pigmentation, liver enlargement, loss of hair, and very poor resistance to infection of the skin, lungs, and intestinal tract, often resulting in death. Physical stunting is so general among the survivors that it is often mistaken for a genetic character. Permanent impairment of the central nervous system has also been reported. This is not surprising since the most acute forms of malnutrition commonly occur during the most active period of brain development. In certain areas, up to 50 per cent of the children may die of protein malnutrition before school age.

In brief, protein deficiency and especially amino acid imbalance are responsible for high death rates among the young, poor physical development in the adults, and chronic illness that persists throughout life. This form of malnutrition (not necessarily undernutrition) probably constitutes the most important cause of disease in the underprivileged parts of the world.

b. Breast-feeding vs. bottle-feeding

In general, breast-fed children suffer little up to weaning time even in the areas where protein malnutrition is prevalent; their growth rate and resistance to infection are adequate as long as breast-feeding is continued. However, nutritional misery commonly occurs immediately after weaning and results in an enormous death rate from diarrhea and infection. To a very large extent, the frequency and severity of infantile diarrhea in tropical areas are due to malnutrition following discontinuance of breast-feeding. For this reason, it seems scientifically justified to call this disease "weanling diarrhea" as has been recently suggested.

In the past, many children in the tropics and in other low-income areas did manage to overcome the early environmental stresses, thanks to the fact that breast-feeding up to two years of age was the general practice. However, the dangers of the post-weaning period are now rapidly increasing because breast-feeding is becoming less common and less prolonged in the countries undergoing industrialization and urbanization. Bottle-feeding is thus responsible at the present time for some of the most serious health problems among children in these areas.

The severity of infantile diarrhea and other enteric disorders in bottle-fed children is commonly thought to result from the failure to observe sanitary practices, and from the consequent heavy contamination of the bottle contents. Almost certainly, however, the poor nutritional quality of the food is often as important as microbial contamination in constituting the primary cause of the disease. A typical culture pattern in low-income groups is to feed the children, immediately after weaning, starchy foods that may or may not be supplemented with miscellaneous vegetables, fruits, and legumes, but in any case do not provide an adequate source of protein. This situation is very common in tropical countries afflicted with kwashiorkor, especially in Latin America, the Caribbean islands, Africa, the Middle East, India, and Southeast Asia.

c. *Meat substitutes*

The largest difficulties that stand in the way of nutritional improvement in the underdeveloped countries are of an economic nature, but others have their origin in cultural patterns. Food habits, taboos, and religious scruples often interfere with the acceptance of products or nutritional practices that are nutritionally desirable. But even if these difficulties could be overcome, the fact would remain that the production of animal proteins cannot be accelerated rapidly enough to meet the world's demands. Attempts must be made therefore to extend the use of proteins obtained from plants and other sources.

Fortunately, meat is not essential for growth and health—witness the fact that many individual persons and whole populations live on vegetarian diets, supplemented or not by dairy products. This is true even for children. Shortly after World War II, for example, it was found that children in German orphanages grew well and remained in excellent health without milk and meat, provided their diet contained a good mixture of vegetable proteins. Studies aimed at the development of dietary regimens not based on meat or dairy products are in progress along several different lines:

1. Production of comestible microorganisms under industrial conditions. The most extensive experience so far is with the production and use of algae. Yeasts and bacteria can also be produced at small cost on a large scale by growing these organisms on low-grade petroleum products. Algae, yeasts, and bacteria provide the amino acids necessary for animal growth, but these materials have not yet been adequately tested in long-range experiments with regard to their possible toxic effects; furthermore, their acceptance by man is still questionable.

2. Separation of the protein constituents from the leaves of forage plants. The fractions thus obtained have a high nutritional value and seem to be accepted by farm animals and to a limited extent also by human beings. It is not yet certain that leaf proteins can be prepared and purified at a cost low enough to make them acceptable and of practical use in the poor countries where they are most needed.

3. Preparation of plant mixtures having the proper balance of amino acids. This approach is based on the knowledge that the storage organs of plants are usually deficient in one or several of the amino acids, but differ sufficiently in percentage composition to make it possible to prepare mixtures having an amino acid balance suitable for human growth. From a practical point of view, the protein sources of greatest interest seem to be presscake from oil seeds, e.g. cottonseed flour, soybean products, and peanut flour, and to a lesser extent, beans, peas, sesame, sunflower, and coconut. These materials can be usefully mixed with fish products, casein, and grain.

A product called Incaparina has been produced on an industrial scale in Central America from locally available materials. The formula INCAP 9B contains only seed meal, corn, sorghum, yeast, synthetic vitamin A, and calcium carbonate, yet provides all the requirements of a child at weaning, except for calories and vita-

TABLE 4. Composition of INCAP Vegetable Mixture 8 *

Ingredients	(per cent)	Composition	(per cent)
Dried corn masa	50	Protein	25.1
Sesame meal	35	Fat	13.7
Cottonseed presscake	9	Carbohydrate	46.3
Torula yeast	3	Crude fiber	3.2
Kiluyu-leaf meal	3	Ash	6.6

* Adapted from N. S. Scrimshaw, R. L. Squibb, R. Bressani, M. Béhar, F. Viteri, and G. Arroyave in *Amino Acid Malnutrition*, W. Cole, editor, Rutgers University Press, 1957.

min C. The cost of the ingredients is said to be only 2 or 3 cents for a daily ration. Nutritional field studies carried out under the sponsorship of the Pan American Health Organization indicate that Incaparina has a protein quality comparable to that of milk; there is evidence that the native populations in Central America can be taught to accept it. The product has proven effective for the treatment of severe kwashiorkor even when used as sole source food.

4. Supplementation of plant proteins with purified amino acids. This approach is based on the theory that since most cereal and tuber proteins are deficient in several amino acids, in particular lysine, methionine, and threonine, their growth-promoting property can be much improved by the proper kind of amino acid supplementation. In practice, supplementation has become possible through the recent development of industrial methods for the large-scale production of certain amino acids, by chemical synthesis and by fermentation. The manufacturing costs are sufficiently low that amino acid supplementation of certain animal feeds has proven economically profitable in animal husbandry, especially for chicken production. A few large-scale tests are being conducted at present in several parts of the world to determine the practicability and nutritional usefulness of adding lysine to human of protein.

5. Fish meal. This source of protein is obtained by treating otherwise poorly acceptable sea fish with organic solvent which eliminates objectionable oils and fats. The fish powder thus obtained is white, almost tasteless, and of very high nutritive value. Its cost is low, of the order of 3 cents a pound dry weight, according to the manufacturer, and the potential world supply appears almost unlimited at the present time.

Needless to say, protein content and amino acid composition are not the only factors to be considered in comparing foods of animal and plant origin. Recent studies on several groups of vegans (vegetarians who consume no animal foods or dairy products whatever) have shown that these people are usually deficient in iron and in vitamin B_{12}. Their B_{12} serum level is approximately 120 μg. as compared with 360 for meat-eaters and 330 for ordinary vegetarians; the latter have a B_{12} serum level higher than that of vegans because they supplement their vegetable diets with eggs, milk, and cheese. As pathological conditions are likely to develop when the serum level of B_{12} falls below 200 μg./ml., regimens consisting only of plant products should be supplemented with a source of this vitamin. The fact that vitamin B_{12} is synthesized by intestinal bacteria probably explains why certain strict vegans do not exhibit the deficiency.

As already mentioned, the largest nutritional problem in the world today is not so much undernutrition as malnutrition. In fact, the debility caused by protein malnutrition has consequences that transcend its purely medical aspects. It constitutes a social problem of immense importance because it restrains economic and social progress. Almost half the world's population lives in areas where this form of malnutrition prevails, clearly a shackle to progress and a goad to revolutions.

d. *Overnutrition*

It is only during recent years that overnutrition has been recognized as a form of malnutrition and that obesity has come to be considered as the most common nutritional disease in the prosperous countries of the Western world. To a large extent, concern with obesity has emerged from the actuarial evidence that the obese have a reduced expectancy of life and from the observation that ischemic heart disease is often associated with dietary excesses. Death on the expense account is a characteristic feature of the affluent society.

Experiments in animals are also beginning to provide evidence that too abundant nutrition is not an unmixed blessing. Rats and mice fed ad lib. have a shorter average life span than animals prevented from gaining weight by a diet restricted in quantity but well balanced in quality. Limitation of food intake to approximately two-thirds of what the animal would eat if unrestricted decreases the incidence of pulmonary, cardiac, renal, and vascular diseases as well as cancers. In brief, the animals fed diets providing the fastest growth rate tend to have the lowest life expectancy and the highest incidence of age-associated diseases. The results of experimentation in animals are thus compatible with the actuarial evidence that lean persons fare better than their overweight contemporaries with regard to both expectancy of life and general state of health.

Most emphasized at present is the effect of intake of saturated fats on cardiovascular disease, but despite so much publicity this relationship is still *sub judice*. In any case, there is no doubt that the larger the income the greater the tendency to replace starchy foods in the diet with fat, protein, and refined sugar. During the discussions on the genesis of cardiovascular diseases at the Sixth International Congress of Nutrition held in Edinburgh in 1963, emphasis was laid not only on the possible effects of too much animal fats but also of too much protein. Surfeit of steak was considered to be as dangerous as surfeit of butter. Other very recent studies have revealed a striking correlation between sucrose consumption and cardiovascular accidents; this correlation has received some support from experiments in animals.

Recent epidemiological surveys, on the other hand, point to the probable importance of factors other than nutrition in the causation of cardiovascular disorders. Thus, an Italian community in Pennsylvania was found to enjoy an unusually low rate of these disorders despite the high prevalence of obesity among both men and women, caused by a diet generous in calories, fat, and wine. The Masai tribesmen mentioned earlier with their meat- and fat-loaded diets and the Samburn warriors who drink at least six liters of milk a day also are said to be essentially free of ischemic heart disease. The attempts to explain away these anomalies by invoking genetic factors, intense physical exercise, happy and relaxed ways of life, and freedom from competition serve only to emphasize the obscurities attending this important problem.

Other aspects of overnutrition may also be of great importance even though less dramatic. The fact that human breast milk contains smaller concentrations of several dietary constituents, especially protein, than does cow's milk has led certain pediatricians to regard the rapid growth of the bottle-fed baby as a curse rather than a blessing. Indeed, there has long been a belief among certain clinicians that the fat infant tolerates gastrointestinal infections poorly, and that weight reduction often has a salutary effect on infantile eczema. It has been claimed also that a slight degree of undernutrition is associated with a higher resistance of adults to certain viral infections—a view expressed as early as 1552 by John Caius in his famous essay on the sweating sickness.

Overnutrition is not a phenomenon peculiar to our times. Human beings always have been prone to overcompensate for fear of food shortages by engaging in nutritional excesses whenever they could afford it. As a result, overnutrition has been a feature of all periods of abundance or material prosperity. "In the old days," wrote Lucretius in the fifth book of *De Rerum Natura,* "lack of food gave languishing limbs to Lethe; contrariwise today surfeit of things stifles us." Thus, history repeats itself. Like the prosperous Romans of 2,000 years ago, many prosperous citizens of the Western world today dig their own graves through overeating.

3. Nutritional state and development

a. Nutrition and stature

The dimensions of the armor and costumes displayed in museums make it obvious that the average American male of today is much taller than were the medieval and Renaissance knights, the members of the aristocracy that dominated Europe in the succeeding centuries, and the average nineteenth-century workingman or member of the bourgeois classes. Likewise, today's debutante is a much enlarged version of the eighteenth-century belle. These impressions can be documented by actual measurements, which show that during recent decades an increase in stature affects each chronological age down to the preschool years. The gain in height and weight is still continuing today, and there is even some evidence that the final heights may be increasing as well. Physiological characteristics are also changing. Boys and girls are becoming physically and sexually mature six months to two years earlier than did their parents; on the average, in fact, sexual maturity is now reached almost three years earlier than it was a century ago.

The change in size became noticeable first in the prosperous industrial countries, particularly in the United States. A few decades ago, American children looked like young giants compared with their European counterparts, and the contrast with Asiatic children was even more striking. These differences were thought at first to be of genetic origin, but recent observations indicate that they are caused in most cases by environmental factors. Several comparative studies made during the present century of young people whose families had emigrated to the United States and of their kin who had stayed in Europe or in Oriental countries revealed an astonishing difference in size between the groups; in all cases, the American-born children were by far the tallest. The situation, however, is rapidly changing. Increase in size and advance in sexual maturity are now taking place in all countries of Western civilization, as well as in some Asiatic countries.

A survey recently made in Glasgow showed that Scottish boys in 1960 were taller by almost 4 inches at age 13 than were the 13-year-olds of 40 years ago; the corresponding gains for girls has been 3¼ inches. The boys weighed 14½ pounds and the girls 16½ pounds heavier than the children of 40 years earlier. Similar findings have been made in other parts of the Western world. In London, choirmasters became aware of the accelerated maturation of children when, according to an editorial in *Lancet,* they began experiencing difficulties in finding choirboys with soprano voices! The Chinese, Japanese, and other Oriental people who live in the Hawaiian Islands are also growing faster than did people of the same nationalities two generations ago. In Honolulu the Oriental teenagers are now able to engage in basketball games with their Occidental contemporaries!

The factors that have brought about such an acceleration in the growth of children apparently do not affect the size of the neonate. In fact, birth weights have not changed appreciably during the past 300 years. Today as in the past, the range is from 3,000 to 3,500 gm. for boys and 2,900 to 3,400 gm. for girls, except under conditions of acute food shortage.

The acceleration of growth after birth is probably the result of several independent causes. It can be assumed that more sensible clothing and control of infectious diseases have played their part. Genetic changes may also have been of some importance, for the reason that more convenient and more rapid means of transportation facilitated mating among groups of people who in the past had only an extremely narrow range of human contacts. More interbreeding may thus have resulted in greater hybrid vigor. Most important, certainly, were the changes in nutrition that have occurred almost everywhere in the Western world during the past century. Under normal conditions, the growth of the modern child in the Western world is not interrupted by seasonal nutritional deficiencies, as was the case in the past and as still occurs among most nonindustrialized people. At the present time in the United States, seasonal changes in the rate of growth have practically disappeared, except for the small number of Indian children who live in parts of the reservations where the ways of life have not yet been completely Westernized. In all other population groups, the general availability of milk, of vitamin sources, as well as of a great variety of food permits uninterrupted and rapid growth throughout the year.

Whatever the reasons for the acceleration of gain in weight and height, there is as yet no proof that the biggest baby, vitamin- and protein-stuffed throughout the year, is necessarily the healthiest baby. Nor is it certain that enjoying three square meals every day, with a constant supply of all nutritional factors at all seasons, constitutes an unmixed blessing for the adult. As we have seen (Chapter II.2), many physiological mechanisms of man and animals are geared to cyclical changes in nature. A nutritional pattern designed for continuous growth at a maximum rate may therefore not be ideal from the physiological point of view, since the genetic constitution of man evolved as an adaptive response to seasonal fluctuations in food supply. It may thus turn out that religious fasts and dieting fads have some physiological justification after all, by providing the human body with an opportunity to give play to the physiological emergency mechanisms built into it during evolutionary development.

b. Nutrition and general resistance

The experience of everyday life makes it obvious that the response of man to almost any kind of environmental stimulus or insult is conditioned by the nutritional state. Yet it is very difficult to establish convincingly a causal relationship between any type of response and any particular dietary factor, the reason being of course that malnutrition decreases general resistance through several different mechanisms.

For example, malnutrition interferes with the production of new cells in the liver, spleen, and bone marrow. Even more important, perhaps, it reduces or abolishes the detoxifying mechanisms of the body and thereby may increase indirectly the damaging effect of certain substances. Thus, a short period of fasting is sufficient to prolong the sleep induced by hexobarbital in mice, probably by slowing the rate at which the liver oxidizes the drug. Fasting likewise slows down the dealkylation of pyramidon, the hydroxylation of acetanilid, the oxidation of chlorpromazine, and may prolong thereby the action of these drugs as well as of various toxins. The LD50 of bacterial endotoxin for guinea pigs can be reduced 100 to 1,000 times by starving the animals or treating them first with carbon tetrachloride. Evidence has been adduced recently that toxic pregnant women exhibit a decreased ability to detoxify anisic acid, probably as a result of change in liver function. In these women, clinical improvement of toxemia resulted from administration of a high protein diet, coupled with treatment with neomycin. As shown in Chapter VI.1, amino acid malnutrition in mice retards the recovery from loss of weight caused by endotoxin.

In many cases, probably, the loss of resistance resulting from acute malnutrition is the indirect outcome of inadequate production of the pituitary or adrenocortical hormones needed in increased amounts during stress. The possibility exists, furthermore, that nutritional factors as yet unrecognized or not essential under normal conditions can become critical when needed for recovery from certain types of insults. In the case of at least four substances (thyroid extract, cortisone, promine, and atabrine) there seems to exist in the liver a protective factor that is retained in the water-insoluble fraction and is apparently distinct from any of the known nutrients. This liver fraction also prolongs the survival of rats which have received multiple doses of X irradiation.

c. Nutrition, behavior, and learning

It is a very ancient belief that nutrition can modify behavior. Porridge was suitable for the farmer, meat for the warrior, and the chief was given the additional privilege of eating raw liver. The illustrious German chemist Justius von Liebig had no doubts on the matter. In the thirty-fifth of his *Letters on Chemistry* he wrote of a bear in Giessen which was very gentle when fed bread and vegetables, but ferocious and dangerous after being fed a meat diet for a few days. Liebig accepted without question the folk belief that swine fed meat can become so aggressive as to attack man. Recent observations with ferrets seem to provide some support for this belief. Initially fierce and intractable, ferrets became so tame that they could be handled without gloves after being fed regularly on a soft diet; they reverted to their original wild behavior after being fed the usual dry diet.

Experiments with laboratory animals have provided some information on the effect of diet on behavior. In rats, food deprivation reduces random activity but increases voluntary activity. On the other hand, rats made to grow rapidly by a superabundance of food in the first two weeks of life are more active and inquisitive at every chronological age than are their litter mates made to grow more slowly by being suckled in large litters.

In man, mental activity and learning are certainly affected by protein and amino acid intake. Thus, mental lethargy is constantly present and usually severe in children suffering from kwashiorkor. Inadequate dietary protein, especially at a critical period early in life, impairs or at least slows down mental development. As is well known, the lack of hydroxylase activity for phenylalanine is associated with severe mental retardation in patients with phenylketonuria. Perhaps through an analogous mechanism, a very high dietary intake of phenylalanine seems to interfere with the development of conditioned avoidance behavior in the rat. In a more general way, it is very likely that disturbances of amino acid metabolism in infants and young children can result in gross mental defects. It has also been claimed that administration of large doses of certain l-amino acids to schizophrenic patients can affect their mental condition.

d. Nutrition and resistance to infection

This topic is discussed in Chapter VI.

4. Nutritional adaptations

In theory, all human beings have essentially the same nutritional requirements and suffer in a similar way from the various kinds of deficiencies. In practice, however, immense numbers of human beings all over the world remain healthy and vigorous on diets that are considered grossly deficient by nutritionists and would appear close to a famine ration for the American public. This paradoxical state of affairs can be explained in part by more efficient use of foodstuffs in underprivileged populations. As we have seen, many primitive people have developed empirical practices that permit them to find fairly adequate subsistence under conditions where food supplies are restricted (Chapter III.1). In addition to these social adaptations of each community to its food supplies, there are biological adaptations such as the biosynthesis of essential growth factors. For example, certain vitamins and amino acids can be produced by the flora of the digestive tract. Of greater interest and importance probably is the fact that man and animals can develop truly metabolic adaptations to a wide range of nutritional changes, both qualitative and quantitative.

Thus, rats on a restricted but constant food intake at first lose weight, but eventually regain it. When their food intake is so limited that they can just maintain their weight, their needs gradually decline; the ultimate caloric requirement is determined by the composition of the diet. Likewise, rats fed diets containing the amino acid antagonist ethionine lose weight initially, but start growing again after 6 or 7 days, at a rate comparable to that of the control animals.

Even more remarkable are some observations made on young pigs given only water without food. Under such conditions, the animals fed a high protein diet prior to the starvation test did not survive as long as pigs of the same age which had been fed a poorer diet and therefore had smaller protein stores. The average daily loss of protein during starvation was much less pronounced in the latter group of animals than in the former group. Moreover, protein catabolism persisted in the animals with the greatest protein stores throughout the period of stress, as judged by the concentration of urea in their blood. It would seem as if the body, once accustomed to a high protein intake, is unable to "gear down" its metabolic demands. These findings lend credence to the sailors' tales according to which the most robust members of shipwrecked crews are likely to die sooner of starvation than do their less muscular companions. It appears, therefore, that the ability to utilize food can undergo marked changes and is a function of the body's adaptive state, within the limits imposed of course by genetic factors. The direction of the metabolic shifts is toward the maintenance of the more vital at the expense of the less vital functions.

Many observations indicate that feeding habits in man are established early during infancy. Underfed babies come to accept smaller food intakes and therefore grow slowly, whereas babies who have been pressed to take more food become habituated to this regimen and maintain a higher rate of growth. Even prenatal nutrition is probably of importance; thus, the weights of newborn infants were below normal in the countries suffering from food shortages during the war. This is consonant with the results of animal studies which indicate that large placental size is associated with large offspring; furthermore, ill-nourished animals produce smaller offspring.

An extensive and prolonged study was made of a large group of beggars who had long been on a diet deficient in total calories, proteins of animal origin, vitamins (A, B, C, and D), and minerals (iron, calcium, phosphorous). Clinical examination revealed the surprising fact that only 4 per cent of these beggars showed signs of nutritional deficiency; on radiologic examination most of them proved to have a normally calcified skeleton. In the women, moreover, the low nutritional intakes did not have an obviously deleterious influence on pregnancy or lactation; even the growth rate of children was almost normal. It appeared, in other words, as if chronic food shortage had caused the group to develop a state of physiological adaptation to low nutritional intake.

Men placed on an emergency pemmican ration providing only 1,000 calories a day developed an increased resistance to starvation that persisted for several weeks. A second period of starvation caused physiological disturbances that were less pronounced than those experienced during the first period of limited food intake. For example, the excretion of nitrogen and ketone bodies was smaller during the second than during the first starvation tests.

When calcium intake is stepped up, some of the increment usually is retained for the first few days, after which equilibrium is again reached. On the other hand, a lower intake usually is compensated by an increase in the percentage of calcium absorbed by the intestine; absorption is also increased when greater amounts of calcium are required for purposes of repair or growth. The mechanism by which the intestine is "instructed" as to the amounts of calcium needed by the body is usually so effective that the amount deposited in the bones of the normal adult in a state of equilibrium is equal to that being reabsorbed during the continuous dynamic process of bone modeling.

One of the strangest nutritional adaptations on record has been observed in a large town of North Africa. The only ground water used by the inhabitants of this town is extremely high in sodium chloride (3,000 parts per million). This water is unacceptable to outsiders, but the inhabitants are so accustomed to it that they take salt with them on their travels to mix with their coffee! In contrast, there have been and still are many groups of men all over the world who do not add sodium chloride to their diet, whatever the nature of their food; indeed, some tribes in Africa use wood ashes, rich in potassium, as a condiment instead of salt.

Adjustment to an inadequate diet is not of course without danger, even when it appears to be successful. Thus, while men in primitive societies seem to become adjusted to the recurrence of what is called in Africa the "hunger season," this period of deprivation between periods of plenty may have serious consequences for the organism in the long run. The necessity to respond repeatedly to intake of one or another substance in amounts larger than the tissues are adapted to tolerate is likely to result in physiological disturbances. Experimental animals which have gone repeatedly through cycles of low and high feedings, water being given ad lib., exhibit lesions in the liver and pancreas much more severe than do animals constantly maintained on restricted diets. Furthermore, a tremendous increase of liver fat occurs when rats previously kept on a low protein diet are suddenly allowed a much higher protein intake. In view of these facts, the great seasonal variations that are common in primitive and underprivileged communities, with regard to both quantity and quality of food, may have some deleterious effects on growth and health, especially among children.

Granted these dangers, evidence is accumulating that the minimum nutritional and especially protein needs of the human body may not be as great as commonly assumed in the United States. In addition to its theoretical interest, this fact renders more hopeful the efforts that are being made to improve nutrition in the underprivileged countries. There is no way to provide for the world as a whole a level of protein intake that would even approach that generally considered ideal in the industrialized countries of the Western world and that is indeed commonly achieved in the United States. This high level of protein nutrition probably accelerates growth in the early years of life, advances sexual maturity, and leads to larger adult size. However, such rapid development is not necessarily a biological advantage; it may correspond rather to temporary social tastes and conventions. Obesity has been considered in certain civilizations and at certain times as an enviable attribute, whereas the tendency today is to regard it as a disfigurement, even though it is so common in our communities.

For the sake of world nutrition, and of health in this country, it is urgent to determine the extent to which the human body can become adapted to intakes of the various food constituents lower than those commonly recommended today. Man apparently can adapt to overfeeding as well as to underfeeding, but little is known of his tolerance levels in either one of these extreme nutritional situations.

5. The Enlargement of Nutritional Science

Nutritional science developed at a time when the worst effects of malnutrition were caused by shortages of food, resulting in gross deficiency states. The most obvious manifestation of these nutritional defects was limitation of growth. It was therefore natural and indeed essential to emphasize at first the need to accelerate growth rates, since this goal provided a convenient index of improvement in the nutritional status. The knowledge acquired from the study of growth rates had the further practical importance of being applicable to the problems of animal husbandry; it helped make the production of meat and dairy products more economical and more efficient. However, while measurement of growth rate constitutes an excellent and probably the most useful guide for improvement of nutrition in animal husbandry, this index is not sufficient for an evaluation of nutritional regimens in man.

The dietary factors most effective in the production of meat, milk, and eggs also bring about a rapid growth of children; but growth rates are not the most significant values of human life. Life span, resistance to disease, intellectual performance, emotional responsiveness and perceptiveness, etc. are characteristics that cannot be measured on a weight scale and are of little relevance to the production of market pigs or chickens, yet should be of paramount importance in judging the value of a diet for man.

It is clear, furthermore, that the nutritional requirements of man are not as completely understood as was once believed. One reason, as we have seen, is that man has a very wide range of metabolic adaptabilities that allow him to survive, function, and multiply on a great variety of nutritional regimens (Chapter III.4). Another reason is that when the essential requirements were determined around the turn of the century, the criterion used was maximum growth of young adults living a kind of existence very different from that which is most common today in modern communities. The values thus obtained are not entirely applicable to the needs of modern man, wheel-borne and living in the sheltered and air-conditioned environment created by industrial technology. What was adequate nutrition yesterday is probably overnutrition and also malnutrition today. The American teenager, tall and big because fed such an abundant and rich diet early in life and continuously thereafter, develops appetites that probably greatly exceed his true physiological needs. As a result, he may find it increasingly difficult in the future to compete with his contemporaries who live on more spartan diets in less prosperous parts of the world.

There are many obscurities also with regard to the qualitative composition of nutritional regimens. In practice, diets are made up of natural products containing many substances other than those listed in textbooks of nutrition. Some of these substances have biological activities that do not necessarily express themselves in growth rates, but may nevertheless result in important physiological effects, either beneficial or injurious. Thus, it has been known for almost 40 years that several fodder plants, in particular alfalfa and clover, contain substances capable of producing oestrus in animals which eat them. Some of these substances resemble natural estrogens, and others have an entirely different structure, such as genistein 4-methyl ether. The amounts of such substances present in pasture plants can be sufficiently great in certain cases to cause pathological effects in grazing animals; in Western Australia, for example, sheep put on pastures rich in subterranean clover ceased to reproduce normally because this legume plant contains large amounts of a substance with estrogenlike activity. In any case, the estrogenic substances that occur normally in certain foodstuffs probably affect the regulation not only of fertility but also of processes, such as lactation, that are under endocrine control.

It is not unlikely that in man also many components of natural foodstuffs have effects that are not expressed in growth rates. Examples that have come to light recently are the goitrogenic activity of substances in certain plants of the cabbage family, the opiumlike property of lettuce, the corticosteroidlike effect of licorice.

Man is more than an animal and does not live by bread alone; and for this reason the purely biochemical and physiological points of view are not sufficient to formulate dietary regimens adequate for his welfare. The relation of food to human existence involves factors that go above and beyond the relation of biochemical nutrition to physical life. The following statement, written by a woman anthropologist, may help to convey why the chemical knowledge of nutrition should be supplemented by an awareness of the emotional state in which normal human beings approach every article of food and perform the act of eating as a symbolic rite.

> Food can engage me completely as a person. It can bring to life my keen anticipation, my impatient or happy waiting; it can evoke my memory in all its pain and joy; it can pierce me with nostalgia; it can revolt me to the point of vomiting. Its preparation can be an act of relatedness, of obligation, of self-fulfillment, of creativity, of love; its eating can be participating and communion. Let us hope it also nourishes me physically, supplying me with the proteins, vitamins, carbohydrates and other nutrients I need to stay alive and healthy so that I can experience the fullness of human existence. . . .
>
> In some societies, no new relationship can be initiated or no transaction carried on without gifts of food. This gift has no function other than to be a medium of relatedness. One man will give to the other a basket of yams and be given a basket of yams in return. The yams are of the same size, kind, and degree of ripeness. The baskets are the same size. All that has happened is that food has been given as a gift, back and forth. . . .
>
> For many corn-growing Indians, the cornfield was traditionally an altar and the act of cultivation was an act of worship. If we were to describe the agriculture of the Hopi Indians as it was in traditional times, we would have to describe the entire culture: the ceremonials, the ball games, the dancing, the mask-making, the scrambling up to the aeries of eagles, the silent retreats to the kivas, the long journeys to the Pacific Ocean in search of turtle-shell, the joyous cries of the playing children (yes, even these) and the smiles of the parents. All these found their meaning and their function in the growing of corn, which was ultimately an act of worship. Corn itself meant life. . . . When I take away corn from such people, I take away not only nutrition, not just a loved food, I take away an entire way of life and the meaning of life [Lee, 1962].

IV. The Living World

1. The interplay between living things

a. Competition, peaceful coexistence, and symbiosis

Man is wont to claim that he has "mastered" his physical environment, even though in reality he escapes from natural forces rather than masters them when he takes shelter in rooms heated during the winter and cooled during the summer. With more justification, he can boast of having "conquered" most members of his biological environment. He has indeed destroyed a few kinds of animals and plants, put others behind walls in zoological or botanical gardens, and enslaved some in his homes and farms. During the past few decades, he has come to believe that he can also conquer insects and microbes. Some epidemiologists go so far as to assert that it is possible, and desirable, to eradicate several species, i.e. to eliminate them completely from the surface of the earth (Chapter XIV.1).

The words mastery, conquest, and eradication have unbiological or perhaps even antibiological overtones. They symbolize an attitude that tends to place man above and indeed outside the natural world and implies that he can best assure his survival through his destructive powers. Yet, it is far from certain that trying to eliminate living species is a wiser and more effective course than learning to achieve peaceful coexistence with them through an understanding of their habits.

Admittedly, infectious diseases and the objectionable properties or activities of numerous animals and plants constitute the manifestations of the interplay between man and the rest of the living world which are the most obvious and which have been the most emphasized. But these negative aspects of biological interrelationships are not the most common in nature, nor are they the most important in the long run. Under natural conditions, a state of equilibrium is commonly reached between various living forms when they have long been associated in a given area. This equilibrium is achieved through biological and social mechanisms that permit the survival and perpetuation of all species involved.

The favorable results and vast potentialities of peaceful coexistence among different species are apparent in the countless examples of symbiosis throughout nature. With regard to man, peaceful coexistence and symbiosis will be illustrated later in some detail by the example of the indigenous microbiota and its contribution to the well-being of its host (Chapter V.2, 3, and 4) and also by the phenomenon of microbial persistence (Chapter VII.3). Symbiotic and commensal relationships constitute aspects of biological science that have been much neglected, even though the welfare of man certainly depends on the understanding of these relationships as much as it does on the knowledge of infectious diseases and other forms of destructive competition.

Ultimately, in fact, the outcome of biological interrelationships is rarely determined by man's intent to conquer or destroy. In the vast majority of cases, the determinant factors of health and disease are genetic forces over which man has little control and environmental conditions that he understands but superficially. While man cannot do much as yet to modify his own genetic constitution, he is learning to manipulate his internal environment, even though within narrow limits, and thus can control to some extent the responses of his body and his mind to environmental stimuli and insults. Furthermore, the range of his adaptive potentialities is so wide that he can make a satisfactory response to many kinds of threats, whether these originate from physicochemical forces, from microbes, or from his fellow men.

In the case of microbial agents, it is the failure of the adaptive response, namely microbial disease, which naturally captures attention. But more remarkable than the occurrence of disease is the fact that man so often achieves biological equilibrum with many of the potentially pathogenic microorganisms with which he comes into contact. Furthermore, when biological equilibrium cannot be reached, he can often modify the ecological situation in such a manner that his contacts with the pathogen are extremely rare. Malaria control was achieved in many parts of western Europe during medieval times, and in the Ohio Valley during our times, not by pursuing the elusive goal of complete eradication of malaria parasites but more simply by keeping the mosquitoes at bay through drainage and farming.

Living is everywhere a collective enterprise. However primitive or complex, all organisms in nature spend much of their existence in the company of several representatives of their own kind. Even more interestingly, they always occur in intimate and lasting associations with other forms of life not genetically related to them. In fact, most living things soon die of starvation or disease when separated from the other species with which they exist in partnership under natural conditions.

Countless examples of biological interdependence have been described in great detail since the subject became popular around the turn of the century through the publication of the famous book *Mutual Aid, A Factor of Evolution,* by the Russian naturalist and sociologist Prince Kropotkin. During his travels in Siberia and Manchuria, Kropotkin had observed the frequent occurrence of cooperative endeavors in the animal world; moreover, the historical study of social institutions had convinced him that a desire for mutual aid was natural to man. The kind of association and interdependence that he observed among higher animals has now been shown to exist at all levels of development in the living world.

Intimate and probably indispensable partnerships occur even among unicellular organisms. One of the examples most recently described may be worth mentioning, in part for its picturesque quality and also because it is so unexpected. In the gut of an Australian termite lives a unicellular organism, the protozoon *Mixotricha paradoxa,* which is literally taken for a ride by thousands of spirochetes that cover its surface. The spirochetes are attached to brackets on the outer surface of the protozoon. When they move, *M. paradoxa* is propelled smoothly along; its own flagella play no part in the movement but are used for steering.

A more classical example of association is provided by the protozoon *Paramecia bursaria,* which owes its green coloration to the fact that it normally harbors microscopic algae of the *Chlorella* genus. The protozoon can be rid of the algae by treatment with certain drugs, but then it loses its green pigmentation and fails to develop full size. What might be regarded as an infection constitutes therefore in this case, a successful and almost indispensable association. As we shall see, such associations are also very common between bacteria and viruses.

Despite the fact that no organism ever exists in nature except in association with others of different genetic constitution, many biologists tend to limit their studies to individual specimens or to populations they assume to represent "pure" biological entities.

Microbiologists have based their science on the premise that they can obtain "pure" cultures of bacteria, molds, protozoa, or other microorganisms. Tissue-culturists struggle to establish and maintain "pure" lines of cells. Gnotobiologists are now extending this approach by developing colonies of germ-free animals.[1]

It is probable, however, that biological purity is as elusive a concept as chemical purity. Experience has taught chemists that the purity of a given preparation is always relative, even in the case of the cleanest, most beautiful crystals. Similarly, biologists are now realizing that natural populations are always made up of mixtures of several biological entities. Indeed, many organisms long thought to constitute well-defined species in reality are made up of two or more constituents that are genetically unrelated and yet are so intimately associated that it is difficult to separate them. The protozoon *Paramecia bursaria* is a case in point, and a few other examples will be discussed in the following pages. Suffice it to mention here in addition that electron micrographs have recently revealed the presence of innumerable structures that appear to be viruses in the thymus and other organs of so-called germ-free mice.

While it would be out of place and indeed impossible to describe here or even to merely list all the known examples of biological associations, it is essential to emphasize again that such associations occur throughout the living world. Some of them eventually manifest themselves in the form of disease. More commonly and interestingly, however, many biological associations are of nutritional advantage to both partners and may even be found to express themselves in the emergence of new and unexpected properties.

b. Beneficial effects of biological associations

To devour was long regarded as the only alternative to being devoured. But this oversimplified formulation of the law of the jungle accounts for only a small aspect of the interplay between living things. There are throughout the living world countless types of symbiotic associations, in which both partners derive advantage from their nutritional complementariness and often require each other's presence to complete their development. (For a general discussion of symbiosis, see Nutman and Mosse, 1963.)

1. It has been suggested during recent years that the expression "germ free" be replaced by "axenic" or "gnotobiotic." I shall continue in the present volume to designate as "germ free" animals raised under such conditions that they do not carry any microorganisms culturable or detectable by the usual laboratory techniques. Whether such animals harbor virsues or other living agents remains *sub judice.*

The word symbiosis was coined by deBary for the special case of lichens, organisms made up of an alga and a fungus. While it is easy to isolate in pure culture the individual species of alga and fungus in any given lichen and to cultivate them separately free from each other on artificial culture media, it has proved very difficult to regenerate the lichen from its two separate constituents. Under ordinary conditions of culture, the alga and the fungus either develop independently side by side without uniting to form a lichen, or more commonly one acts as a parasite for the other and destroys it.

The first step in true lichen synthesis in the laboratory was recently achieved by the simple artifice of using a culture medium so deficient nutritionally that neither the alga nor the fungus could develop on it. When the two organisms were placed side by side on this deficient medium, association took place between them and resulted in lichen development. Restating this finding in teleological terms, the lichen is formed when its two constituents *need* to become associated in order to develop. In the nutrition of lichens, the photosynthetic activities of the alga are complemented by the ability of the fungus to assimilate various minerals and other nutrients.

Another classical example of nutritional symbiosis is provided by the association of leguminous plants with bacteria of the species *Rhizobium leguminosarum*. In this case, the bacteria live in root nodules and fix atmospheric nitrogen, which they supply to the plant; in return the plant supplies them with carbohydrates, minerals, and water. The various species of orchids likewise exhibit different degrees of dependence on fungi that penetrate their cells and enhance their development.

Nutritional symbiosis is very widespread among insects. For example, the utilization of wood by termites is made possible by the activity of cellulose-decomposing protozoa which these insects constantly harbor in their intestines and which are highly adapted to this environment. Similarly, the common roach, like many other insects, is dependent for its survival in nature on rickettsialike symbionts that are located in specialized organs of its body. These symbionts can be eliminated by treating the roach with the proper antimicrobial drugs, but then the insect fails to reach complete development or may not develop at all. Supplementation of the roach's diet with yeast and certain mixtures of vitamins and minerals corrects in part the nutritional deficiencies resulting from elimination of the symbionts, but in part only. Either the symbionts provide the roach with other nutritional factors that have not yet been identified, or more likely they exert metabolic and morphogenetic effects that are essential for complete development.

The microbiota of the gastrointestinal tract in animals and man unquestionably contributes to the nutrition of the host by synthesizing certain essential growth factors. Because of its importance for human welfare, this subject will be discussed in further detail in Chapter V.

Among the most unexpected and picturesque beneficial manifestations that give a high survival value to biological partnerships are those resulting in disease control. The sanitary effect of association between two different species is seen in a striking form in the colonies of ants that cultivate fungi and feed on them. These fungi, which have never been found growing anywhere except in the ant nests, thrive and remain uncontaminated in their gardens, but are rapidly overwhelmed by common microorganisms under usual laboratory conditions. In the gardens, the ant workers continuously move over the fungus growth and discard unwanted material from it; they also deposit on its surface saliva and feces, which favor continued fungal development. Interestingly enough, the ants themselves have become almost completely dependent for their survival on the particular species of fungi they cultivate in their gardens.

Even more striking is the phenomenon recently designated "cleaning symbiosis," which has been observed in many animal species. Skin divers have discovered, for example, that a surprisingly large number of marine organisms specialize in cleaning others. As will be seen from the following description, this peculiar behavior promotes the well-being of the cleaned animals and provides food for those which do the cleaning.

> One finds in the Bahamas the highly organized relationship between the Pederson shrimp (*Periclimenes pedersoni*) and its numerous clients. The transparent body of this tiny animal is striped with white and spotted with violet, and its conspicuous antennae are considerably longer than its body. It establishes its station in quiet water where fishes congregate or frequently pass, always in association with the sea anemone *Bartholomea annulata*, usually clinging to it or occupying the same hole.
>
> When a fish approaches, the shrimp will whip its long antennae and sway its body back and forth. If the fish is inter-

> ested, it will swim directly to the shrimp and stop an inch or two away. The fish usually presents its head or a gill cover for cleaning, but if it is bothered by something out of the ordinary such as an injury near its tail, it presents itself tail first. The shrimp swims or crawls forward, climbs aboard and walks rapidly over the fish, checking irregularities, tugging at parasites with its claws and cleaning injured areas. The fish remains almost motionless during this inspection and allows the shrimp to make minor incisions in order to get at subcutaneous parasites. As the shrimp approaches the gill covers, the fish opens each one in turn and allows the shrimp to enter and forage among the gills. The shrimp is even permitted to enter and leave the fish's mouth cavity. Local fishes quickly learn the location of these shrimp. They line up or crowd around for their turn and often wait to be cleaned when the shrimp has retired into the hole beside the anemone [Limbaugh, 1961].

Experiments in which the cleaners had been removed from a certain area have demonstrated that "cleaning symbiosis" is of great importance in maintaining the health of the marine population. "Within a few days [after removal of the cleaner] the number of fish was drastically reduced; within two weeks almost all except the territorial fishes had disappeared. Many of the fish remaining developed fuzzy white blotches, swelling, ulcerated sores, and frayed fins" (Limbaugh, 1961). When a "cleaner" shrimp was then introduced, it began immediately to clean the infected fishes and their health promptly improved.

A somewhat related mechanism of disease control has been observed among mice. These animals commonly treat each other's wounds by licking, which is particularly important when the wounds are in the head region since these can be treated only by another mouse. In a group of mice that had been caged singly, such head lesions failed to heal. But healing occurred rapidly after the mice had been regrouped and the wounds were licked by the new mates of the wounded animals.

c. Morphological and physiological changes produced by biological associations

The plant tumors known as crown galls can be induced at will by inoculating certain plants with pure cultures of the proper strain of the bacterium *Agrobacterium tumefaciens.* Although the secondary tumors that develop on the same plant at sites removed from the initial infection are usually free of bacteria, they can be transferred in series to new plants or propagated in tissue culture as self-reproducing entities. All these derivative neoplastic growths remain free of the initiating bacterial agent, *A. tumefaciens.* Furthermore, this bacterium can be eliminated from the primary tumor tissue by controlled heating, without affecting the power of autonomous growth of the tumor. Thus, the crown gall can be propagated independently of *A. tumefaciens* even though this bacterium constitutes the only etiological agent so far known to be capable of producing the gall experimentally.

In order to be transformed by the bacterium into tumor tissue and to produce galls, plant cells must first be conditioned by certain stimuli associated with wound healing. The physiological state of the host cells is therefore an essential determinant of crown gall. Although normal plant tissue requires indole acetic acid and the coconut-milk factor for its growth, the self-producing tumor tissue does not need these growth factors. Indeed, it can synthesize them, and perhaps others. The acquisition of this new synthetic power probably accounts for the ability of the cancer cells to grow profusely and escape from the restrictive forces in the normal plant.

In addition to the ordinary type of crown gall tumors characterized by a complete loss of ability to organize tissues and organs, there exists in certain plant species, in tobacco for example, another morphological type of tumor, made up of cells which possess a pronounced capacity to organize highly abnormal leaves and buds and which retain this characteristic indefinitely in culture. These teratomata are transplantable, like those of the unorganized type. Furthermore, cells isolated from them grow indefinitely, as do cells obtained from an unorganized tumor, on a simple culture medium containing only inorganic salts and sucrose, which cannot support the continued growth of normal plant cells from which the tumor cells were originally derived.

A most remarkable advance in the study of these plant tumors has come from the discovery of techniques to revert the teratoma cells back to the state of normal cells identical with those in the normal plant before exposure to the bacterium. This back transformation can be brought about by forcing shoots derived from tumor buds to grow rapidly through a series of graftings to healthy plants. To this end, the single teratoma cells are planted on agar medium, and the tissues of single-cell origin are grafted to cut stem ends of tobacco plants from which the auxiliary buds have been removed, thus giving rise to overgrowth of considerable size. The organized structures that arise from the tumor tissue at the periphery of the expanding tumor mass can then be grafted to the cut stem ends of healthy tobacco plants.

After several such graftings, the growth becomes progressively more rapid and more normal in appearance, and when the tips of such shoots are grafted to healthy plants, they develop rapidly and appear normal in every respect, ultimately flowering and setting seeds that yield normal *Nicotiana tabacum* plants. It is of special interest that the cells of such recovered scions fail to grow on a basic culture medium that supported the continued rapid growth of crown gall tumor cells (reviewed in Braun, 1954, 1959).

The fact that the cells from the teratomata gradually recover and become normal in every respect under the conditions just described indicates that the cellular alteration in crown gall does not involve a true mutation, since heritable changes of that type would not be lost by rapid growth. The findings suggest rather that there exists in the tumor an autonomous entity that can be eliminated by dilution in very rapidly dividing cells. This entity would then be responsible for the continuity of the tumorous properties from one cell generation to the next.

It is tempting of course to postulate an analogy between the genesis of plant galls and of mammalian cancers. But this hypothesis must remain a matter for speculation at the present time, even though it has received support from the discovery that several types of viral infections can produce tumors in experimental animals. Even more suggestive is the fact that some viruses are capable of inducing in vitro lasting cellular alterations by forming nonlethal and transmissible associations with tissue cultures (reviewed in Koprowski, 1963; Paul, 1963; Shein et al., 1963).

The effects of biological associations thus constitute a vast and continued spectrum ranging from destructive pathology to lasting cellular and organismic alterations, reaching into the multifarious forms of interdependence and symbiosis.

The most extensively studied manifestations of symbiosis, because the most obvious, are those which result in nutritional advantages for one or both symbionts. But interesting and important as they are, these nutritional effects constitute only one of the manifestations that give such a high survival value to symbiotic partnerships. More remarkable is the fact that symbiosis commonly results in the production of substances and structures that neither one of the two components of the system produces when growing alone. (Examples of such creative effects are reviewed in Allen and Allen, 1954; Dubos, 1961; Humm, 1944; Jacob and Wollman, 1961; Nutman and Mosse, 1963.)

The creativeness of biological associations will be illustrated in this chapter by examples taken from various forms of life. Those provided by the bacterial world have probably the greatest telling value because bacteriologists have been so exacting in the development of techniques for securing and maintaining cultures of bacteria in what they have thought to be a "pure" state. The very austerity of their attitude toward the pure culture doctrine gives particular significance to the recent discovery that toxin production occurs in certain bacterial cultures, assumed in the past to be always toxigenic, only when the bacteria harbor the proper kind of bacteriophage. Thus, the classical diphtheria bacilli are associated with a temperate bacteriophage that endows them with the ability to produce the diphtheria toxin. Toxigenic cultures of *Corynebacterium diphtheriae,* considered pure since their discovery in 1883, provide therefore an example of the fact that association between a bacterium and a virus can create a new biological characteristic, namely toxigenicity. A similar phenomenon has been established with regard to streptococcal erythrogenic toxin. Here again, the toxin is produced only when streptococci of group A are made to carry the proper strain of temperate bacteriophage.

In an analogous manner, the serological identity of strains of salmonellae is determined by the presence of bacteriophages that are incorporated in the genetic apparatus of these organisms and that induce them to produce the characteristic somatic antigens. More generally, it can be said that most strains of bacteria, perhaps all, normally harbor bacteriophages that play a very important part in the structural make-up and in the biochemical activities of the so-called "pure" bacterial cultures.

The "killer" strains of *Paramecia aurelia* are rendered able to kill other strains of this species through the possession of the so-called Kappa factor, an intracellular organelle genetically unrelated to the strains that harbor it. This organelle has similarities to rickettsia and can be eliminated from its protozoon host by the proper techniques of cultivation. Elimination of the Kappa factor deprives the paramecia of their "killer" property, but does not apparently affect their viability or other biological characteristics (reviewed in Sonneborn, 1961).

Chloroplasts likewise can be eliminated from *Chlorella* by treatment with streptomycin. Although the alga naturally becomes unable to derive energy from photosynthesis after the loss of its chloroplasts, it can nevertheless survive and continue to grow in their absence if it is supplied with sugar. The chloroplast is an organelle genetically unrelated to and independent of *Chlorella,* but intimately associated with it; the association is so intimate indeed that the choroplasts behave as cellular constituents indispensable for the life of the alga under natural conditions (reviewed by Gibor and Granick, 1964).

Many other organisms of greater complexity that had long been regarded as pure species have also been found to constitute highly integrated biological associations. Thus, a certain strain of *Drosophila* normally carries an organelle that is not one of its genetic constituents but nevertheless confers upon it the peculiar physiological characteristic of being highly susceptible to CO_2. A great variety of morphogenetic effects have been found to be exerted by mycorrhiza fungi associated with the roots of trees, by bacteria in the roots or leaf nodules of many plant species, by viruses in the leaves and flowers of several kinds of ornamental plants. It is clear, in other words, that higher plants are commonly associated with other living things which determine some of their morphological and physiological characters, just as is the case with animals.

The generalization seems therefore warranted that, in the natural state, most unicellular as well as multicellular organisms exist in intimate association with other self-duplicating structures that profoundly affect their phenotypic characters. The mechanisms of such partnerships can range all the way from incorporation of the foreign self-duplicating structure in the chromosomal organization of the host cells, as occurs in the phenomenon of bacterial lysogeny, to the many various forms of highly involved symbiosis as occurs in lichens, in root nodules, or in insects. What matters in the present discussion, however, is not the precise mechanism through which one organism becomes associated with another, but rather the large biological significance of the morphogenetic effects that result from the association. The example of lichens is convenient to illustrate some of the far-reaching effects of such biological partnerships.

In addition to its nutritional usefulness to both members of the partnership, the association between the alga and the fungus in lichens has a number of biological consequences that could not have been predicted from the performance of the constituent microorganisms when they live independent of each other. Lichens produce organic compounds, some of them apparently unique in nature, that are not produced by either algae or fungi growing alone. Furthermore, lichens exhibit complex anatomical structures and reproductive bodies which are of no apparent use for the survival or multiplication of their isolated fungal and algal constituents, but which are essential for the maintenance and reproduction of lichens as a biological entity. The biological association of the alga and the fungus corresponds therefore not merely to an additive phenomenon but to a process creative of new forms and new properties.

Examples of the biological creativeness resulting from the integrated activities of two different organisms are found in all classes of life. In plants, one might mention the production of a hemoglobinlike substance and the fixation of atmospheric gaseous nitrogen when *Rhizobium leguminosarum* becomes associated with legume plants in the root nodules; or the deep-seated morphogenetic effects of the microorganisms (unidentified) responsible for the leaf nodules of several tropical species; or the variegated patterns produced in the petals and leaves of tulips and other plants carrying the proper viral strains. Many insects harbor in specialized organs bacteria or rickettsialike structures that also exert complex morphogenetic and nutritional effects. In certain cases, as we have seen, this association is indispensable for complete growth and development of the insect. (The profound morphogenetic effects of the microbiota on the histological development of birds and mammals will be considered in Chapter IV.3.)

As already mentioned, there is no reason to believe that the varied effects of biological associations which have been considered in the preceding pages depend on similar physiological and biochemical mechanisms. Each particular case will probably be found to involve highly specialized reactions. But the findings have large implications essentially independent of the particular mechanisms involved in each type of association.

The fact of general importance is that it is possible by several different techniques to remove from many cells various types of organelles which had been considered essential components of the cellular structure. Not only can the organelles be removed without killing the cell but, more interestingly, cell division can continue indefinitely in their absence. It is sufficient that the cell be placed in a medium that allows it to call into play alternate metabolic pathways, as has been shown with yeast and *Euglena*. Furthermore, as we have seen, integration of viruses into cells can bring about lasting physiological and even morphological alterations. (Examples illustrating these statements are quoted by Dubos, 1961; Gibor and Granick, 1964; Lederberg, 1952; Salser, 1961.)

In the light of these facts, it is clear that many if not all types of cells possess several mechanisms of information storage that are interdependent, yet remain genetically unrelated. It might be useful therefore to think of the cell as made up of an assembly of several different genetic entities that have become thoroughly integrated. Such integration, however, is likely to break down when environmental conditions become unfavorable or merely change. For example, when lysogenic bacteria are irradiated or placed in a medium unbalanced with regard to amino acid composition, the result is likely to be proliferation of the bacteriophage and cell destruction. In animals and man also irradiation and many other forms of stimuli can upset the equilibrium that normally exists between cells and their indigenous microbiota, often with the production of disease (Chapter VII.5). Ill defined as they are, these facts clearly point to the possibility that environmental stimuli can indirectly be responsible for many types of pathological processes, including cancers, by upsetting the integration between healthy cells and the unrelated genetic entities they normally harbor.

2. The social environment

a. Physiological responses to population density

The word crowd has unpleasant connotations. It evokes disease, pestilence, and group-generated attitudes often irrational and either too submissive or too aggressive. Congested cities call to mind unhealthy complexions and harassed behavior; city crowds are accused of accepting despotic power and of blindly engaging in acts of violence. In contrast, rural areas and small towns are thought to foster health and freedom. The legendary Arcadia and the Utopias of all times are imagined as comfortably populated by human beings enjoying vast horizons. The nature and history of man are far too complex, of course, to justify such generalizations, but there is some truth nevertheless in the belief that crowding generates problems of disease and behavior. However, these problems are poorly understood and their formulation is rendered even more difficult by a number of oversimplified and erroneous concepts inherited from the late nineteenth century.

During the Industrial Revolution, the crowding in tenements, factories, and offices was associated with tremendous increases in morbidity and mortality rates. Along with malnutrition, the various "fevers" were the most obvious causes of ill health. Epidemic outbreaks and chronic forms of microbial disease constituted the largest medical problems of the late nineteenth century because they were extremely prevalent, not only among the economically destitute but also among the more favored classes. The new science of microbiology that developed during that period provided a theory that appeared sufficient at first sight to explain the explosive spread of infection. The germ theory made it obvious that crowding facilitates the transfer of microbes from one person to another, and this led to the reasonable conclusion that the newly industrialized communities had been caught in a web of infection, resulting from the increase in human contacts.

The expression "crowd diseases" thus became, and has remained ever since, identified with a state of affairs conducive to the rapid spread of infective agents, particularly under unsanitary conditions. Epidemiologists have built their science on the hypothesis that the pattern of microbial diseases in a given community of animals or men is determined by the channels available for the spread of microbes. In reality, however, the rise and fall of animal populations, both in confined environments and in the field, present aspects that cannot be entirely accounted for by these classical concepts of epidemiology. The reason, as we shall now see, is that crowding has several independent effects. On the one hand, it facilitates the spread of infective agents; on the other hand, it also modifies the manner in which men and animals respond to the presence of these agents and thereby increases indirectly the prevalence and severity of microbial disease. In fact, crowding affects the response of the individual and social body, not only to infection, but also to most of life's stresses.

In many species, the numbers of animals increase continuously from year to year until a maximum population density is reached; then suddenly an enormous mortality descends. This phenomenon, known as "population crash," has long been assumed to be caused by epidemics corresponding to those which have been so destructive in the course of human history, for example plague or yellow fever. Indeed, several different kinds of pathogens have been found to attack animal populations at the time of the crash. Pasteurellae and salmonellae are among the bacterial organisms that used to be most commonly incriminated; two decades ago a particular strain of *Mycobacterium muris* (the vole bacillus), isolated from field mice in England, was thought for a while to be responsible for population crashes in these rodents. Now that viruses have taken the limelight from bacteria, they in turn have been made responsible for occurrences of widespread mortality in several animal species.

It has become apparent, however, that the relation between population crashes and microbial diseases is far less clear than was once thought. On the one hand, several different types of pathogens can be associated with crashes in a given animal species. On the other hand, there are certain crashes for which no pathogen has been found to account for the pathological picture. These puzzling observations have led to the theory that the microbial diseases associated with population crashes are but secondary phenomena, and that the primary cause is a metabolic disturbance. (This subject is further discussed from another point of view in Chapter XI.2.)

Food shortages, or at least nutritional deficiencies, were long considered as a probable cause of drastic population decline. It is well known, in fact, that when wild animals multipy without check under natural conditions they exhaust their food supply, lose weight, and bear fewer young; this occurs for example when their predators are eliminated. However, a poor nutritional state can hardly account alone for population crashes. Its effect is rather to limit reproduction, either by failure of conception or by abortion; the overall result is an automatic adjustment of population size to the food supply instead of a massive crash. In fact, drastic population declines commonly occur even when the food supply is abundant (Chapter XI.2).

The trend during recent years has been to explain population crashes by a "shock disease" related in some obscure way to overactivity of the adrenopituitary system. A notorious example of this type of crowd disease is the mass migration of the Norwegian lemmings from the mountaintops of Scandinavia. According to an ancient Norwegian belief, the lemmings periodically experience an irresistable "collective urge" either to commit suicide or to search for their ancestral home on the lost Atlantic Continent, and consequently they march unswervingly into the sea. In reality, such migrations take place whenever the lemmings become overcrowded, a situation that occurs every third or fourth year, as each mating pair produces 13 to 16 young annually. The migration of Norwegian lemmings was so massive in 1960–61 that a steamer entering the Trondheim Fjord took one hour to pass through a two-mile-long pack of swimming and sinking rodents!

Although the nature of the initial stimulus that prompts the lemmings to migrate is not understood, crowding is almost certainly one of its aspects. As the rodents become more and more crowded they fall victims to a kind of mass psychosis. This results in a wild scrambling about that, contrary to legend, is not necessarily a march toward the sea but merely random movement. The animals die, not by drowning, but from metabolic derangements associated with stress; lesions are commonly found in the brain and the adrenals.

Profound changes have also been observed to occur at more or less regular intervals in the population of snowshoe hares. According to a classical description, these animals observed in Minnesota during periods of crash

> characteristically died in convulsive seizures with sudden onset, running movements, hind leg extension, retraction of the head and neck, and sudden leaps with clonic seizures upon alighting. Other animals were typically lethargic or comatose. . . . This syndrome was characterized primarily by decrease in liver glycogen and a hypoglycemia preceding death. Petechial or ecchymotic brain hemorrhages, and congestion and hemorrhage of the adrenals, thyroid, and kidneys were frequent findings [Deevey, 1960].

Interestingly enough, many of the signs and symptoms observed in wild animals dying during population crashes have been reproduced in the laboratory by subjecting experimental animals to crowding and other forms of stress. Voles placed for a few hours a day during a month in cages containing another pair of aggressive voles eventually died, but not of wounds. The main finding at necropsy was a marked increase in the weight of their adrenals and spleen and a decrease in the weight of the thymus. Similar findings have been made on captive and wild rats.

Crowding can act as a form of stress in most species of experimental animals. In chickens, mice, rats, and voles, it causes an enlargement of the adrenals chiefly through cellular hyperplasia in the cortical areas; in addition it interferes with both growth and reproductive function.

Crowding affects many other biological characteristics of animal population; for example, the reproducibility of the response to various abnormal states, such as barbiturate anaesthesia, is affected by population density. The toxicity of central nervous system stimulants such as amphetamine is remarkably enhanced when the animals are placed in a crowded environment; central depressants protect to some degree against this aggregation effect. The experimental hypertension produced in rats bearing regenerating adrenals is increased by crowding, and coronary arteriosclerosis develops more rapidly and more intensely in chickens that are grouped than in animals kept isolated.

Field studies of voles in England have revealed the puzzling fact that their population continues to fall the year after the crash. It would appear, therefore, that the reduced viability responsible for the crash is transmitted from one generation to another. This finding is compatible with other observations which indicate that crowding of the mother affects the physical development and behavior of the offspring (Chapter I.2).

The response to almost any kind of stimulus can be modified by crowding, as is illustrated by the production of experimental granuloma. Cotton pellets impregnated with turpentine were introduced subcutaneously into groups of mice that were then either caged individually or in groups. The granulomas that developed in the grouped mice weighed 19 per cent less than in the other animals, a result probably due to the fact that the greater adrenocortical activity in the grouped mice had exerted a suppressive effect on the inflammatory reaction.

It is probable that the effect of crowding on tissue response accounts for the decrease in resistance to infection. In order to put this hypothesis to the test, mice were infected with a standardized dose of *Trichinella* and then were either isolated in individual jars or caged in groups immediately after infection. When these mice were sacrificed 15 days later, it was found that all the grouped animals had large numbers of worms (15 to 51) in their intestines, whereas only 3 out of 12 of the isolated animals showed any sign of infection. Although exposure to infection had been identical, crowding had therefore increased the ability of trichinella to invade the intestinal wall, probably by decreasing the inflammatory response to the parasite. Analogous observations have been made with regard to infantile diarrhea of mice. The incidence of clinical signs of this disease remains small or is nil when the population density in the animal room is low, but it increases as the colony approaches peak production. The infection is endemic in most colonies, but the disease does not become overt until the animals are crowded.

The grouping of several organisms of one given species has certainly many physiological consequences more subtle than those mentioned above. One such curious effect has been observed in male ducks kept constantly either in the dark or exposed to artificial light for periods of over two years. In both cases, these abnormal conditions of light exposure resulted in marked disturbances of the sexual cycles, which were no longer in phase with the seasonal rhythms. However, the animals within each group exhibited a remarkable synchronism of testicular evolution, thus revealing a "group effect" on sexual activity that was independent of light, of season, and of the presence of animals of the opposite sex.

b. Territoriality, dominance, and adaptation to crowding

As we have just seen, the epidemiology of "crowd" diseases involves factors other than those affecting the spread of infectious agents. Association with other living things modifies the total response of the organism to the various environmental forces and thereby affects susceptibility to a multiplicity of noxious influences, including infection.

A quantitative statement of population density is not sufficient, however, to forecast the effects of crowding on human beings or animals. Even more important than numbers of specimens of a given species per unit area is the manner in which each particular person or animal responds to the other members of the group under a given set of conditions. The response to population density is determined in large part by the history of the group and of its individual members; furthermore, it may be favorable or unfavorable, depending upon the circumstances.

Many types of rodents, such as laboratory rats and mice, prefer to be somewhat crowded. In fact, individually housed rats and mice usually behave in a more "emotional" or "frightened" manner than their group-housed counterparts; they are also less able to adapt to a variety of experimental procedures such as food restriction, food selection, or cold stress. Isolated mice are less able than grouped mice to overcome the disturbances in intestinal ecology caused by antimicrobial drugs and other physiological disturbances (unpublished observations). As mentioned earlier, the practice of mutual cleaning accelerates wound healing in many animal species (Chapter IV.1), and isolation has unfavorable effects on the behavior and personality structure of animals and man (Chapter I.5).

In most animal species, probably in all, each group develops a complex social organization based on territoriality and on a social hierarchy comprising subordinate and dominant members, the so-called pecking order. The place of each animal in the hierarchy is probably determined in part by anatomical and physiological endowments and in part by the history of the group. In any case, the behavioral differences that result from the pecking order eventually bring about anatomical and physiological differences far more profound than those initially present. For example, the dominant animals usually have larger adrenals than the subordinates and they grow more rapidly because they have more ready access to food. It appears also that in rhesus monkeys the young males issued from females with a high social rank have a better chance than other males to become dominant in the colony.

Under a given set of conditions, the relative rank of each individual animal is fairly predictable. Social competition is often restricted to the male sex, the reproductive fortunes of the female being determined by the status of the male which selects her. Females associated with subordinate males in experimental populations may entirely fail to reproduce. However, the pecking order is valid only for well-defined environmental conditions. For example, each canary bird is dominant in the region near its nest; and similarly chickens in their home yard win more combats than strangers to that yard. The successes of animals on their own territorial grounds bring to mind the better performance of baseball teams on their home fields.

Successful competition within the group naturally confers advantages. The despot has first choice with regard to food and mates, and its position may even increase its resistance to certain forms of stress such as infection. In a particular experiment involving tenches, one fish in the group was found to dominate the whole territory and to be the first one to feed. This dominance had such profound physiological consequences that when all the tenches were infected with trypanosomes, the infection disappeared first from the dominant fish. When this fish was removed from the tank, fighting started among those remaining; the fish that became dominant in the new grouping in its turn had first access to the food, and soon got rid of its trypanosome infection.

The phenomenon of dominance has a social meaning which transcends the advantages that it gives to the dominant individuals. Acceptance of the hierarchical order reduces fighting and other forms of social tensions and thus provides a stability that is beneficial to the group as a whole. In an undisturbed organized flock of chickens, for example, the individual animals peck each other less frequently and less violently, eat more, maintain weight better, and lay more eggs than do chickens in flocks undergoing social reorganization through removal of some animals or addition of new ones. Furthermore, the subordinate animals do not suffer as much as could be expected from their low rank in the pecking order. There is no direct competition for food or for mates in the well-organized group; the subordinates readily yield their place to the dominants at the feeding box; they exhibit no sexual interest, often behaving as if they were "socially castrated" (Chapter XI.2). Thus, the establishment of an accepted hierarchy in a stable group of animals almost eliminates the stresses of social tension and results in a kind of social homeostasis.

Needless to say, there are limits to the protective efficacy social organization can provide against the dangers created by high population density. Excessive crowding has deleterious effects even in the most gregarious rodents. When laboratory rats are allowed to multiply without restriction in a confined space, an excess of food being available at all times, they develop abnormal behavior with regard to mating, nest building, and care of the young as soon as the population becomes too dense. However, such conditions of life are extremely artificial. Under the usual conditions of rodent life in the wild, animals migrate or are killed when the population becomes too large for the amount of food available.

Although man is a gregarious animal, sudden increases in population density can be as dangerous for him as they are for animals. The biological disturbances created during the Industrial Revolution by lack of sanitation and by crowding in tenements and factories were aggravated by the fact that most members of the new labor class had immigrated from rural areas and were totally unadapted to urban life. In contrast, the world is now becoming more and more urbanized. Constant and intimate contact with hordes of human beings has come to constitute the "normal" way of life, and men have eagerly adjusted to it. This change has certainly brought about all kinds of phenotypic adaptations that are making it easier for urban man to respond successfully to situations that in the past constituted biological and emotional threats (Chapter X).

There may be here an analogy with the fact that domesticated animals do not respond to various types of threatening situations in the laboratory as do wild animals of the same or related species. In any case, the effects of crowding on modern urban man are certainly very different from those experienced by the farmer and his family when they were first and suddenly exposed a century ago to the city environment of industrialized societies.

The readiness with which man adapts to potentially dangerous situations makes it unwise to apply directly to human life the results of experiments designed to test the acute effects of crowding on animals. Under normal circumstances, the dangerous consequences of crowding are mollified by a multiplicity of biological and social adaptations. In fact, crowding per se, i.e. population density, is probably far less important in the long run even in animals than is the intensity of the social conflicts, or the relative peace achieved after social adjustments have been made. As already mentioned, animal populations in which status differences are clearly established are likely to reach a greater size than those in which differences in rank are less well defined.

Little is known concerning the density of population or the intensity of stimulation that is optimum in the long run for the body and the mind of man. Crowding is a relative term. The biological significance of population density must be evaluated in the light of the past experience of the group concerned, because this experience conditions the manner in which each of its members responds to the others as well as to environmental stimuli and trauma.

Laying claim to a territory and maintaining a certain distance from one's fellow are probably as real biological needs in man as they are in animals, but their expressions are culturally conditioned. The proper distance between persons in a group varies from culture to culture. People reared in cultures where the proper distance is short appear "pushy" to those coming from social groups where propriety demands greater physical separation. In contrast, the latter will appear to the former as behaving in a cold, aloof, withdrawn, and standoffish manner. Although social anthropologists have not yet adequately explained the origin of these differences, they have provided evidence that ignorance of them in human relations or in the design of dwellings and hospitals can have serious social and pathological consequences.

The problems posed by crowding in human populations are thus more complex than those which exist in animal populations because they are so profoundly conditioned by social and cultural determinants. Indeed, there is probably no aspect of human life for which it is easier to agree with Ortega y Gasset that "man has no nature. What he has is a history" (Chapter I.1). Most experimental biologists are inclined to scorn discussions of mob psychology and related problems because they feel that the time is not yet ripe for scientific studies on the mechanisms of collective behavior. Yet the phrase "mob psychology" serves at least to emphasize that the response of human beings to any situation is profoundly influenced by the structure of the social environment.

The numerous outbreaks of dancing manias that occurred in Europe from the fourteenth to sixteenth century constitute a picturesque illustration of abnormal collective behavior; such an event was witnessed by P. Breughel the Elder and became the subject of one of his most famous paintings, "The Saint Vitus Dancers," now in Vienna. Even today, revivalists, tremblers and shakers often outdo the feats of the medieval performers during the dancing manias. And millions of people can still be collectively bewitched by the antics of a Hitler or other self-proclaimed prophet, to whom they yield body and soul. What happens in the mind of man is always reflected in the diseases of his body. The epidemiology of crowd diseases cannot be completely understood without knowledge of mob psychology.

V. The Indigenous Microbiota

1. The composition of the indigenous microbiota

a. Indigenous, autochthonous, and normal microbiota

Animals and human beings have evolved in intimate and constant association with a complex microflora and microfauna. Under natural conditions, the development and functions of their tissues are influenced by countless microorganisms that are always present in the digestive and respiratory tracts, and probably also in other organs. It is to be expected, therefore, that anatomical structures and physiological needs have been determined in part by the microbiota that prevailed during evolutionary development, and that many manifestations of the body at any given time are influenced by the microbiota now present. The microbiota is part of the environment to which man has had to become adapted.

Paradoxically, observations on germ-free animals have provided the most compelling evidence that higher living forms are dependent on microbial activities for the completion of their development and for the performance of their normal activities. Although animals of many different species can grow and reproduce in the complete absence of detectable microorganisms,[1] the nutritional requirements of so-called germ-free animals are more exacting than those of conventional animals; their histological development is incomplete; their susceptibility to infection and to various forms of stress is greater. Such peculiarities would preclude their survival under normal conditions. It is certain, in other words, that several kinds of microbes play an essential role in the development and physiological activities of normal animals and man.

The following discussion will be focused on the composition and activities of the indigenous microbiota, i.e. on the microbiota present in the healthy state. No mention will be made of the

1. As mentioned earlier, it remains to be proven that so-called germ-free animals do not harbor filterable viruses (see Chapter IV.2).

microbial species usually regarded as pathogenic. In reality, however, the contrast between the two groups is not as sharp as used to be thought. Indeed, it is so difficult to formulate criteria which differentiate clearly the indigenous from the pathogenic biota that the distinction is arbitrary and usually meaningless. On the one hand, most microorganisms commonly harbored by the body in the state of health are capable of exerting a wide range of pathological effects under special conditions. On the other hand, many of the microorganisms classified as pathogens, indeed probably all of them, often persist in vivo without causing overt disease (see Chapter VII.3 and 5).

At first sight, the statement that the indigenous microbiota is responsible for a great variety of disease processes seems to be incompatible with the view expressed earlier than animals and man have become dependent on this microbiota during evolutionary development. The following facts may help in accounting for this apparent anomaly.

The nefarious effects of the indigenous microbiota become manifest chiefly, and perhaps only, when animals or human beings are placed under conditions that differ from those under which the evolutionary equilibrium between host and microbes became established. Nutritional deficiencies, exposure to toxic agents, and certain kinds of physiological stress are among the many causes of disturbances associated with disease processes caused by the indigenous microbiota. In other words, some form of pathological or at least abnormal state must exist before the indigenous biota can multiply to such an extent that it causes detectable deleterious effects. As in so many other cases, Pasteur had reached very early a remarkable prescience of the pathogenic activities of the indigenous flora, and clearly suggested that a physiological disturbance might be in certain cases the primary cause of the infectious process, rather than its consequence. He made the prophetic remark that when a common microorganism is found in association with a pathological state, it is difficult to decide, as he put it, "which came first, the chicken or the egg? . . . It is possible on logical grounds to take the view that the disease itself, whatever its unknown original cause, was the condition which made possible the development of the microorganism." Half a century later, George Bernard Shaw was to make this idea his own in the preface of *The Doctor's Dilemma:* "The characteristic microbe of a disease might be a symptom instead of a cause."

Another factor to be taken into consideration in relating the indigenous microbiota to pathological processes is that this microbiota includes not only the flora and fauna acquired in the course of evolutionary development, but, in addition, a great variety of other microorganisms that find their way into the body of each individual person or animal as a result of the accidents of life.

The indigenous biota is thus probably made up of several classes of microorganisms, which are completely different in origin. Some of the microorganisms have achieved a symbiotic status with their host through a long period of evolutionary association; it seems appropriate to refer to them as constituting the *autochthonous* biota. The others possess some degree of pathogenicity and therefore are capable of becoming established in the tissues, but can nevertheless achieve a temporary state of biological equilibrium with the host under the proper conditions. In fact, it has now been established beyond doubt that even the most virulent pathogens can persist in the tissues, often without manifesting their presence by obvious pathological manifestations. (See Chapter VII.3 for discussion of "persisters.")

The expression "indigenous microbiota" as commonly used thus includes true pathogens. It would seem useful to reserve the expression *normal* microbiota for designating those microorganisms so common in a given community that they become established in practically all its members. The normal microbiota of the New Guinea aborigines includes malaria parasites, whereas most if not all New York City residents harbor a variety of ill-defined viruses potentially capable of causing respiratory disease. For example, more than 150 different viruses have been recovered from human beings whereas only 50 different viral diseases of man have been recognized so far. In summary, the indigenous microbiota of each individual person is made up of microorganisms present during evolution (the autochthonous microbiota), of those which are ubiquitous in his community (the normal microbiota), and of true pathogens which have been accidentally acquired and are capable of persisting in the tissues.

Studies of the bacterial components of the indigenous microbiota have been concerned chiefly with the common Gram-negative enterobacilli, the enterococci, the Gram-positive aerobic sporeformers, and the clostridia, all organisms that grow rapidly on artificial culture media and therefore can be readily isolated from the intestinal tract. Emphasis on these organisms is explainable in large part, if not completely, by the fact that they lend themselves to in-vitro studies. Yet it is certain that they represent but a very small part of the total indigenous microbiota, and not the most important. A large percentage of indigenous microorganisms cannot be identified, let alone enumerated, for lack of adequate cultural techniques; their biological characteristics and role in vivo are consequently unknown. The most that can be done at present, therefore, is to report a number of facts derived from the detailed study of a few selected cases, in the hope that such information will help in creating a background from which knowledge of the true composition of the autochthonous microbiota will eventually emerge. The following discussion will be limited to the bacterial flora of the gastrointestinal tract.

b. The bacterial flora of the gastrointestinal tract

The fetus seems to be essentially free of bacteria before birth; this view is supported by the experience gained from the study of germ-free animals. There are some indications, however, that protoplastlike forms can be transferred from the mother to the fetus in utero. But on the whole, the indigenous bacteria are derived first from the mother at the time of or immediately after birth, then from the external world.

In general, most of the microorganisms acquired from the environment with which the infant or newborn animal comes into contact are rapidly destroyed or eliminated. The ones that persist in the body are those which are adapted to a symbiotic life with their host, or which are capable of becoming established in its tissues by virtue of some pathogenic property. On very slim evidence we shall attempt to differentiate here the former group, designated above as constituting the autochthonous microbiota, from the latter, which has thus an accidental origin even though it includes microorganisms commonly present in the tissues because they are ubiquitous in the community. Most of the published information refers to the intestinal tract, but many of the findings apply probably just as well to the rest of the gastrointestinal tract, including the buccal cavity.

As has been known ever since Tissier's pioneering observations, the bacterial flora of the feces in breast-fed human infants is almost exclusively Gram-positive and consists chiefly of lactobacilli. Other bacterial species, even the coliform bacilli, become prominent only after the infant begins eating other food, or if he is bottle-fed from the beginning. (Further details concerning Tissier's findings and the biological characteristics of lactobacilli will be considered later.)

Many studies carried out during the past 50 years have confirmed that lactobacilli of several species constitute one of the dominant microbial groups in the intestine and the mouth, not only of the infant, but also of children and adults. This is true also in certain animal species, but not in all. Numbers exceeding 10^9 or even 10^{10} viable lactobacilli per gm. of stool or of intestinal contents have been frequently reported for man, pigs, rats, mice, and other animals. There is also general agreement that Gram-negative, nonsporulating, anaerobic bacilli, loosely classified under the genus *Bacteroides,* are extremely numerous in the buccal and intestinal flora of man and of certain animals. In addition, stained preparations reveal the presence in the mouth and the intestine of large numbers of fusiform bacteria, but these organisms have not been satisfactorily classified or enumerated, for lack of quantitative cultural techniques. (Detailed accounts of this classical bacteriological knowledge are presented in the reviews by Haenel, 1960; Petuely, 1962; and Rosebury, 1962.) The indigenous microbiota certainly contains in addition many other species which have not yet been identified, cultured, or even recognized. There is suggestive evidence, for example, that certain microorganisms are present in the linings of the intestine in the rabbit, some of them perhaps intracellularly; however, nothing is known of their nature.

Studies now in progress in our laboratory have revealed that, contrary to general belief, there normally exists a very large bacterial population in the stomach and in the small intestine. Experiments with mice, white rats, and hogs have given consistent results, which are summarized in Table 5. It is seen in this table that organisms of the bacteroides group are extremely abundant in the cecum, but much less numerous or absent from the stomach and intestine of the mouse. In marked contrast, lactobacilli and anaerobic streptococci (group N) are normally present in very large numbers not only in the cecum but also in the stomach and intestine of healthy animals; indeed, they are at least as numerous in the stomach as in the cecum. It is worth emphasizing that none of these organisms have been recovered from the lungs, the spleen, the kidneys, or the heart of healthy mice.

The lactobacilli and anaerobic streptococci (group N) become

TABLE 5. Digestive Flora of NCS Mice *

	Lacto-bacilli [a]	*Strept. (N)* [a]	*Bacter-oides* [a]	*Coliforms & enterococci*	*Clostridia*
Stomach	10^9 [b]	10^8	0	0	0
Small intestine	10^8	10^6	0	0	0
Cecum	10^9	10^8	10^9	10^3	0

* Tests carried out on animals ranging from 4 weeks to 8 months of age.

a. Numbers of viable organisms determined by plating on selective media incubated under anaerobic conditions.

b. The figures indicate the approximate numbers of colonies recovered per gm. of organ homogenate.

established very early in all parts of the gastrointestinal tract and persist uniformly throughout life. Moreover, they are very closely associated with the walls of the organs in which they occur and seem to be located in the mucous layer. Very consistently, much larger numbers of these organisms could be recovered from the homogenates of washed organs than from either the washings or even the contents.

Tests carried out in germ-free animals have confirmed the findings made in conventional animals. Germ-free mice were fed cultures of lactobacilli, anaerobic streptococci (group N), and bacteroides isolated from conventional mice, and they were sacrificed at various times thereafter. As illustrated in Table 6, the lactobacilli and streptococci rapidly became established in all organs of the gastrointestinal tract and were found in close association with the walls, whereas the bacteroides multiplied only in the cecum. In the light of all these facts, the possibility seems worth considering that lactobacilli, anaerobic streptococci, organisms of the bacteroides group, and perhaps also the ill-defined fusiform bacteria constitute an important part of the autochthonous microbiota of the gastrointestinal tract in man and several animal species.

It may seem surprising that the enterobacteria (especially *Escherichia coli* and the *Proteus* and *Pseudomonas* groups), the enterococci, and the clostridia so commonly found in the intestine have not been mentioned as part of the autochthonous microbiota. The decision not to include them here is based on two different kinds of facts.

It appears from a review of the literature that the organisms

TABLE 6. Colonies Recovered 3 Weeks after Feeding Bacterial Cultures to Germ-Free Mice *

		Lacto-bacilli	*Strept. (N)*	*Bacter-oides*
Stomach	Content	80 [a]	25	<1
	Wash [b]	7	1	
	Homogenate [c]	50	10	0
Small intestine	Content	25	8	<1
	Wash	3	1	
	Homogenate	11	5	0
Cecum	Content	200	50	10,000
	Wash	25	1	
	Homogenate	200	40	1,000

* The cultures of lactobacilli, streptococci (Group N), and bacteroides recovered from normal mice were fed to germ-free mice approximately 4 weeks old. (Adapted from R. Dubos and R. Schaedler, 1964.)

a. The figures refer to the numbers of colonies recovered from the animals 3 weeks after feeding cultures. Figures must be multiplied by 10^6 to give number of colonies per gm. of organ.

b. "Wash" refers to colonies recovered from third washing (from 1 gm. of organ).

c. "Homogenate" refers to colonies recovered per gm. of washed organ homogenized in Teflon grinder.

of the groups just listed are usually far less numerous in healthy men and animals than are the lactobacilli, anaerobic streptococci, and organisms of the bacteroides group. More importantly perhaps, their numbers are extremely erratic. The findings for mice reported in Table 7 illustrate the range of the differences in the populations of the various bacterial groups recovered from fecal material. It has been found in a particular study, for example, that the coliform count per gram of human feces ranged from 10^4 to 2.6×10^8, and the enterococcus count from 10^3 to 2.7×10^8. In our experience, such large variability is never observed in healthy animals with regard to the populations of lactobacilli, anaerobic streptococci, or bacteroides. As already mentioned, the emphasis on common enterococci and coliform bacilli is to a large extent a consequence of the fact that these organisms can be readily cultivated and enumerated, rather than an expression of their significance as members of the indigenous flora.

Another reason for regarding organisms such as *E. coli*, the *Proteus* and *Pseudomonas* groups, enterococci, and the clostridia as foreign to the autochthonous biota is provided by the experience gained with the NCS colony of albino mice that has been maintained essentially free of these organisms for the past five years in our laboratory and for the past three years at the Pasteur Institute in Garches, France. As other results obtained with the NCS colony will be discussed later (Chapter VI.1), it will be useful to present here a few facts concerning its development and characteristics.

The NCS colony is made up of the progeny of seven males and five females obtained by Caesarian section from three animals of the standard Swiss [2] colony (SS) of albino mice. The young were foster-fed by mice of the so called "Princeton" colony. They were treated (especially with piperazine) to rid them of ecto and endo parasites. When the animals so obtained are maintained under properly sheltered conditions, their stools remain free of detectable *Eperythrozoon coccoides, E. coli, Ps. aeruginosa, Proteus sp.*, hemolytic streptococci, clostridia, and PPLO. The animals also seem to be free of the viruses of ectromelia, epizootic bronchiectasis, epidemic diarrhea of infant mice, and lymphocytic choriomeningitis.

Table 7 presents some of the differences in intestinal bacterial

TABLE 7. Comparative Fecal Flora of Two Strains of "Swiss" Mice

	SS mice [a]	*NCS mice* [b]
Lactobacilli	10^8 [c]	10^9
Enterococci	10^6	10^3
Coliform bacilli	10^9	10^4
Lactose fermenters	10^7	0
Pseudomonas	present	0
Proteus	"	0

a. 7 other strains of mice obtained from commerical breeding farms gave similar results.

b. Raised in our laboratories at the Rockefeller Institute.

c. Numbers of viable bacteria per gm. stool.

flora between animals of the NCS colony and of the SS colony, from which the former was derived. The table shows that, as already mentioned, NCS animals differ from those of the parent SS colony in the fact that cultures of their intestinal contents do not

2. The so-called "Swiss" colony of albino mice was probably established at the Pasteur Institute in Paris by A. Borrel, then transported by him to Strasbourg. From there, a subcolony was established at the Institut du Cancer in Lausanne, Switzerland. In 1926, Dr. C. B. Lynch brought some of the animals from Lausanne to the Rockefeller Institute in New York (hence the name "Swiss" mice). Ever since that time, the colony has been maintained by brother-sister matings not only at the Rockefeller Institute, but also in many other laboratories all over the world.

yield any *E. coli, Proteus, Pseudomonas,* or clostridia, and only small numbers of enterococci and enterobacteria (nonlactose-fermenters). In contrast, they consistently yield very large populations of lactobacilli (at least two different types), of anaerobic streptococci, and also of Gram-negative nonsporulating anaerobic bacilli (at least two different types of bacteroides).

The most unexpected finding has been that, with the proper diet, NCS mice retain their characteristic fecal flora, with very small numbers of enterococci and of coliform bacilli, even when placed for many months after weaning in a general animal room under conditions that would seem to allow contamination with other bacterial species. Of special interest is the fact that, under these conditions, their fecal flora remains entirely free of lactose-fermenting coliform bacilli, of *Ps. aeruginosa,* and of *Proteus;* their enterococcus population remains small. In fact, it has proved difficult to establish *E. coli* in normal adult NCS mice, even after feeding them for several days large doses of a culture of this organism freshly isolated from mice of the SS colony. This finding presents similarities to the recent discovery that, whereas young pigs suffer from severe enteritis when infected with the proper strain of *E. coli,* they fail to acquire this infection or even to become carriers of the organism when their exposure to it is delayed until later in life.

By sacrificing mice at daily intervals after birth, it has been possible to determine the approximate time at which some components of their flora become established in the intestinal tract. In brief, it was found that the lactobacilli and anaerobic streptococci begin to multiply immediately after birth and reach a high level around the fourth to seventh day of life; their population remains consistently high thereafter. In contrast, the enterococci and enterobacteria (nonlactose-fermenters) begin to multiply somewhat later. Their population reaches a maximum after two weeks, then falls precipitously. Thereafter it remains at a low level as long as the conditions of husbandry are good. In conventional mice, the population of *E. coli* in the intestinal tract is already abundant when the animals are one week of age; it reaches a maximum at two weeks, then falls to a lower level.

In all colonies of mice so far tested, the changes in population size of enterobacteria and enterococci suggest that these organisms cause an intestinal infection, but one from which the animals recover. In this light, the persistence of the organisms in the feces would correspond to a carrier state, or to a low-grade continuous infectious process. The fact that the populations of lactobacilli, anaerobic streptococci, and bacteroides remain at a high level throughout the lives of healthy animals suggests in contrast that these organisms do not evoke a protective immunological response; this is one of the facts which supports the hypothesis that they are part of a truly symbiotic biota.

Of even greater significance, probably, are the results obtained by breeding ordinary mice under sheltered conditions. Such mice, which had the usual type of complex intestinal flora, were housed and allowed to breed in cages protected from outside contamination by a filter of glass wool. Food, water, and bedding were sterilized; piperazine was added to the drinking water in an attempt to reduce helminth and other parasitic infections, but no antibacterial drug was used at any time. Progressively, the intestinal flora became simpler in this protected environment. After a few generations *E. coli* all but disappeared from the feces of some colonies; the numbers of enterobacteria and enterococci fell to low levels; whereas the populations of lactobacilli, anaerobic streptococci, and bacteroides remained as high as in NCS mice.

The few bacteriological studies available for the human intestinal flora provide a picture similar to that derived from the study of mice. Many recent authors have confirmed Tissier's statements that the *Lactobacillus* population reaches a very high level during the first few days of life in the breast-fed infant. These organisms remain extremely numerous in bottle-fed infants, but then large numbers of other microorganisms also invade the intestinal tract. *Clostridium perfringens, E. coli,* and enterococci tend to reach high levels during early life, then to decrease more or less rapidly. This trend, which has been referred to as "a general feature of aging," has been observed in several animal species as well as in the human infant (Table 8). In agreement with the hypothesis outlined above, these findings might signify that the normal body has mechanisms to eliminate, or at least to keep under control, microorganisms different from those which became symbiotic during evolutionary development.

c. *Influence of diet and drugs on the bacterial flora of the gastrointestinal tract*

The nutritional state of the host influences profoundly the body response to certain pathogens and may even determine whether or not these can multiply in the tissues; dietary factors probably

TABLE 8. Viable Fecal Flora of Infants *

Milk formula	Intestinal pH	Colony counts [a] Lactob. Bifidus	Enterococci	Coliforms
Unheated human milk	5.8	1,100	5	10
Boiled human milk (full-term infants)	5.8	120	55	35
Boiled human milk (premature infants)	5.9	600	300	440
⅔ cow's milk (extra sucrose)	7.4	1,200	580	1,450
⅔ cow's milk (extra lactose hexametaphosphate)	7.2	3,600	500	2,000

* Adapted from Gyllenberg and Roine, 1957.
a. Multiply by 10^6 to obtain values per gm. of feces.

condition also the ability of the oral bacteria to cause dental caries (Chapter VI.4). Unfortunately, the mechanisms of the interrelationships between nutritional state and microbial disease are still poorly understood. Somewhat more direct information is available concerning the effect of the dietary regimen on the composition of the bacterial flora of the gastrointestinal tract.

Observations in animals and man indicate that the composition of the intestinal flora in each particular animal or person remains approximately constant under stable conditions. The organisms of one host may fail to establish themselves in another host of the same species even when contact is intimate and frequent. It is certain, on the other hand, that the flora can be profoundly altered, both qualitatively and quantitatively, by a variety of environmental and physiological influences.

The role of the nutritional regimen in intestinal physiology became a popular topic around the turn of the century through Metchnikoff's writings. For various reasons, entertaining if not convincing, Metchnikoff came to believe that many of the ills of old age are caused by toxic products originating from intestinal fermentation. He believed also that peasants in Bulgaria and in Southern Russia lived to a ripe old age because they consumed large amounts of milk acidified by the growth of lactobacilli. From these unproven premises, he concluded that intestinal intoxication could be prevented, or at least minimized, by establishing in the intestine the proper kind of lactobacillus flora capable of inhibiting the growth of putrefying organisms.

Although Metchnikoff's imaginings were not supported by experimental or clinical evidence, belief in the noxious effects of intestinal intoxication became so firmly established and so widespread that it stimulated a fashion for colostomy. More profitably, perhaps, it also led bacteriologists to isolate and cultivate lactobacilli from the human intestine in the hope that these organisms could be used to control intestinal intoxication in human beings. A number of clinicians came to believe that feeding cultures of lactobacilli to their patients had beneficial effects. The popular interest in this field generated the yogurt and kumiss fads among the general public; it also inspired Aldous Huxley to write a novel entitled *After Many a Summer Dies the Swan,* in which he made human beings achieve some sort of immortality by altering the composition of their intestinal flora through the consumption of the raw contents of carp gut!

Progressively, however, scientific and public interest in the subject began to wane; by the second quarter of the present century, only a few pediatricians, medical historians, food faddists, and manufacturers of acidified milk were still aware of Metchnikoff's claims. The renewal of interest in this field dates from two unrelated findings, each of great practical importance. One was the frequent occurrence of intestinal disorders following the administration of antimicrobial drugs; the other was the acceleration of growth rates achieved by adding small amounts of these drugs to the feed of farm animals. (These two topics will be considered in greater detail in Chapter V.2.)

The first convincing discovery inspired by Metchnikoff's writings was the demonstration by Tissier that the fecal flora of breastfed human infants consists chiefly of Gram-positive lactobacilli, and that other bacterial species, especially Gram-negative enterobacilli, become numerous in it only after initiation of bottle-feeding. The strain that dominates the intestinal flora until the time of weaning was designated by Tissier *Lactobacillus bifidus.* More recently, it has been reclassified as *Actinomyces bifidus.* Granted the validity of this taxonomic differentiation, it will be more convenient even though perhaps less accurate to refer here to the whole group by the designation *Lactobacillus,* without attempting to distinguish between species.

Ever since Tissier's classical observations, studies on man and animals have confirmed beyond doubt that the nutritional regimen can effect profoundly the composition of the intestinal flora.[3]

3. For the American medical public, it is of interest to note that the illustrious neurosurgeon Harvey Cushing devoted an extensive experimental study to this problem as early as 1900.

(The early literature is reviewed in the monograph by Rettger et al., 1935.) Rats being the animals most commonly used for nutritional experiments, it is not surprising that the effect of diet on their intestinal flora was extensively studied half a century ago. The composition of this flora was found to change rapidly when the animals were transferred from a regimen of ordinary mixed food to a better-defined mixture consisting of "starch, lard, protein-free milk, and a pure protein." With this simpler regimen, the Gram-positive microbial population increased from 35 per cent to 85–100 per cent. In particular the *Lactobacillus* population increased markedly whereas that of *E. coli* became much less abundant.

It is now certain that, in addition to the strain of *L. bifidus* first recognized by Tissier, many other strains of the same species occur in the stools of man and certain animals. Most of these strains have exacting and rather unusual growth requirements. Several different growth factors required by various strains of *Lactobacillus* have been identified at the present time: Bifidus factor 1, an amino sugar; Bifidus factor 2, a peptide related to strepogenin; and hypoxanthine. Other nutritional factors certainly affect the comparative proliferation of the various strains of lactobacilli. For example, the growth of the strain initially isolated from human infants by Tissier is inhibited by adenine, guanine, and xanthine, but this inhibitory effect is neutralized by hypoxanthine. Furthermore, lactulose is said to increase the multiplication in vivo of the strain of *L. bifidus* characteristic of breast-fed infants, although it does not seem to act as a growth factor for this organism in vitro.

In human infants, as already mentioned, the substitution of breast-feeding by bottle-feeding results in a dramatic increase in the numbers of Gram-negative bacilli and of clostridia in the stools (Table 8). However, the numbers of lactobacilli do not necessarily decrease at the time of weaning. In fact, this bacterial group apparently remains extremely abundant in the human fecal flora even during adult life under ordinary conditions. There is strong evidence, however, that the types of lactobacilli change with the diet and with age; in other words, the lactobacilli present in the adult may be different from those present in the infant.

In mice also the composition of the diet affects the numbers and types of lactobacilli present in the gastrointestinal tract. In brief, the organs of mice contain many more lactobacilli and anaerobic streptococci when the animals are fed certain brands of commercial pellets than when they are fed a semisynthetic diet, containing casein supplemented with cystine and all known growth factors. The differences in the numbers of lactobacilli in the two groups of animals are of the order of 100- to 1,000-fold. Moreover, the rhizoid type of *Lactobacillus* that is abundant in the mice fed pellets is consistently absent from those fed the casein diet. These differences, first recognized in stool cultures, have now been confirmed and much extended by studies of the bacterial population of the stomach. The nutritional factors responsible for this consistent and striking effect of the diet on the gastrointestinal flora of mice have not yet been identified. But there is no doubt that, in mice fed the proper kind of mixed food as provided by certain commercial pellets, the mucous layer of the stomach contains an immense population of rhizoid lactobacilli and anaerobic streptococci which do not appear in cultures recovered from mice fed a semisynthetic casein diet.

The influence of the nutritional regimen on the flora of the gastrointestinal tract is even more striking in mice given penicillin. As will be shown later (Table 9), administration of this antimicrobial drug brings about the elimination of lactobacilli, anaerobic bacilli, and bacteroides from the gastrointestinal tract and the proliferation of clostridia, enterococci, and coliform bacilli. Whereas the flora returns to its original pre-penicillin state within a few days or at most a few weeks after discontinuance of the drug if the animals are fed pellets, a much longer time is required for this change back to normal if they are fed the semisynthetic casein diet. On this latter regimen, indeed, the animals may fail altogether to overcome the disturbances in intestinal ecology caused by penicillin treatment.

It has long been known that the various utilizable carbohydrates differ in their effects on the nutritional state. In pigs, for example, lactose is far superior to cornstarch with respect to both weight gain and efficiency in food utilization. The type of carbohydrate used in the diet also affects profoundly the composition of the intestinal flora. Coliform bacilli, streptococci, staphylococci, molds, and yeasts are less numerous in each section of the intestinal tract when the pigs are fed lactose than when they receive starch; in contrast, the numbers of lactobacilli and anaerobes are high when the diet contains lactose.

The nature of the carbohydrate source in the diet seems to affect most animal species, and it influences many different physiological and even anatomical characteristics. In rats, lactose feeding brings about a decrease in the percentage of carcass fat and an increase in cecum weight. In this case again, lactose tends to encourage a predominantly Gram-positive bacterial flora, a fact which may be of significance in accounting for the enlargement of the cecum in these animals (Chapter V.3). In a particular experiment, the pH of the cecum was 6.5 in animals fed lactose, whereas it was 7.9 in animals fed either glucose, or glucose and galactose. As we shall see later, an increase in the concentration of certain organic acids in the intestinal tract is correlated with higher resistance to certain infections (Chapter V.4).

There have been several reports that the composition of the digestive flora is affected by various physiological and environmental factors other than composition of the diet. For example, it is claimed in the early literature on the subject that the lactobacillus flora of infants decreases during periods of föhn wind; more recently it has been found that the biochemical activities of the intestinal flora in the pig exhibit a seasonal fluctuation (Chapter III.2). For obvious reasons, extensive studies have recently been made of the effect exerted on the indigenous flora by antimicrobial drugs. Disturbances of intestinal ecology caused by an antimicrobial drug were probably first observed in white rats fed a diet containing 0.5 per cent succinylsulfathiazole; this treatment brought about a lasting depression in the numbers of lactobacilli followed by an increase in enterococci and yeastlike forms.

Because the indigenous flora of NCS mice is relatively simple under undisturbed conditions (Chapter V.1), these animals lend themselves admirably to the demonstration that antimicrobial drugs can upset the ecology of the intestine. As seen in Figures 3 and 4, administration *per os* of even small amounts of penicillin or tetracycline to NCS mice causes a sudden disappearance of lactobacilli and anaerobic streptococci from the fecal flora, accompanied by an explosive and lasting increase in enterococci and in Gram-negative enterobacilli. It is of special interest that, whereas lactose-fermenting coliforms cannot be detected in the stool cultures of the untreated animals, these bacteria seem to appear "spontaneously" and soon multiply extensively after administration of penicillin. Many similar experiments with several drugs used in various concentrations have established a striking cor-

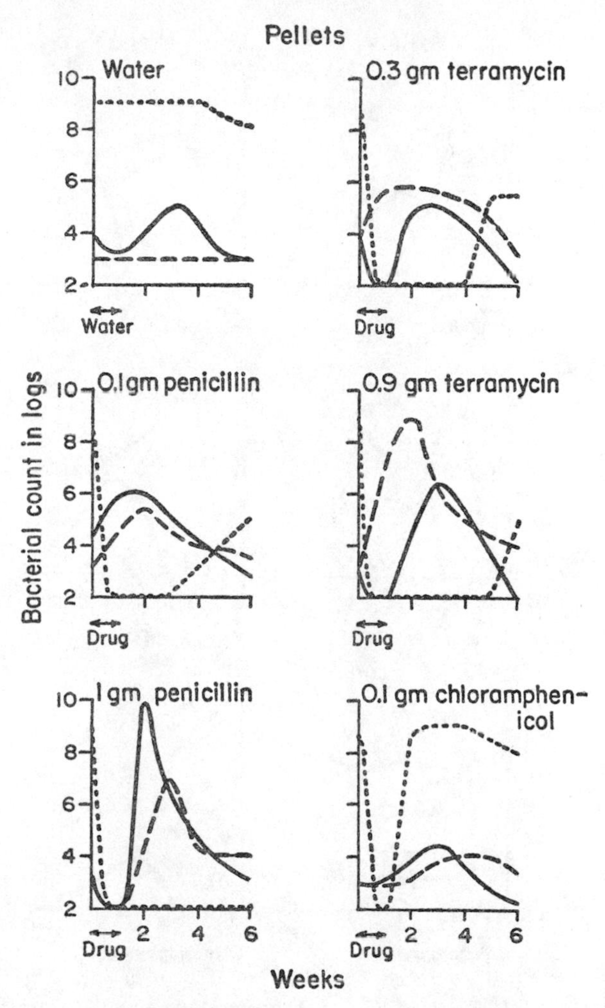

FIGURE 3. Effect of various antimicrobial drugs on the fecal flora of NCS mice. All animals were fed D&G pellets throughout the experiment. The drugs were added to the drinking water during the first week of the experiment, then discontinued. The following concentrations were used: penicillin G, 1 or 0.1 gm. per liter; terramycin, 0.9 or 0.3 gm. per liter; chloramphenicol, 0.1 gm. per liter.

Lactobacilli -----

Enterococci ———

Gram-negative bacilli ————; • indicates the presence of lactose fermenters.

Results: All drugs caused a rapid disappearance of the lactobacilli from the fecal flora, and an increase in enterococci and coliform bacilli. The fecal flora progressively returned to its initial state after discontinuance of the drugs. The results are given as logarithms of numbers of colonies obtained per gm. of stool (collected between 9 A.M. and 12 A.M. (Reproduced from R. Dubos, R. Schaedler, and M. Stephens, 1963.)

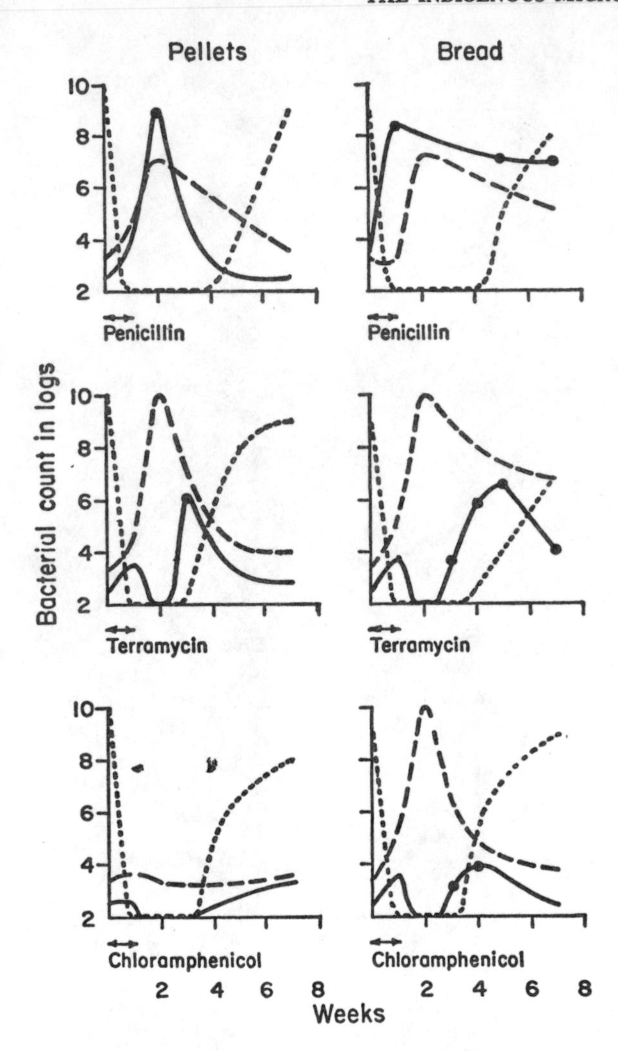

FIGURE 4. Effect of various antimicrobial drugs and diet on the fecal flora of NCS mice. Mice were fed either D&G pellets or enriched bread throughout the experiment. Penicillin G, terramycin, or chloramphenicol was added in a concentration of 0.3 gm. per liter to the drinking water during the first week, then discontinued.

Results: Administration of penicillin or terramycin was followed by disappearance of lactobacilli and by a marked increase in the numbers of Gram-negative bacilli and enterococci. Lactose-fermenters were abundant in mice fed bread and treated with penicillin. The disturbances in fecal flora were least pronounced with chloramphenicol. In all mice, the fecal flora progressively returned to its initial state following discontinuance of the drugs; this occurred more rapidly in animals fed pellets. For further details see Figure 3. (Reproduced from R. Dubos, R. Schaedler, and M. Stephens, 1963.)

relation between the disappearance of lactobacilli, anaerobic streptococci, and bacteroides and the increase in numbers of certain other components of the intestinal flora. Experiments in progress at the time of writing show that the findings in stools are the expression of even more striking changes in the flora of the stomach and intestine (unpublished observations by R. Dubos, R. Schaedler, and R. Costello).

Following discontinuance of drug therapy, lactobacilli, anaerobic streptococci, and bacteroides slowly reappear in the gastrointestinal tract of mice and eventually return to their initial level when the animals are fed the proper diet; the populations of enterococci and Gram-negative enterobacilli progressively decrease at the same time. Needless to say, the fact that the two trends are more or less simultaneous does not mean that they are causally related. Drug treatment may bring about the elimination of components of the flora of the gastrointestinal tract that have not yet been identified. Furthermore, the change goes pari passu with extensive multiplication of certain anaerobic species, in particular clostridia. Unfortunately, knowledge of the anaerobic indigenous flora is so inadequate that an exact description of its evolution under drug therapy is impossible at the present time.

Incomplete as it is, the knowledge concerning the effects of antimicrobial drugs on the microbial ecology of the gastrointestinal tract is sufficient to provide an explanation for the well-documented clinical experience that antimicrobial therapy often results in infections of various organs caused by bacteria, yeasts, fungi, and other microorganisms usually considered of low pathogenicity. These pathological phenomena occur not only in man

TABLE 9. Alteration of the Fecal Flora by Penicillin

Time (days)		*Lacto-bacilli*	*Bacter-oides*	*Coliforms*	*Clostridia*
0		10^9*	10^7	10^3	0
1		±	±	±	0
2	oral penicillin	0	0	0	0
6		0	0	0	0
9		0	0	10^8	10^8
15		10^7	10^6	10^6	10^5

* The figures indicate numbers of colonies recovered per gm. of feces.

but also in animals. In guinea pigs, administration of penicillin commonly brings about such an intense invasion of the intestinal tract by various bacteria that a fatal outcome usually ensues. This fact explains why penicillin, a drug otherwise so remarkably safe, is so highly toxic for these animals.

Fatal accidents caused by antimicrobial drugs have occurred also under practical conditions of animal husbandry. A few years ago, diarrheal disease suddenly struck down a large number of animals on a chinchilla farm. This catastrophe was apparently caused by staphylococci overgrowing the indigenous intestinal biota of the chinchilla after one of the tetracycline drugs had been added to their diet in order to enhance growth (Chapter V.2). Such accidents are often observed, of course, in human patients treated with antibacterial drugs. Their most plausible interpretation is that treatment with these drugs eliminates certain components of the autochthonous microbiota that normally hold in check potentially pathogenic members of the indigenous microbiota.

2. Effects of the Microbiota on Host Nutrition

As we have seen, many kinds of microorganisms live in a state of symbiotic relationship with plant or animal hosts, supplying them with a variety of nutritional factors essential for growth and development (Chapter IV.1). The flora of the gastrointestinal tract in man and animals probably contributes to host nutrition by synthesizing certain vitamins and amino acids.

In sheep, the disease known as bush sickness provides a striking example of the nutritional effects exerted by the indigenous flora. This disease is essentially due to a vitamin B_{12} deficiency that occurs when the sheep are pastured on fields deficient in cobalt. Under these conditions, their intestinal bacteria cannot synthesize the vitamin. Another line of evidence for the nutritional importance of the indigenous flora comes from the observation that germ-free rats develop severe signs of folic acid and vitamin K deficiencies shortly after being placed on diets lacking these vitamins, whereas ordinary rats remain free of symptoms on the same regimens. Contamination of germ-free animals with various bacterial cultures protects them against folic acid deficiency.

The deleterious effects on growth resulting from the prevention of coprophagy in rats indicate that, in these animals at least, the nutritional role of the intestinal biota is probably more complex than commonly assumed. Rats fed diets deficient in thiamine, riboflavin, pyridoxine, pantothenic acid, vitamin K, and vitamin B_{12} develops symptoms of deficiency much sooner if coprophagy is prevented by the use of tail cups. Indeed, prevention of coprophagy by cupping reduces the growth rate by some 20 per cent even on complete diets. Cupping also changes markedly the composition of the rat's intestinal flora, decreasing significantly for example its content in lactobacilli. However, lactobacilli cannot account alone, if at all, for the beneficial effects of coprophagy since adding these organisms to the diet does not correct the inhibition of growth caused by cupping.

A curious and unexplained fact is that coprophagy is effective only if the animals take the feces directly from the anus. This suggests that either some microorganisms present on the surface of the perianal areas are involved, or that the active organisms, or materials, are extremely unstable and are rapidly inactivated outside the body. The most likely explanation of the need for a "closed circuit" in rat coprophagy is that an organism with exacting anaerobic requirements is an essential part of the system.

It is difficult to prove that the microbiota of the gastrointestinal tract plays an important nutritional role in man. Its usefulness is suggested, however, by the fact that certain persons fail to show the expected deficiency symptoms when fed diets that are known to be nutritionally inadequate. Thus, some vegans who live on a strictly vegetarian diet remain healthy even though their B_{12} intake is extremely low (Chapter III.2). A plausible explanation of this finding is that bacterial synthesis of the vitamin takes place in the intestine of these vegans, much as it does in sheep and other animals. Since many types of bacteria synthesize essential amino acids, it is possible also that the gastrointestinal flora can in certain cases supplement diets consisting primarily of plant products in which the protein composition is inadequate for human growth.

Granted that certain bacteria of the gastrointestinal tract can synthesize various vitamins and amino acids, it is also true that other components of this biota can aggravate nutritional deficiencies either by competing with their host for available nutrients, or by increasing nutritional needs through the physiological disturbances they cause.

The most obvious examples of nutritional competition are provided by infestation with helminths, since these parasites derive from the intestinal tract all the nutrients they require for maintenance and growth. Thus, *Diphylobothrium* can cause megaloblastic anemia by storing large amounts of vitamin B_{12} and thereby competing with its human host for this essential nutrient. In such cases, the anemia can be cured by eliminating the worms or by increasing vitamin intake. Other types of nutritional competition are probably just as important, even though less obvious. For example, certain forms of megaloblastic anemia that occur in Africa can be cured by penicillin treatment, presumably because this drug eliminates bacteria with high requirements for B_{12}. The incidence of anemias of this type varies with the season and seems to be related to the diet, an indication that some dietary constituents may favor the growth of the bacteria that compete with the human organism for the hematopoietic substance.

The ability of antibacterial drugs to promote the growth of farm animals, and to increase their efficiency in food utilization, constitutes another line of evidence supporting the hypothesis that some components of the microbiota have unfavorable effects on the nutritional state. The phenomenon itself is so well documented that supplementation of agricultural feeds with antimicrobial drugs has now become a general practice in animal husbandry. Organic arsenicals and copper sulfate have also been used to the same end.

While most of the information regarding the growth-promoting effect of antimicrobial drugs applies to farm animals, similar findings have been made in laboratory animals, especially in mice. Suggestive evidence of a growth-promoting effect has also been obtained in young human adults. Thus, healthy Navy recruits receiving chemoprophylactic doses of chlortetracycline or penicillin daily for seven weeks gained an average of 4.8 or 4.1 pounds during that period, as against 2.7 pounds for the group receiving placebo. Improvement in weight gain and in protein utilization has also been observed in human beings treated with antimicrobial drugs after partial gastrectomy, particularly in patients with a histamine-fast achlorhydria.

Under the proper conditions, increase in weight gain of farm animals can be achieved by adding any one of the common antimicrobial drugs to the feed in proportions as small as 10 parts per million or even less. However, the mechanism of this growth enhancement is so poorly understood that at least some fifty different hypotheses have been formulated to account for the effect (reviewed in Luckey, 1959).

The simplest and most significant fact so far established is that the growth-promoting effect does not occur when the antimicrobial drugs are administered to germ-free animals or to farm animals maintained under very sanitary conditions. It is therefore extremely probable that the drugs exert their beneficial influence by inhibiting certain microorganisms that interfere with growth. Clostridia have been most frequently incriminated and there is indeed suggestive evidence that their presence in the intestinal tract of chickens is correlated with a slower growth rate. Furthermore, contamination of germ-free birds with *Clostridium welchii* represses their rate of growth, and this effect can be corrected by adding to the diet antibacterial drugs active against this microbial species. Observations made in our laboratory show that intestinal clostridia retard weight gain also in mice. It is unlikely, however, that clostridia are the only intestinal microorganisms that can depress growth, or that the same microbial species are involved in all animal species. In a particular experiment, for example, the growth of chickens was markedly retarded by adding enterococci to their diets, and this deleterious effect was completely corrected by treating the birds with chlortetracyline.

The more unsanitary the quarters in which the animals are kept, the greater, in general, the enhancement of growth by dietary antimicrobials. But any particular drug progressively loses its growth-promoting ability when it is administered for a long period of time in the same unsanitary area. This finding provides further support for the hypothesis that enhancement of growth is due to the control of some microbial activity. It is probable that the beneficial effect disappears after prolonged drug intake because a drug-resistant flora eventually becomes established in the animals, thus rendering the drug ineffective.

Two hypotheses have been repeatedly mentioned to account for the retardation of host growth by the indigenous microbiota. One is that certain bacteria compete with the host by utilizing or destroying ingested nutrients. Another is that bacteria produce substances that poison the mucosal cells and thus interfere with its absorptive functions. In this regard it is of interest that addition of antibacterial drugs to the diet usually brings about a reduction in the thickness of the intestinal wall. The intestines of animals so treated resemble somewhat those of germ-free animals (Chapter V.3). Furthermore, the feeding of antimicrobial drugs increases the rate of absorption, at least as tested with radiolysine. It is possible, therefore, that more efficient absorption is indeed partly responsible for the faster growth rate and the better food utilization that result from such treatment.

The growth-enhancing effect of antimicrobial drugs may be due in part to the fact that they prevent or at least minimize low-grade toxemias caused by certain bacteria present in the digestive tract. Ammonia, a variety of amines, and several phenolic compounds that are released into the intestine by bacterial decomposition of amino acids are among the substances which might play a role in such low-grade toxemias. For example, all amino acids are deaminated when incubated in vitro with the washed contents of the intestinal tract of swine—asparagine and glutamic acid being the most rapidly attacked. The ammonia arises from direct deamination since the bacterial mass does not give the Stickland reaction. Deamination of arginine can be brought about by pure cultures of enterobacilli, enterococci, and heterofermentative lactobacilli, but interestingly enough not by homofermentative lactobacilli.

Tests in vitro reveal the interesting fact that the rate of ammonia production by the total intestinal flora of swine is consistently higher during the summer than during the winter months. A similar seasonal variation occurs in vivo, the ammonia concentration in the portal vein of swine being also higher during the summer. This correlation suggests that the deaminating activity of the flora in vitro is related to the amounts of ammonia it produces in vivo. Other observations support this view. Thus the ammonia content of portal blood in germ-free guinea pigs is barely one-fourth of that found in conventional animals. In swine and guinea pigs, furthermore, the portal blood level of ammonia is considerably lowered by adding small amounts of aureomycin or neomycin to the diet.

The intestinal flora is also capable of decarboxylating most amino acids. In fact, at least seven amines have been detected in various concentrations in the intestinal contents of guinea pigs and swine. Putrescine, histamine, and tyramine are a few of the pharmacologically active substances that have thus been recognized as products of the intestinal flora.

Both deamination and decarboxylation are inhibited by penicillin, tetracylines, neomycin, and other antimicrobial drugs, as well as by certain substances, such as copper sulfate, which are known to enhance the growth of farm animals when small amounts are added to the diet. In contrast, chloramphenicol, although so effective as an antimicrobial agent, is claimed to be inactive with regard to its ability both to inhibit the catabolic activities of bacteria and to enhance the growth of farm animals. On the other hand, it has been observed in our laboratory that chloramphenicol disturbs the autochthonous intestinal flora less profoundly than do other antimicrobial agents, in mice at least.

In swine, and probably also in other animals, microbial deamination and decarboxylation seem to occur all along the digestive tract, from the stomach to the colon, with a maximum of activity in the cecum. It is not surprising, therefore, that the products of bacterial catabolic activity can be absorbed and thus exert a physiological action. The beneficial effect of certain antimicrobial drugs in the treatment of hepatic coma points to the importance of this source of intestinal intoxication. Ammonia produced in the colon is absorbed in the portal circulation, bypasses the liver, reaches the general circulation, and thus may contribute to the encephalopathy in this condition.

There is also some evidence that certain antimicrobial agents can reduce the production of phenols derived from the decomposition of aromatic amino acids. Thus, tetracyclines administered to man decrease the fecal and urinary excretion of phenol and p. cresol originating from tyrosine. A recent observation of some interest in this regard is that the indole compounds that result from bacterial metabolism of tryptophane can depress the respiration of rat brain in vitro.

The flora of the gastrointestinal tract can probably affect the physiological state of the host through a great variety of other mechanisms. For example, one of the effects of bacterial endotoxins is to increase catabolic activities and in particular to disturb amino acid metabolism. As substances having endotoxin activity are constantly being released from many different microbial species, avirulent as well as virulent, and are absorbed from the intestinal tract, it is possible that the intestinal flora exerts a deleterious effect on the physiological and nutritional state through a variety of indirect metabolic channels. Of great potential importance also is the fact that certain intestinal bacteria can attack bile acids or even alter the sterol nucleus and thereby affect the sterol metabolism of their host. Recent studies have shown that the rate of cholesterol catabolism is much greater in conventional than in germ-free rats.

3. Morphogenetic effects of the intestinal microbiota

One of the most important roles of the indigenous microbiota is to exert on its host a morphogenetic stimulus that appears to be essential for the normal development of certain organs. This unexpected phenomenon was first revealed by the fact that germ-free animals exhibit anatomical and physiological abnormalities which would prevent them from competing successfully with their conventional counterparts under natural conditions. For example, their lymphoid tissue is poorly developed, and their serum γ globulin level as well as their properdin titer are abnormally low. Histologically, the intestinal wall of germ-free animals tends to remain in an undifferentiated state; it is thin and elastic. As peristalsis does not force the intestinal contents on to the large intestine, distension of the thin wall allows the cecum to enlarge up to the limits set by other abdominal organs. There is a marked increase in the weight of the cecal contents as well as of the cecal sack. In fact, enlargement of the cecum is one of the most striking abnormalities resulting from germ-free life in many animal species.

The intestinal mucosa remains incomplete also in adult germ-free guinea pigs, and it remains almost in its prenatal state as long as the animals are protected from bacterial contamination. The mucosa differs from that of the conventionally reared animals by a near absence of inflammatory cells in the lamina propria, distinct shallow crypt glands lined by a high proportion of markedly distended goblet cells, absence of degenerative changes in the epithelium lining the villi, taller and more delicately shaped villi in the small intestine, and a villous pattern in the cecum. So-called specific pathogen-free (SPF) rats, which have an intestinal flora simpler than that of ordinary rats, exhibit similar although less pronounced abnormalities in the structure of their intestinal mucosa; their intestinal villi are more "delicate" structures than those in conventional animals.

In general, the anatomical abnormalities of germ-free animals are rapidly corrected when these animals are brought into contact with the proper kind of bacteria. In the germ-free rat or mouse, for example, the cecum may occupy 30 to 90 per cent of the abdominal cavity, its wall is thin, and its contents are fluid; but reduction of cecum size and change in the consistency of its contents can be achieved rapidly, in certain cases within 24 hours, by administering orally cultures of *Clostridium difficile,* of *Bacteroides,* of of certain streptococci, isolated from the fecal contents of conventional mice. Lactobacilli and other bacteria isolated from the same animals are much less active in this respect. The

FIGURE 5. (a) *Left:* Digestive tract of germ-free female mouse (20 weeks old). *Right:* Conventional female mouse (20 weeks old). (Courtesy of Dr. Helmut Gordon, Department of Pharmacology, University of Kentucky, Lexington.)

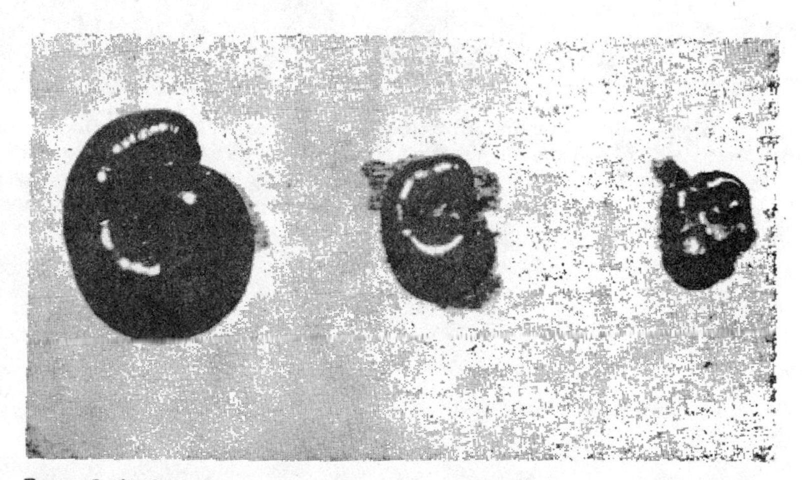

FIGURE 5. (b) *Left:* Cecum of germ-free mouse (19% of total body weight). *Right:* Cecum of conventional mouse (1.6% of total body weight). *Middle:* Cecum of germ-free mouse reassociated with lactobacilli and other bacilli (4.5% of total body weight).

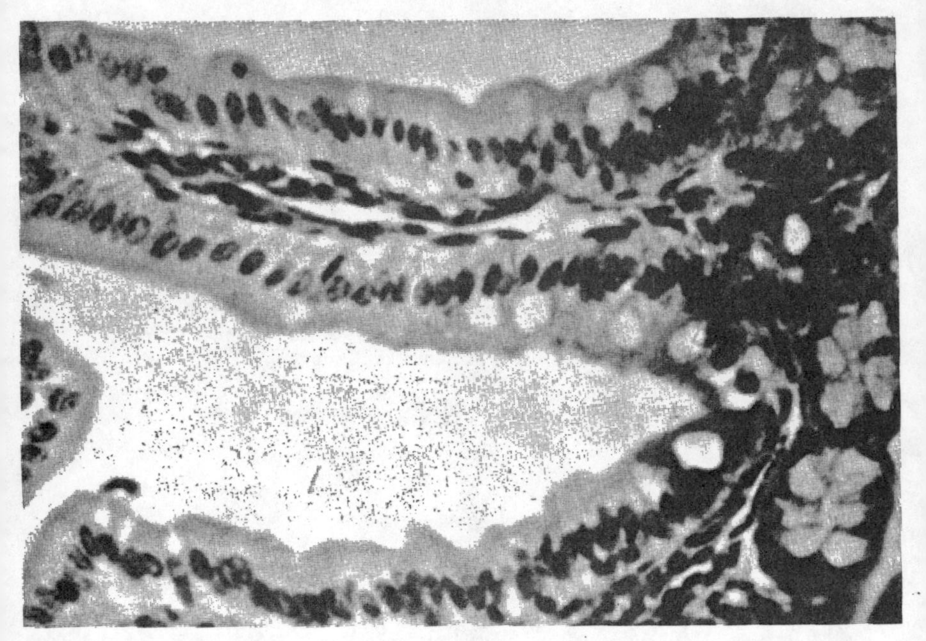

FIGURE 6. (a) Ileum of a germ-free guinea pig. Note sparsity of cells in the lamina propria, the relatively tall slender villi in comparison to the relatively narrow non-villus portion of the mucosa and the fairly large goblet cells. These goblet cells are so well filled with mucus that they appear to comprise the majority of cells lining the shallow crypt glands. Note also the uniformly shaped absorptive cells lining the villi.

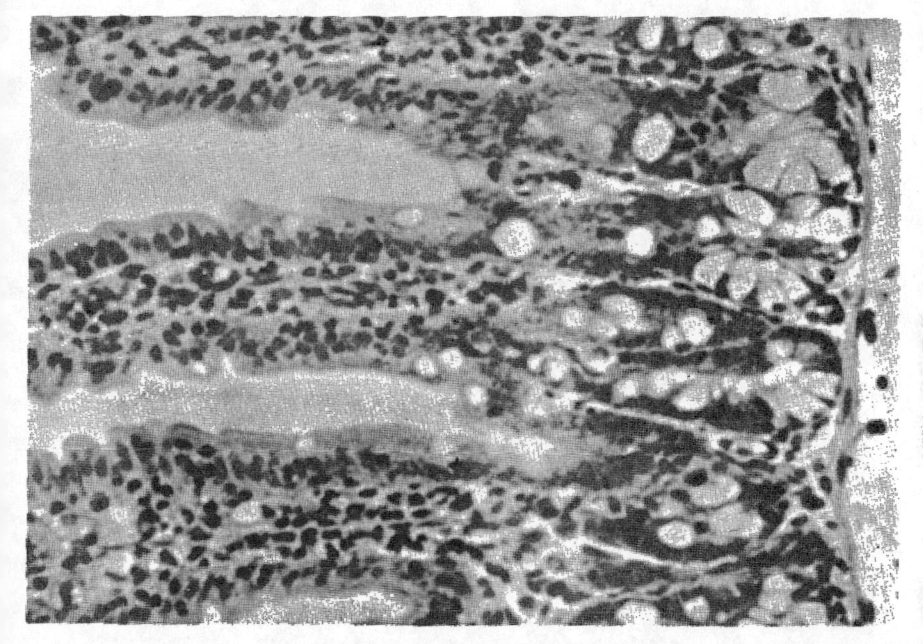

FIGURE 6. (b) Ileum of conventionally raised guinea pig—the animal comes from a particularly "clean" colony. Note the expansion of the non-villus portion of the mucosa, the crypt glands are now erect and much taller. Note also the inflammatory cells in the tunica propria. (Courtesy of Colonel Helmuth Sprinz, Chief, Department of Experimental Pathology, Walter Reed Army Institute of Research.)

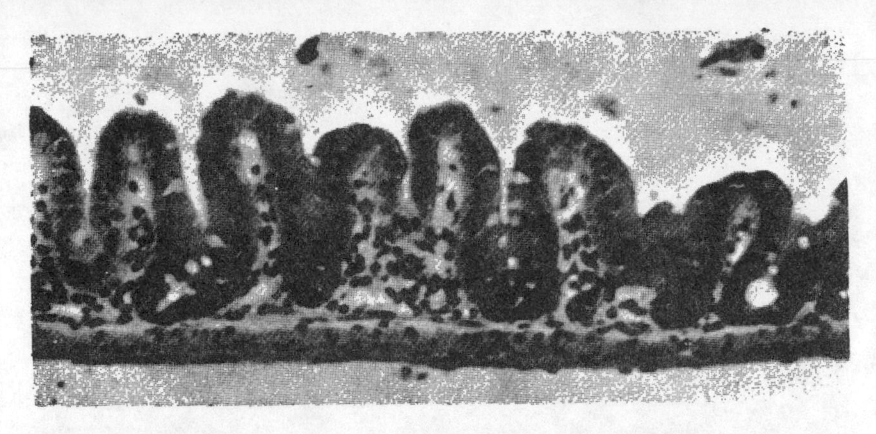

FIGURE 7. Walls of cecum of 7-8 week-old "Swiss" mice (X200). (Courtesy of Dr. Dwayne C. Savage, The Rockefeller University, New York, New York.)

Above: Animal raised under germ-free conditions: the wall is thin, the villi prominent, and the epithelium composed almost exclusively of columnar cells. The lamina propria is relatively acellular and the muscularis mucosa, submucosa, and muscularis externa are all very thin. *Below:* Animal raised under conventional conditions: the thickness is due to an extensive cellular lamina propria, a prominent muscularis mucosa and submucosa, and an extremely thick muscularis externa.

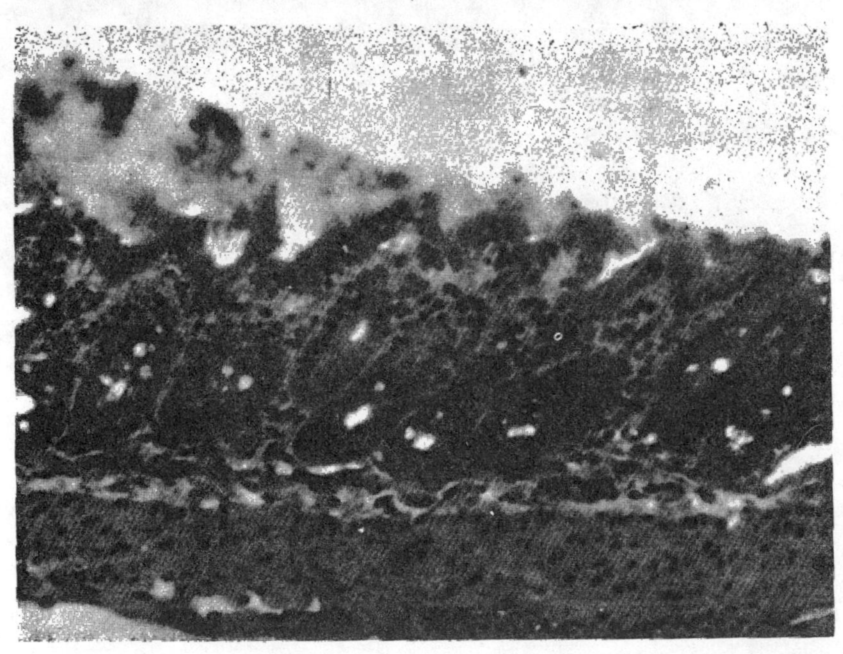

cecal apical patch which does not exist in germ-free rats or mice rapidly develops following association with the proper kind of bacteria; the quantity of Peyer's patches increases within 7 hours and a fivefold enlargement of ileocecal nodes occurs within one week. Unpublished experiments from our laboratory have revealed that the size and the contents of the cecum are normal when mice issued from germ-free animals associated with lactobacilli and anaerobic streptococci are maintained from then on in a sterile environment.

Oral introduction of *E. coli* also brings about a rapid change in the architecture and histological pattern of the bowel in germ-free guinea pigs; within a few weeks after contamination, it approaches that seen in conventionally raised animals. It is worth mentioning in passing that addition of antibacterial drugs to the diet of conventional rats occasionally causes enlargement of the cecum, probably because the drugs alter the composition of the intestinal flora and thus indirectly the histological response of intestinal tissue.

Thus, several independent lines of evidence indicate that the bacteria in the intestinal tract exert a distinct morphogenetic effect. In fact, it has been suggested that the structure of the so-called normal intestinal mucosa corresponds in reality to a sort of physiological inflammation. The normal physiological range of this response has not yet been clearly defined, but pathological studies in human beings suggest that the intensity of the inflammatory reaction is affected by the socioeconomic status. For example, the intestinal mucosa of "normal" Thai persons of low socioeconomic status presents a histologic picture similar to that of European and American patients with nontropical sprue and idiopathic steatorrhea. It seems likely that these morphological alterations are produced by irritants to which the Thai population is chronically exposed, such as highly spiced foods and enteric pathogens.

4. Physiological effects of the indigenous microbiota

Animals raised under germ-free conditions are generally much more susceptible to various forms of stress than are conventional animals. It has been reported, for example, that germ-free guinea pigs do not heal wounds readily and that germ-free mice support food deprivation and nutritional deficiencies less well than do their conventional counterparts. In general also, germ-free animals show little resistance when first exposed to experimental and accidental infections. This inadequacy of initial response can probably be traced to the fact that the general defense mechanisms do not develop normally under the conditions of germ-free life. Among the deficiencies that come to mind are the poor development of the lymphoid system; the low γ globulin content of the serum; the limited clearing activity of the reticuloendothelial system; the absence of nutritional or protective substances normally produced by the indigenous microbiota.

While the mechanisms through which the indigenous microbiota exerts its protective effects are ill understood, the effectiveness of the protection it elicits has been experimentally proved repeatedly. In brief, the general resistance of germ-free animals can be increased by merely associating them with the proper microbial agents. Indeed monocontamination is often sufficient to stimulate a variety of defense mechanisms. For example, monocontamination of germ-free mice with a nonpathogenic strain of staphylococcus enables these animals to survive doses of Coxsackie B virus which would otherwise have been lethal. Even the resistance of mice to food deprivation can be increased by monocontamination with *E. coli.*

In contrast, the resistance of conventional animals to various stresses can be decreased at will by manipulations that disturb their indigenous microbiota. Thus, exposure to large doses of ionizing radiation often brings about an overwhelming septicemia caused by Gram-negative organisms commonly present in the intestinal tract, and similar effects have been observed in man. Several mechanisms have been invoked to account for the failure of the body to hold enterobacteria in check after irradiation. On the one hand, pathological alterations of the intestinal mucosa probably facilitate the passage of microorganisms from the lumen into the general circulation; on the other hand, interference with the detoxifying activity of the liver may prevent the destruction of bacterial endotoxins and secondarily increase susceptibility to any form of stress. Furthermore, doses of X irradiation that cause an increase of the coliform population in the intestine have been reported to bring about a decrease in the number of lactobacilli. It seems worth postulating therefore that one of the mechanisms by which radiation decreases resistance to infection is by disturbing the normal microbial ecology of the intestine.

The disturbing effect of antimicrobial drugs on intestinal ecology has provided a convenient experimental technique for decreasing the resistance of guinea pigs and mice to *Vibrio cholera, Salmonella, Shigella, Staphylococcus,* and enteropathogenic strains of *E. coli.* Treatment with streptomycin, neomycin, or other antibacterial drugs eliminates for a while many types of bacteria from the intestinal tract and thereby usually makes it possible to establish infection with drug-resistant forms of the desired pathogens introduced at the critical time. Under these conditions, an oral dose containing very few infective organisms is often sufficient to establish intestinal infection, especially if steps are taken to inhibit the motility of the small intestine, as with opium or morphine. Of special interest is the fact that when infective microorganisms are introduced along with drugs, they tend to colonize not only the lower region of the intestine but also the upper areas. A few clinical reports indicate that in man also administration of antimicrobial drugs facilitates the establishment of certain pathogenic agents, for example staphylococci.

Many efforts have been made to correct the disturbances resulting from treatment with antimicrobial drugs, by administering lactobacilli to animals or human patients. This practice naturally follows from the emphasis placed by Metchnikoff and his followers on the beneficial role of the *Lactobacillus* flora (see Chapter V.2). On the whole, however, the results of this type of microbial substitution therapy have often been disappointing. In contrast, mice that have received streptomycin can be made to recover some or all of their normal resistance to salmonellosis by feeding them fresh fecal material obtained from untreated animals. The protective effect of the fecal material is destroyed by heat or by treatment with germicides. It seems likely, therefore, that the increase in resistance induced by the fecal material is due to certain components of the microbiota that normally protect the animal against infection. In unpublished experiments carried out in our laboratory, similar results have been obtained with mice by the following technique.

As mentioned earlier, large populations of coliform bacilli, enterococci, and clostridia develop in the intestinal tract of NCS mice following the administration of penicillin (Chapter V.1). Normally, the disturbance in the intestinal ecology persists for many weeks after discontinuance of the drug. However, the ecological disturbance produced by penicillin is more rapidly corrected if the animals are fed the intestinal contents of normal NCS mice shortly after the drug has been discontinued. Following one single feeding of normal intestinal content, the usual populations of *Lactobacillus* and *Bacteroides* reappear, and after 2 to 3 weeks, the coliform bacilli, enterococci, and clostridia disappear more or less completely. In this case again, the corrective activity of the intestinal contents is completely abolished by heating to 80° C for 30 minutes; moreover, the material rapidly loses its activity when manipulated for the purpose of purifying it and isolating the active agents.

Unfortunately, the microbial species responsible for the protective effect exerted by administration of intestinal material have not yet been identified. In fact, none of the pure cultures isolated from active intestinal content has been able so far, when tested alone or in association, either to restore resistance to salmonellosis after streptomycin treatment or to accelerate the elimination of coliform bacilli, enterococci, or clostridia following discontinuance of penicillin treatment. The most likely explanation of these negative results is that the protective activity is exerted by microbial inhabitants of the intestine which readily lose their viability and which have not yet been obtained in culture.

The biochemical conditions prevailing in the intestinal tract are certainly relevant to the mechanism by which the indigenous microbiota exerts its protective effect against infection. In general, the pH of the cecum and transverse colon is slightly acidic, especially when the diet contains lactose. In normal mice, for example, the intestine contains large amounts of volatile acids (especially acetic, butyric, and propionic acids), and the Eh is —0.2 to —0.4 V; such an environment is inhibitory to *Salmonella typhimurium* and is even bactericidal for this species in vitro. When mice are treated with streptomycin, however, the concentration of volatile fatty acids falls, and the Eh rises to +0.2 V. The volatile acids, the acidity, and the reducing environment provide conditions highly unfavorable to the growth of *Salmonella* and of many other pathogenic microbial species.

The digestive tract of penguins contains large concentrations of acrylic acid, probably produced by the plankton ingested by these animals. Acrylic acid has a powerful antibacterial effect at the low pH of the intestine, and may be instrumental in preventing *E. coli* from becoming established in the penguins' intestines. In addition to its intrinsic interest, this observation illustrates that the components of the normal microbiota which constitute an ecologic barrier to ordinary pathogens probably differ from one animal species to another. Thus, lactobacilli and organisms of the bacteroides group are very abundant in man, rats, mice, and swine, but there are indications that they do not occur normally in certain other mammalian species or in birds. Likewise, it can be expected that differences in the conditions of evolutionary development are reflected in the presence or absence of other members of the autochthonous microbiota. It is probable, on the other hand, that strict anaerobes are abundant in the intestine of all animal species. Unfortunately, so little is known of their characteristics as well as of their physiological requirements and activities that the role of the anaerobic microbiota must remain hypothetical for the present.

Taken together, the facts outlined in the preceding pages make it seem likely that certain components of the indigenous microbiota play a useful role by increasing resistance not only against pathogens but also against less virulent but nevertheless potentially dangerous microbial species which would otherwise multiply extensively in the intestinal tract. The nature of the organisms responsible for this beneficial effect, the conditions that favor their development, and the mechanisms of their protective action constitute unsolved problems of great theoretical and practical importance. They point to the need for detailed studies of the biological and biochemical factors that determine the characteristics of the environment prevailing in the various parts of the intestine.

The range of physiological activities of the intestinal microbiota is certainly much wider than has been recognized so far, as illustrated by the following phenomenon, which was entirely unexpected. The cecal contents of germ-free animals contain a toxic fraction that causes contraction of the rat uterus in vitro. This material, which is produced irrespective of the composition of the diet, is much less abundant in the cecal content of conventional animals than in that of germ-free animals; there is evidence, furthermore, that it can be inactivated by their intestinal flora. The contractive material may play a role in the enlargement of the cecum in germ-free rodents as well as in the hemodynamic abnormalities observed in these animals.

Even more recently, it has been shown that the cecal contents of germ-free rats contain up to 40 times as much hexosamine as that of conventional animals. This material is present in a mucoid fraction of high molecular weight which can be degraded by the intestinal flora of conventional rats. Although the mucoid fraction may play some minor role in the cecal enlargement of germ-free animals, it probably does not account alone for this morphological abnormality since it is not degraded in vitro by certain bacteria which are capable of reducing cecum size in vivo.

The nutritional, metabolic, and morphogenetic effects of the indigenous microbiota are thus very complex and not yet completely elucidated. But there cannot be any doubt of their importance for the growth and welfare of animals and man. Observations made with the NCS mouse colony (described in Chapter V.1) illustrate the wide range of these effects.

As will be recalled, the NCS colony was derived from the standard "Swiss" (SS) colony of albino mice; it is maintained under such conditions that the flora of its gastrointestinal tract contains very few enterococci or Gram-negative enterobacilli, and no detectable clostridia, *E. coli, Pseudomonas aeruginosa, Proteus,* or *Salmonella.* Furthermore, the colony is free of many potential pathogens commonly present in mice. This difference in indigenous microbiota between SS and NCS mice is associated with a number of differences in the growth characteristics of the two groups of animals.

As far as can be judged, the average litter size is approximately the same in both the SS and the NCS colonies. Yet the production rate of young animals is much higher in the NCS than in the original colony. Measured in terms of numbers of weaned mice per female per week, the average production rate over a period of two years was 0.9 for the NCS colony as against 0.35 for the SS colony. To a very large extent, and indeed perhaps exclusively, this remarkable difference is due to the very low infant mortality among NCS animals.

Under the proper conditions, NCS mice weigh 2 to 5 gm. more than SS animals when weaned at 3 weeks of age, and they continue to gain weight more rapidly for several months thereafter. (See Figure 8.) Furthermore, they retain throughout their lives some rather unusual characteristics. The size of the liver, lymphoid tissue, and especially the spleen is somewhat below normal, and so are the numbers of blood lymphocytes and eosinophils. The macrophages are extremely abundant in the peritoneal cavity, whereas the poly-

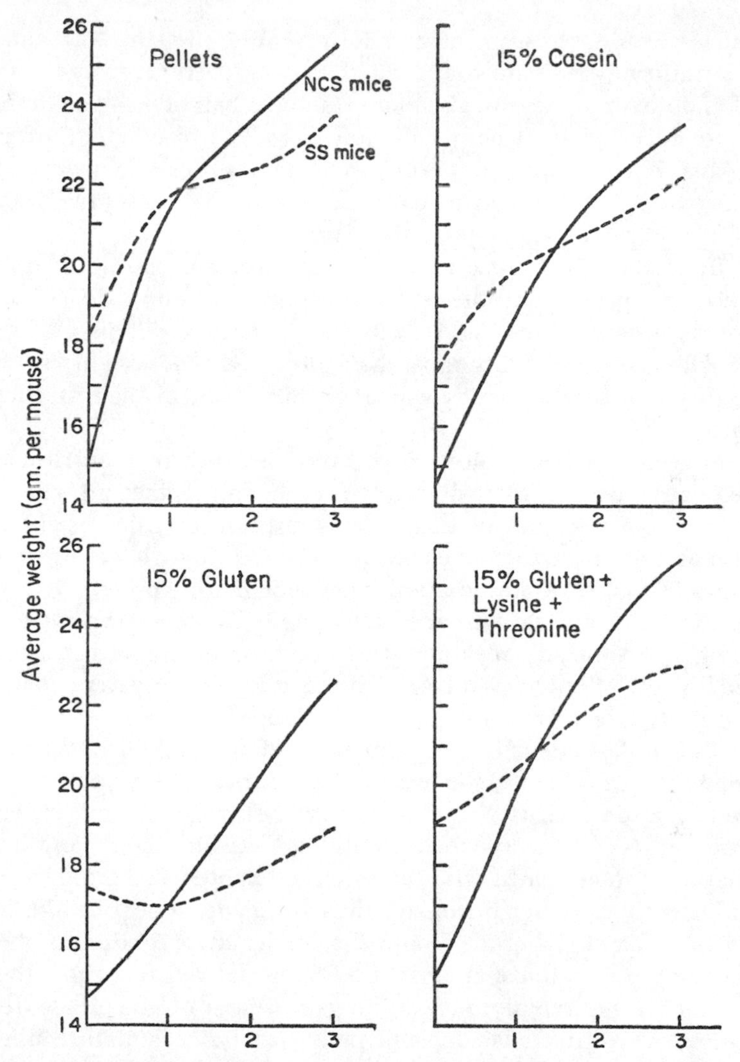

FIGURE 8. Comparative growth rates of NCS and SS mice on four different diets. NCS mice eventually outgrew the SS mice on all diets, even though the animals had been so selected that those of the former group were smaller than those of the latter at the beginning of the experiment. Of special interest is the fact that NCS animals gained weight at a rapid rate when fed a diet in which gluten was the sole source of protein, whereas SS animals failed to gain weight on this diet, as was to be expected. (Reproduced from R. Dubos and R. Schaedler, 1962.)

nuclear eosinophils are absent. Remarkably enough, NCS mice are uniformly resistant to the lethal effect of even very large doses of endotoxins. They are also more resistant than other mice to the acute effects of total body radiation as well as of other injurious agents. For example, no deaths occur among them following injection of 10 mg. of cortisone acetate, a dose that causes the death of a large percentage of mice of other colonies.

In general, NCS mice are somewhat more susceptible than SS mice to experimental infection with *Mycobacterium tuberculosis, Staphylococcus aureus,* and *Klebsiella pneumoniae* (type C). Drs. E. Kilbourne and J. Schulman have observed that certain viruses multiply more abundantly and uniformly in them than in other mice.

Mice of the NCS colony have much less exacting nutritional requirements than those of other mice; moreover, they grow more rapidly on any kind of diet. They continue to gain weight at almost maximum rate when fed diets containing wheat gluten as the sole source of amino acids, even though this protein is very low in lysine and threonine. Mice of the NCS colony can even be maintained for an indefinite period of time on a diet consisting exclusively of autoclaved corn. The SS mice of the parent colony naturally fail to grow when fed these deficient diets.

Two independent lines of experimentation make it clear that the faster growth rate, the less exacting nutritional requirements, and other peculiarities of NCS mice are not of genetic origin. For example, NCS mice lose their peculiar growth characteristics when they are contaminated with the intestinal contents of ordinary SS animals shortly after birth, and thus are made to acquire the intestinal microbiota of the SS animals. Under these conditions, they are smaller than uncontaminated NCS mice at weaning time; they do not grow as rapidly; they fail to gain weight on the gluten diet unless the regimen is supplemented with lysine and threonine. Furthermore, they become sensitive to the lethal effect of endotoxin.

Profound changes can be brought about in NCS mice without artificial contamination simply by treating them with antimicrobial drugs. As we have seen, addition of these drugs to the drinking water for one week causes an immediate change in the composition of their intestinal flora (Chapter V.1). First the lactobacilli, the anaerobic streptococci (group N), and the organisms of the bacteroides group disappear completely and often lastingly. Then discontinuance of the drug is followed by an explosive multiplication of various coliform bacilli, enterococci, and anerobic organisms, including large numbers of clostridia. The populations of several of these bacterial species reach 10^{10} organisms per gm. of fecal material within a few days after administration of the drugs has been discontinued.

This dramatic change in the fecal flora is accompanied by an equally profound change in response to the nutritional regimen. Whereas untreated NCS mice continue to grow on the gluten diet, those having received drugs begin to lose weight immediately after treatment, and the loss continues for several weeks thereafter. The extent and duration of the loss is related to the amount and type of drug administered and therefore to the severity of the change in the composition of the intestinal flora. Weight recovery occurs only when the intestinal flora has approximately returned to its original state.

The findings with NCS mice that have just been described appear at first sight to conflict with those reported earlier showing that administration of antimicrobial drugs to farm animals can enhance the growth rate and the efficiency in the utilization of food. However, the results presented in Figure 9 indicate that both statements are true despite their apparent incompatibility.

Mice of the NCS colony were compared with conventional albino mice of approximately the same age, but obtained from a colony (Ha/ICR) maintained under ordinary conditions and therefore having a complex indigenous biota. Whereas administration of penicillin retarded the growth of NCS mice, it accelerated the growth of the Ha/ICR animals, as seen in Figure 9. The effect was the same whether the animals were fed commercial pellets or a semisynthetic casein diet. The difference between the two groups of Swiss mice with regard to their nutritional response can probably be explained by the theory presented earlier to account for the growth enhancement of farm animals fed antibacterial drugs.

Ordinary mice (Ha/ICR, for example) carry a large variety of microorganisms that probably repress their growth through the various mechanisms discussed in the preceding pages. In such a situation, antimicrobial drugs exert a beneficial effect by inhibiting the objectionable microbial species. In contrast, as we have seen, drug treatment causes an infectious process process in NCS mice; the infection is not sufficiently severe to kill the animals, but the severity of the disturbance is evident from the spectacular and

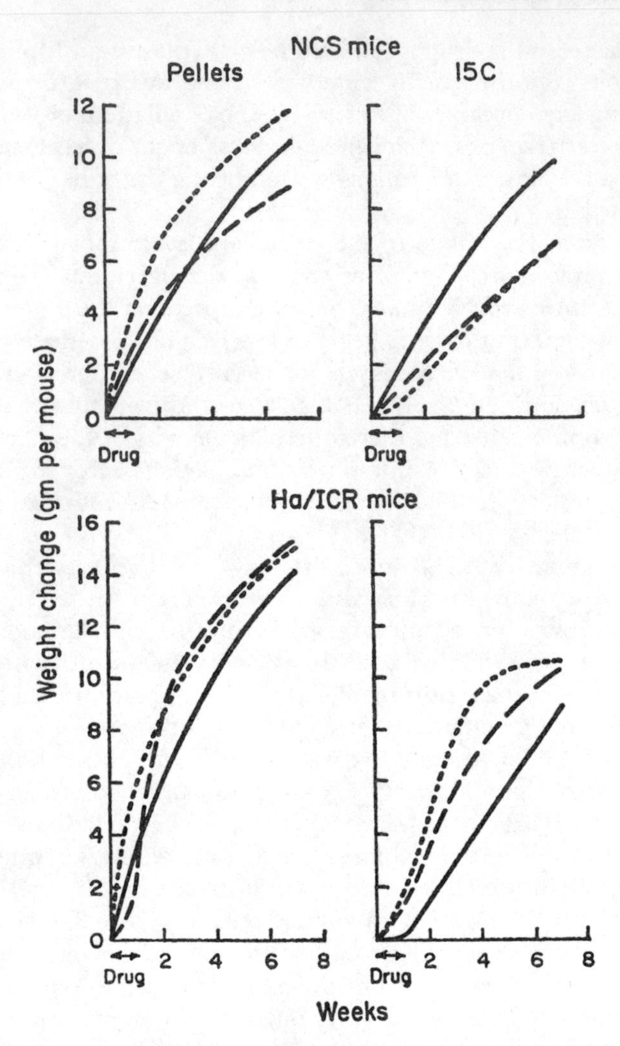

FIGURE 9. Comparative effects of drugs on weights of two strains of Swiss mice. Drugs added to drinking water in concentrations of 0.3 gm. per liter during the first week of experiment, then discontinued. Control ————; penicillin — — —; terramycin - - - - -. In contrast to NCS mice, the Ha/ICR mice had a complex fecal flora, with large numbers of Gram-negative bacilli and enterococci, at the beginning of the experiment. Half the mice of each strain were fed D&G pellets; the other half were fed a synthetic diet containing 15% casein (15C).

Results: Whereas penicillin and terramycin accelerated the weight gain of Ha/ICR mice fed the semisynthetic casein diet, the drugs retarded the weight gain of NCS mice fed the same regimen. The differences were less striking in mice fed D&G pellets. (Reproduced from R. Dubos, R. Schaedler, and R. Costello, 1963.)

sudden increase in the numbers of coliform bacilli, enterococci, and clostridia. The loss of weight and the development of more exacting nutritional requirements in NCS mice treated with antimicrobial drugs are thus expressions of an infectious state. As we shall see, administration of bacterial endotoxin to NCS mice causes them to lose weight and increases their requirements for lysine and threonine.

The rapid growth rate of untreated NCS mice and their ability to fare well on deficient diets acquire particular significance from the fact that similar findings have been made with other animals also raised under conditions designed to control the composition of the intestinal flora. Like NCS mice, the so-called specific pathogen-free rats (SPF) grow more rapidly than their conventional counterparts of the same genetic make-up. Baby pigs obtained by hysterectomy and raised without contact with other swine, but not under germ-free conditions, also grow more rapidly than ordinary pigs and utilize their feed more efficiently; indeed, the difference is so great that the technique presents economic advantages from the point of view of meat production.

Whatever the exact mechanism of the interplay between nutritional state and the microbiota of the digestive tract, it is clear that rate of growth, nutritional requirements, and efficiency in food utilization are characteristics influenced by the presence in the tissues of microorganisms not usually considered to be agents of disease. As mentioned in Chapter III.3, people from Oriental countries and Southern Europe often are of small physical stature in their native environment, but grow more rapidly and larger when raised under the more prosperous conditions prevailing in the Western world. Hybrid vigor and the greater availability of adequate food are certainly of importance in accounting for this phenomenon. But sanitary practices and other aspects of the ways of life may also play an important part by affecting the indigenous microbiota and thereby conditioning the response of human beings to the food they consume. The histological differences observed in the intestinal mucosa depending upon the socioeconomical status may be relevant to the problem, since they probably reflect the intensity of the inflammatory response to the so-called "normal" flora.

Many other facts, not to be described here, have revealed that the indigenous microbiota can influence the morphological and physiological characteristics of its host to such an extent that traits assumed to be the unavoidable consequence of the genetic endowment are determined in reality by the microbial environment.

Although little is known of the comparative biochemical activities of the different bacterial species present in the digestive tract, a few facts may serve as guideposts for the development of working hypotheses in this field. The species most numerous under normal conditions, such as the lactobacilli, anaerobic streptococci, and bacteroides, have a much narrower range of biochemical activities than the clostridia, coliform bacilli, or enterococci, which multiply abundantly as a result of various types of disturbances. It is likely, therefore, that the chemical transformation of metabolites in the intestine is less pronounced when the autochthonous organisms predominate than when other species gain the upper hand.

We have seen that the flora of the gastrointestinal tract can exert dramatic morphogenetic effects essential for normal histological development and function. Furthermore, this flora can attack many substances such as amino acids and bile acids, and the products of this attack differ according to the type of microbial species that predominate. The composition of the microbial flora, which depends upon environmental and internal factors, consequently determines the kind of biologically active substances that are released from the digestive tract into the general circulation. Thus, interference with any one of these components is likely to disturb the equilibrium between them and to affect thereby the functions of the system as a whole. In other words, the different parts of the digestive tract, the microorganisms it harbors, and the conditions that govern the interplay between these various components constitute a highly integrated ecosystem.

VI. Nutrition and Infection[1]

1. Nutritional factors and resistance to infection

a. Correlations between nutritional state and infection

Famine and pestilence are mentioned together on several occasions in the Bible, and these two horsemen of the Apocalypse still ride in association today. Despite all the euphoric talk about sanitation, miracle drugs, and vaccines, infectious diseases continue to take a heavy toll wherever undernutrition or malnutrition are prevalent, in the less privileged social classes and in the underdeveloped countries. Infection is also commonly associated with the kind of malnutrition that often results from trauma or surgery and occurs as a secondary consequence of certain metabolic disorders, such as uncontrolled diabetes or cystic fibrosis of the pancreas.

Suggestive as it is, the evidence provided by such epidemiological and clinical data is not sufficient to prove that nutrition per se plays a determinant role in resistance to infection. In social groups as well as in individual patients, undernutrition or malnutrition usually occurs simultaneously with other disturbances, which may, on their own account, increase the frequency and severity of disease. Where food is lacking, the social conditions usually lead to crowding and unsanitary practices and thereby facilitate the spread of infectious agents. Metabolic defects that interfere with the utilization of food may decrease resistance through mechanisms which do not have a nutritional basis. In other words, inadequacies in the supply or in the utilization of food are usually associated with other social or physiological disturbances which may constitute the primary cause of susceptibility to infection. Because of these complexities, the role the nutritional state plays in resistance to infec-

1. The facts to be discussed under this title could have been considered in Chapter III, "Man's Food," and Chapter VII, "The Evolution of Microbial Diseases." They have been grouped in a separate chapter to emphasize that many phenomena of infection are influenced more decisively by the nutritional state of the host than by the specific nature of the causative microbial agent.

tion either in man or in experimental animals has not yet been clearly defined. The conviction that good food makes for greater resistance has become established in the public and scientific mind more by reiteration than by demonstration.

The relation between nutritional state and infectious disease is in fact much too complex to be stated in the form of a simple rule. The following generalizations quoted from a recent review bring out the fact that nutritional deficiencies can either aggravate or minimize the manifestations of infectious processes.

> a) Bacteria, rickettsiae and helminths are regularly synergistic with nutritional deficiencies; b) protozoa are as often antagonistic as synergistic; c) viruses are more often antagonistic than synergistic.
>
> Patterns of interaction according to deficiencies in nutrients have the following characteristics: a) General inanition is regularly synergistic with most infections, but antagonism has been found with viruses and protozoa; b) protein deficiencies produce synergistic effects; rare instances of antagonism result from need of an infectious agent for a specific amino-acid; c) vitamin A deficiency is regularly synergistic; d) vitamin D deficiency commonly fails to interact with infections, but synergism has been demonstrated; e) vitamin B deficiencies result in synergism or antagonism depending upon agent and host; they are responsible for most known instances of antagonism; f) vitamin C deficiencies are usually synergistic, but antagonism has been demonstrated; g) specific minerals are either synergistic or antagonistic [Scrimshaw, Taylor, and Gordon, 1959].

A few specialized aspects of these complex relationships will be discussed here in further detail. It seems useful, however, first to consider the effect of nutrition on infection from a more general point of view.

b. Effect of nutritional state on the response of the body to infection

The first mechanism that comes to mind to account for the enhancing effect of malnutrition on infection is that protective antibodies are less readily formed or more rapidly lost in malnourished individuals. Antibody depletion has been repeatedly demonstrated in animals fed diets deficient either in amino acids, in pantothenic acid, in pyridoxine, or in other essential growth factors. In reality, however, the serum levels of antibodies are appreciably lowered only in the case of profound deficiencies, rarely encountered under usual conditions. Negative nitrogen balance with depletion of serum proteins and therefore of antibodies occurs also very rapidly when amino acid antagonists are fed to experimental animals; but despite its theoretical interest, this finding is not likely to have its counterpart in nature. In fact, loss of antibodies or interference with their production is not a common expression of nutritional deficiencies. On the other hand, there is indirect but nevertheless strong evidence that malnutrition can interfere with resistance to infection through mechanisms other than those affecting serum antibody levels.

It was reported many years ago that rats fed a diet deficient in some unidentified water-soluble vitamin (then designated vitamin H, and probably corresponding to biotin) commonly died with the pathological changes of pseudotuberculosis caused by the proliferation of a murine strain of *Corynebacterium*. Corynebacterial pseudotuberculosis also appears spontaneously in animals maintained on pantothenate-deficient diets. More recent studies have revealed that while healthy rodents generally harbor corynebacteria, such latent infection rarely evolves into overt disease under usual conditions. These findings clearly point to the fact that nutritional deficiencies can activate latent infections and in this manner cause microbial disease. This is well known, of course, in the case of human pulmonary tuberculosis.

There is evidence also that the development of immunity against tuberculosis in experimental animals can be influenced by changing the concentration of protein or the type of fatty acids in the diet. Although the mechanism of these effects is not yet understood, some light has been thrown on the subject by studies on the pathogenesis of intracellular infections.

It has been established recently that acquired immunity to intracellular infections depends upon the simultaneous operation of two entirely different factors. One consists in the immunological response per se and is therefore specific for the particular microbial agent under consideration. The other is nonspecific and consists in a change in the metabolic activities of the macrophages and in their ability to create an intracellular environment unfavorable to the survival or multiplication of the pathogen. There is reason to believe that this metabolic change is profoundly affected by the general state of the host. Nutritional factors could therefore affect the immune state indirectly through their influence on macrophage metabolism. In the light of this theory, it becomes possible to understand why reactivation of arrested tuberculosis is so frequent among malnourished persons and also why BCG vaccination against tuberculosis so commonly fails in underprivileged areas where malnutrition prevails (reviewed in Elberg, 1960; Mackaness, 1964; Dubos, 1964). More generally, disturbances in macrophage metabolism may account for the historical fact that famine paves the way for pestilence.

The effect of the nutritional state on the inflammatory reaction and on the histologic tissue response is also of direct relevance to the natural history of infectious processes. While little can be said concerning the effect of dietary factors on the inflammatory response, it is obvious that the walling-off reaction, involving as it does the deposition of acidic polysaccharides and fibrous proteins (fibrin and collagen), imposes great metabolic demands on the infected organism for synthesis of these substances. Although the detailed steps of this histological response are not well understood, it is known that the synthesis of mucopolysaccharides and of collagen is greatly retarded when the diet is deficient in certain amino acids, particularly lysine and the sulfur amino acids, or in ascorbic acid. There might be some justification, consequently, in regarding ascorbic acid as an "anti-infectious" vitamin.

Lysine deserves further emphasis in this connection, since it is required for the synthesis of β hydroxylysine, which is one of the essential constituents of collagen. For this reason, the possibility is worth considering that plant proteins deficient in lysine are unsatisfactory from the point of view of resistance to infection because they are not adequate for collagen synthesis. Witness, for example, the failure of tuberculous patients to develop a productive type of tissue response when fed diets deficient in "animal" proteins.

Unquestionably, the nutritional state affects susceptibility to infection through many other indirect channels. For example, lysine deficiency in rats markedly decreases the activity of the reticuloendothelial system (as measured by clearance of colloidal chromic phosphate) and seems to increase thereby susceptibility to anthrax. It is known also that starvation can activate the pituitary-adrenal cortical system, whereas chronic undernutrition decreases adrenal activity. In view of the important role played by pituitary and adrenal hormones in infectious processes, it can be surmised that some of the effects associated with starvation and chronic undernutrition are indirectly the result of disturbances in hormonal activity.

It is generally assumed that a good nutritional state increases resistance to infection by rendering the body more effective in destroying pathogens or in slowing down their rate of multiplication. However, this view is not supported by bacteriological evidence. In tests carried out with animals infected experimentally with tubercle bacilli, staphylococci, or Friedländer bacilli, the numbers of organisms recoverable by culture from blood, liver, lung, spleen, and kidney were essentially the same during the 48 hours that followed infection, irrespective of the composition of the diet and of its effect on eventual survival time. It seems unlikely, therefore, that poor diets depress resistance merely by interfering with the bactericidal power of the organs or by favoring the multiplication of the pathogens, at least during the early phase of infection. Another possibility is that inadequate diets decrease resistance to the toxic effects produced by the infectious state. As we shall now see, amino acid deficiencies can indeed interfere with recovery from certain forms of toxemia.

c. *Amino acid intake and infection*

As stated earlier, protein deficiency and amino acid imbalance constitute forms of malnutrition which seem clearly to be associated with lower resistance to infection. It has long been known, for example, that the exudative type of tuberculosis is one of the most common and striking pathological features in populations deprived of "animal" proteins (reviewed in Dubos and Dubos, 1952). Throughout history, indeed, death rates of this disease have been high wherever and whenever protein nutrition has been quantitatively or qualitatively inadequate.

Children suffering from kwashiorkor are extremely susceptible to most types of microbial diseases. Indeed, they often die of intercurrent infections. The following facts, taken from a study in one of the Central American countries, are typical of the findings in areas where protein malnutrition prevails. Of 109 deaths of children one to four years of age, 40 were in association with kwashiorkor. Most of them were attributed on the death certificates to parasitic infestations, infectious diarrhea, and various other infectious processes. It is likely, however, that these infections would not have been fatal in well-nourished children.

While it is certain that microbial diseases are commonly very severe in malnourished persons, the epidemiological facts do not point to any specific relation between protein deficiency and any particular infectious agent. True enough, pulmonary phthisis was many times more prevalent among the vegetarian Africans of the Akikuyu tribe than among their meat-eating neighbors of the Masai tribe at the time of a comparative survey in 1931, but similar differences were also found with regard to other infectious disorders. For example, inadequate protein intake increases susceptibility to parasitic infections, such as those caused by *Leishmania donovani, Trypanosoma cruzi,* or hookworms, just as it does to bacterial infections.

There is no evidence, furthermore, that protein malnutrition actually increases the chance of contracting infection. Its effect seems rather to consist in an aggravation and a prolongation of the infectious process. Indeed, malnutrition weakens in a nonspecific manner the body's response to any form of stress (Chapter III.3). Children on a low protein diet commonly show protracted depression of growth following the usual respiratory diseases, whereas well-nourished children are more likely to regain weight rapidly. Since infection, whatever its causative agent, naturally acts by itself as a form of stress, it thereby tends to aggravate the effects of any underlying quantitative or qualitative protein deficiency.

Experiments with mice have provided a striking demonstration of the influence of amino acid intake on ability to recover from infectious stress. When these animals are made to lose weight by the administration of a small dose of bacterial endotoxin, they regain their original weight rapidly, within 3 to 4 days, if fed an adequate diet containing a well-balanced mixture of amino acids (see Fig. 10). In contrast, they recover much more slowly, or not at all, if the sole source of amino acid in their diet is wheat gluten, a protein low in lysine and certain other amino acids. Weight recovery is rapid, however, if the diet is supplemented in such a manner as to achieve a proper amino acid balance (see Table 10). These findings call to mind the fact already mentioned that one of the gravest effects of protein malnutrition in human beings is to render infectious disease much more severe and longer lasting.

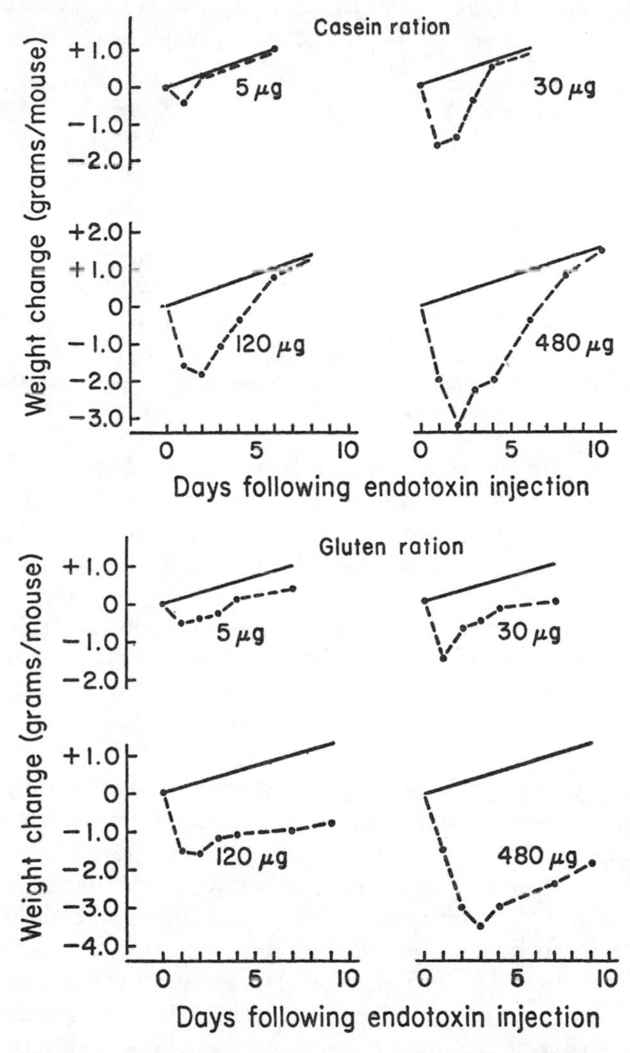

FIGURE 10. Recovery from loss of weight caused by injection of bacterial endotoxin into mice fed either casein or gluten diets. (a) Diet containing 20% casein as sole source of protein. Note that mice regained their original weight within 5 days, irrespective of dose of endotoxin injected. (b) Diet containing 20% gluten as sole source of protein. Note that none of the mice regained their original weight during the period of observation, even when the dose of toxin injected was very small.

It must be acknowledged that attempts to study the effect of protein intake on experimental infections have often given conflicting results in the hands of different investigators; witness the

TABLE 10. Effect of Lysine and Threonine on the Rate of Recovery from Loss of Weight Caused by Injection of Bacterial Endotoxin into Mice.

Diet	*Amino acid supplementation* [a]	*Weight Loss grs./mouse*	*Recovery*	
			7 days (per cent)	*14 days (per cent)*
Casein	0	2.5	100	100
Gluten	0	2.3	0	0
Gluten	lysine & threonine [b]	2.1	100	100
Gluten	threonine	2.1	30–60	30–90
Gluten	lysine	2.1	30–50	30–60

a. All diets are supplemented with 0.1% cystine.

b. Note that when gluten diet was supplemented with lysine and threonine, recovery was as rapid as with the casein diet.

divergencies in the conclusions reached concerning the influence of protein intake on susceptibility of experimental animals to various pathogenic agents. (For reviews of relevant literature, see Dubos and Schaedler, 1959; and Scrimshaw, Taylor, and Gordon, 1959, 1965.)

The very existence of these controversies points to the fact that the relation between nutritional state and infectious process is conditioned by unrecognized or ill-defined factors. The following results obtained with albino mice help somewhat to clarify the situation.

Young mice were fed ad lib. semisynthetic diets that differed in protein and amino acid concentration but otherwise contained optimum amounts of all other known growth factors. The animals were then infected with tubercle bacilli, staphylococci, Friedländer bacilli, or other species of bacteria. Under all conditions of test, mice fed diets containing 20 per cent casein survived the infection longer than did mice receiving diets of the same composition but containing only 8 per cent casein. It is of particular significance that noninfected mice gained weight just as rapidly on diets containing 8 per cent casein as they did with diets containing 20 per cent of the same protein. In other words, the decrease in resistance brought about by the lower protein concentration could not have been predicted from the respective growth curves of the animals on the two diets (for examples see Tables 11, 12, and 13).

The resistance of mice receiving 8 per cent or even only 5 per cent casein could be rendered equal or even superior to that of mice receiving 20 per cent casein by supplementing the low pro-

TABLE 11. Effect of Dietary Protein on Susceptibility of Mice to Tuberculosis [*]

Diet (14 days pre-infection)	*Weight per animal (gm.)*			*Cumulative deaths at indicated days postinfection*			
	Original	*At infection time*	*Gain*	*21*	*28*	*35*	*42*
				(out of 16 mice)			
20 C [a]	26.1	32.0	5.9	1	4	7	9
8 C [b]	26.4	31.1	4.7	5	14	15	16

[*] Male mice infected intravenously with 0.1 ml. *Mycobacterium tuberculosis* (bovine strain). (Reproduced from R. Dubos and R. Schaedler, 1959.)

a. A complete diet containing 20% casein.

b. A complete diet containing 8% casein.

TABLE 12. Effect of Dietary Protein on Susceptibility of Mice to Infection [*]

Protein (per cent)	*Time on diet (days)*	*Average weight change*	*Number of mice*	*Cumulative deaths at indicated days postinfection*		
				3	*5*	*7*
Casein 20 [a]	7	+3.1	7	2	2	7
Casein 5 [b]	7	−2.1	8	6	7	8
Gluten 20 [c]	7	+1.7	8	3	5	8
Casein 20	15	+6.1	8	1	3	6
Casein 5	15	−3.4	8	6	7	8
Gluten 20	15	+3.4	7	4	5	7
Casein 20	21	+7.5	7	3	3	5
Casein 5	21	−3.2	7	7	7	7
Gluten 20	21	+1.8	8	6	6	8

[*] Mice infected with 0.05 ml. *Staphylococcus aureus*. (Reproduced from R. Dubos and R. Schaedler, 1959.)

a. A complete diet containing 20% casein.

b. A complete diet containing 5% casein.

c. A complete diet containing 20% gluten as sole source of protein.

tein diets with adequate mixtures of essential amino acids (Table 13). This increase in resistance was achieved even though the animals fed the amino acid diets ate less avidly than did those receiving 20 per cent casein diet and gained weight much less rapidly. Here again, resistance to infection could not have been predicted from the growth curves of uninfected animals. Other investigators working with more complex and less well-defined diets have also recognized that the regimens that prove best for growth and reproduction are not necessarily those most favorable for resistance to infection.

TABLE 13. Effect of Amino Acids on Susceptibility of Mice to Infection *

Diet (8 days pre-infection)	*Weight per animal (gm.)*			*Cumulative deaths at indicated days postinfection*				
	Original	*At infection time*	*Gain*	*3*	*5*	*14*	*21*	*28*
				(out of 8 mice)				
20 C [a]	14.3	18.1	3.8	0	3	3	3	4
8 C [b]	13.5	17.7	4.2	3	4	5	6	8
8 C + AA [c]	13.7	13.7	0.0	0	0	0	0	0

(16 days pre-infection)	*Original*	*At infection time*	*Gain*	*4*	*7*	*14*
				(out of 16 mice)		
20 C [a]	14.1	?	?	2	4	6
8 C [b]	14.1	22.2	8.1	3	10	12
8 C + AA [c]	13.9	17.9	4.0	0	1	2

* Female mice fed bread and milk for 2 days after weaning and before being placed on experimental diets; infected intravenously with 0.05 ml. *Mycobacterium fortuitum*. (Reproduced from R. Dubos and Schaedler, 1959.)

a. A complete diet containing 20% casein.

b. A complete diet containing 8% casein.

c. A complete diet containing 8% casein, supplemented with 12% of balanced mixture of essential amino acids.

The qualitative aspects of amino acid nutrition influence the course of the infectious processes at least as much as do the quantitative aspects. Even with large concentrations of protein, resistance to infection decreases when the concentration of one or more of the essential amino acids is critically low. Thus, the severity of infection is enhanced when the diets contain either gelatin, gluten, gliadin, or soy bean protein as sole source of amino acids. Resistance is high, however, when the animals are fed a mixture of rice and soy bean in a proportion designed to provide an amino acid pattern similar to that present in casein. In other words, the proper balance of amino acids is probably at least as important as the total amount of these nutrients in determining resistance to infection.

In all published studies of the effects exerted by protein intake on resistance to infection, it has been the practice to keep the animals on the experimental diets throughout the period of experimentation. We have repeatedly observed, however, that short periods of depletion may be sufficient to increase susceptibility. For example, replacement of a complete diet by a protein-deficient diet for a period of just two days in two consecutive weeks during the course of experimental tuberculosis is sufficient to decrease in a dramatic manner the survival time of mice. Many similar experiments have confirmed that transient nutritional disturbances can thus alter profoundly the resistance of animals to infection.

The protein stores of the body deserve some consideration here since they probably influence the rate at which susceptibility to infection increases when the amino acid intake is inadequate. As is well known, the plasma protein level remains remarkably constant under a wide range of conditions. In order to maintain this normal level, however, the body must call on homeostatic mechanisms that mobilize the proteins normally stored in the lymph, in extracellular fluids, and in other reserves. It is possible, therefore, that the potential deleterious effects of amino acid deficiencies are brought out more readily and more rapidly when these protein reserves are low than when they are abundant. In other words, the past nutritional history of the individual as well as the anabolic demands at the time of observation must be kept in mind when one attempts to define the nutritional requirements that are optimal for resistance to infection.

d. Effects of dietary factors other than amino acids on resistance to infection

In addition to proteins and amino acids, many other nutritional factors have been claimed to affect resistance to infection. The "anti-infectious" virtues of vitamins A and C still have many advocates among physicians, and especially among commercial producers of these substances. In reality, however, there is no proof that resistance can be increased by the use of any one of the known vitamins above the intake required for good nutrition. In contrast, dietary constituents not known to promote growth certainly exert a protective effect against a few pathogens under the proper conditions.

A curious and unexplained example of the influence of a dietary factor on infection is the increased resistance of mice to pneumococci and staphylococci when cobalt is added to diets containing excessive amounts of manganese! Probably of more direct relevance to man's problems is the fact that greenstuffs harvested in springtime markedly increase the resistance of guinea pigs to tuberculosis. As far as can be judged, the effective substance is neither vitamin C nor any of the known nutrients. Many other reports also indicate that the resistance of experimental animals to tubercle bacilli, salmonellae, staphylococci, parasites, tapeworms, etc. can be increased by unidentified substances in certain natural foodstuffs which do not seem to affect the well-being of uninfected animals.

Of obvious interest are the claims that breast-fed infants are more resistant than bottle-fed infants to all sorts of respiratory and gastrointestinal infections. If valid, these claims might be related to the influence exerted by human milk, as well as by substances present in certain natural products, on the composition of the intestinal flora in children and in animals. (This topic is discussed in Chapter V.1.)

Among the many odd facts concerning the influence of nutrition on infection is that resistance can be decreased by certain foodstuffs which have high nutritional value. Thus, enrichment of the diet with unsaturated seed oils and particularly with wheat-germ oil aggravates mouse tuberculosis, mouse salmonellosis, and rat leprosy. Paradoxically, nutritional deficiencies can increase resistance under certain circumstances. Examples of this paradox are that riboflavin deficiency in chicks decreases the severity of *Plasmodium lophurae* infections and that ascorbic acid deficiency in monkeys represses the multiplication of *P. knowlesi*. Monkeys and rats do not develop malaria as long as they are fed a diet lacking para-aminobenzoic acid because this vitamin is required for the multiplication of the plasmodia. There is some evidence, likewise, that children exposed to malaria remain free of this disease as long as they are breast-fed because human milk is very low in para-aminobenzoic acid.

It has been repeatedly observed that quantitative limitation of food intake and qualitative deficiencies in essential growth factors (such as various vitamins and amino acids) can retard the progress of viral infection or change the character of the disease they produce. In tissue cultures, certain viruses remain in a dormant state as long as the infected cells are maintained in deficient culture media, the probable reason being that nutritional deficiencies prevent infected cells from carrying on the synthetic processes needed for viral replication. These laboratory findings in vivo and in vitro may account for the clinical experience that viral diseases often exhibit a less acute character in patients who are underfed or malnourished than in well-fed persons.

A most intriguing illustration of the fact that lack of certain dietary factors can exert a protective effect has come recently from investigations on experimental allergic encephalomyelitis. Mice fed a diet consisting of whole wheat and whole milk are more likely to develop this form of encephalomyelitis than are mice fed a semisynthetic diet made up of highly purified constituents; susceptibility to the disease can be increased by adding the proper dietary factors to the semisynthetic diet. In other words, the possibility exists that deleterious responses to the phenomena and products of infection can be enhanced by certain dietary constituents.

There is even evidence that a dietary factor that increases resistance under certain conditions can become deleterious under others. In fact, experimental situations illustrating this paradoxical possibility are easy to imagine. Thus, a dietary factor capable of increasing antibody production, thereby exerting a protective effect, could conceivably intensify potentially dangerous allergic phenomena. Vitamin or amino acid deficiencies capable of lowering resistance to bacterial infections can increase resistance to certain viral diseases by depriving host cells of factors essential for replication of the viruses. Furthermore, one given nutritional factor can either aggravate a viral infection or, on the contrary, retard its course, depending upon the circumstances, as has been shown with pyridoxine in pneumonia virus of mice and with protein in viral influenza. (Reviewed in Dubos and Schaedler, 1959; Scrimshaw and Gordon, 1959, 1965.)

Many other complexities account for some of the conflicts in results that have been reported concerning the effects of dietary constituents on infection. Thus, trauma, with attendant physiologic disturbances and need for tissue repair, may be responsible for periods during which nutritional deficiencies are more critical than in the normal state; this is of particular importance for surgical interventions. Amino acid deficiencies are also more likely to promote infection in young persons during periods of rapid growth than in adults, who have reached a semisteady state. The regularity and rapidity with which diets inadequate in protein content have been found to increase susceptibility to bacterial infections in some of the experiments reported above may be explained by the fact that the tests were carried out with young animals. Relevant to this problem is the finding that the enhancement of infection by nutritional deficiencies is aggravated by submitting the animals to an additional stress, as caused for example by bacterial endotoxins (lipopolysaccharides) at the time of infection. Since lipopolysaccharides cause a pathologic state that has similarities with traumatic shock, one may postulate that the nutritional status is of particular importance in determining susceptibility to infection at the time of trauma.

2. Nutrition and Dental Caries

Studies with germ-free animals have established beyond doubt that dental caries can be produced by certain strains of *Lactobacillus* and especially of *Streptococcus salivarius*. However, production of the disease by these organisms requires that the animals be fed special diets. In man, also, bacterial activity is almost certainly involved in the production of caries, but epidemiological evidence makes it seem likely that dietary factors play a part in the incidence and severity of the disease.

Many surveys carried out all over the world have revealed that dental caries is rare in populations living on a primitive diet, whereas it increases in frequency as the diet becomes more "refined." (Early literature reviewed in Finn, 1952; Price, 1939.) Skulls found in different prehistoric sites show little evidence of dental decay. Examination of the remnants from 1,000 specimens found in Greece revealed 4 to 7 lesions per mouth in the skulls from the Neolithic era, the Early Iron Age, classical times, and the period of Turkish occupation; then a large increase occurred suddenly during the nineteenth century. Whereas excellent teeth are still common today in the poor and isolated areas of Greece, they are rare in the cities.

A similar picture emerges from the study of a large group of complete palates discovered in the vaults of a church in Hythe, England, covering the period between 1250 and 1650. Whereas teeth were sound throughout that period, marked deterioration of dental structures became evident in the population of the seaport at the beginning of the nineteenth century, presumably when food habits began to change. The deterioration has been continuing ever since. As recently as 1962 a survey made in Italy revealed that poor children showed many fewer caries than American children of Italian origin who were fed a richer but not necessarily better diet.

A dramatic correlation between dental decay and adoption of

purified food products is provided by the results of several consecutive dental surveys carried out over a period of 30 years in the island of Tristan da Cunha. When the first survey was made in 1932, the islanders were still living almost exclusively on local products and caries was almost nonexistent among them. By 1955, they had changed largely to imported food, and caries had become as frequent as it is in Western Europe or North America. Similar observations have been made among groups of children in South Africa. The children from families too poor to buy European food had a caries rate of only 11 per cent, whereas the rate was 28 per cent in children of more prosperous families, the same as in European children living in that area. The aborigines of Australia, who live on kangaroo meat, so far have healthy peridontal tissues and teeth capable of transmitting a very powerful biting stress. Similarly, dental caries is rare among the men of the Xavantes tribe of Central Brazil; this Indian population still lives a very primitive and isolated existence, almost without any form of agriculture, in the Brazilian Mato Grosso.

The high incidence of dental disease in all social classes wherever the eating habits of Western civilization prevail strongly suggests that the modern diet is not adequate for maintaining teeth in healthy condition, even though it is highly effective in providing for rapid body growth. Consumption of large amounts of carbohydrates, especially of refined sucrose, is assumed to be an important contributing factor to dental caries, and perhaps its most essential cause. Since certain bacterial species, especially *Streptococcus salivarius,* produce a complex polysaccharide in the presence of sucrose, it has been assumed that the local deposition of this gumlike material facilitates the formation on the teeth of bacterial plaques where organic acids can accumulate and destroy the enamel.

Observations recently made in New Zealand suggest that mineral nutrition may also play a part in the etiology of the disease. It has been noted that caries is uncommon among New Zealand children living in a certain coastal region where local vegetables are grown on recent marine soils and have a high mineral content. Furthermore, the incidence of caries experimentally produced in rats can be reduced by adding the ash of these plants to the diet. The anticaries effect of this ash material is correlated with a high content of molybdenum, possibly in association with other elements.

The probable relation of dietary constituents and habits to dental caries has been singled out for discussion here because, whatever its precise mechanism, it illustrates that the severity of the pathological effects of microbial agents, in this particular case the caries-producing bacteria, is profoundly conditioned by the nutritional state.

As it would not be possible to report in detail all the experimental studies bearing on the effect of the nutritional state on resistance to infection, a few general statements will suffice to illustrate the variety of the mechanisms involved in this relationship. In brief, evidence has been obtained for an effect of nutrition on (1) the multiplication and destruction of the pathogens in vivo; (2) the production of toxins by the pathogens; (3) the rate and effectiveness of immunologic response by the host; (4) the histochemical characters of the lesions; (5) the susceptibility to the various manifestations of toxemia.

VII. The Evolution of Microbial Diseases

1. The so-called conquest of microbial diseases

By the 1950s the most optimistic dreams of the founders of medical microbiology had been essentially fulfilled in several countries of Western civilization. A very large percentage of the microbial agents of disease had been isolated, identified, and cultivated in artificial media or in tissue cultures; bacteriological cleanliness of the food and water supplies had been improved through technological advances; practical procedures had been worked out for the large-scale production of killed or attenuated bacterial and viral vaccines; highly effective drugs had become available for the treatment of bacterial and parasitic infections; a variety of pesticides had been synthesized and had proved their usefulness for the control of insect vectors.

In many places, economic prosperity and social organization have now made it possible to translate into practice the scientific achievements of the microbiological era. As a result, the mortality rates of infectious diseases have been brought down to a very low level, particularly among children and young adults, and the life expectancy at birth has soared to unprecedented high levels.

Most clinicians, public health officers, epidemiologists, and microbiologists felt justified therefore in proclaiming during the 1950s that the conquest of infectious diseases had finally been achieved. The deans of medical schools and their faculties had so little doubt on the subject that they made it a practice (and still continue!) to appoint to the chairs of medical microbiology biochemists or geneticists essentially unconcerned with the mechanisms of infectious processes.

Surprisingly enough, this euphoria has not yet been dampened by the fact that the morbidity rates of infection have not decreased significantly, and in some cases have actually increased. Despite so much oratory on the conquest of microbial diseases, the paradox is that the percentage of hospital beds occupied by patients suffering from infection is now as high as it was fifty years ago. Today, as in the past, moreover, disorders of the respiratory and digestive tracts with a microbial etiology constitute the most frequent causes of absenteeism from school, office, factory, or from training in the armed forces.

There is a widespread tendency to explain away the fact that infection remains a major cause of disease by invoking the emergence of drug-resistant forms of bacteria, the spread of strains of pathogenic agents from one area to another, and the failure to apply with sufficient energy the existing knowledge of prophylaxis and treatment. In reality, however, these explanations deal with only minor aspects of the infectious disease problem in our communities. The more important reason for the stubborn persistence of infection lies in our lack of understanding of the interrelationships between man and his biological environment. As we shall see, there are many forms of infectious diseases that are not prevented or cured by sanitation, vaccines, or drugs, and indeed are probably not amenable to control by these approaches.

The point at issue is that the microbial diseases which account for the greatest morbidity in our communities today are completely different in their origin and manifestations from those which are so effectively dealt with by modern techniques. For this reason, they will require programs of study, as well as methods of prevention and therapy, different from those which have been emphasized so far. The real problem is not how to apply more effectively the control procedures we already possess, or how to improve them, but rather to search for a qualitatively different kind of knowledge.

The sciences concerned with microbial diseases have developed almost exclusively from the study of acute or semi-acute infectious processes caused by virulent microorganisms acquired through exposure to an exogenous source of infection. In contrast, the microbial diseases most common in our communities today arise from the activities of microorganisms that are ubiquitous in the environment, persist in the body without causing any obvious harm under ordinary circumstances, and exert pathological effects only when the infected person is under conditions of physiological stress. In such a type of microbial disease, the event of infection is of less importance than the hidden manifestations of the smouldering infectious process and than the physiological disturbances that convert latent infection into overt symptoms and pathology. This is the reason why the orthodox methods based on the classical doctrines of epidemiology, immunology, and chemotherapy are not sufficient to deal with the problem of endogenous diseases. The need is to develop procedures for reestablishing the state of equilibrium between host and parasite.

The problems posed by microbial diseases of endogenous origin cannot be properly stated without some historical considerations of the changes in virulence that infectious agents have undergone in the course of time. From this point of view, as we shall see, it is not very enlightening to say of a particular microorganism that it has a high or low virulence. A more meaningful statement is that a given pathogen is generally highly destructive in a given population when the pathogen and population first come into contact, and the severity of the infectious process tends to decrease as contact between the two components of the system is continued over several generations. In other words, the problem of virulence cannot be stated without historical perspective.

2. Historical changes in incidence and severity of various microbial diseases

a. Spontaneous changes in the severity of epidemics

Repeatedly all over the world, overwhelming epidemics have arrested invading armies on the march, decimated populations, disorganized the social fabric, changed the pattern of civilizations —but mankind has survived. Human life has proved adaptable enough to triumph over yellow fever, influenza, typhus, plague, cholera, syphilis, and malaria, even without benefit of effective measures to combat these diseases. Less dramatic, but fully as astonishing as the spontaneous and often sudden termination of the great epidemics, is the continuous downward trend of certain diseases in the course of centuries.

Not so long ago tuberculosis was the Great White Plague, the "captain of all men of death." In Boston, New York, Philadelphia, and Charleston—in London, Paris, and Berlin—all available statistics reveal tuberculosis mortality rates higher than 500 per 100,000 inhabitants in the year 1850. Then the mortality rates began to decrease in Europe and North America, and they have continued to decrease ever since, except for brief interruptions in the downward trend associated with the two world wars. In 1947, the tuberculosis mortality rates were below 40 per 100,000 population in several countries, having thus decreased more than tenfold in less than a century. The decrease began before the discovery of the tubercle bacillus, long before specific methods were available for prevention or cure.

Syphilis spread through Europe like a prairie fire during the fifteenth and sixteenth centuries. Its course was then rapid and often fatal, unlike that of the frequently mild and slowly progressing disease of our days. There is still much uncertainty concerning the origin of this epidemic, but there is no doubt concerning its ferocity when it first struck Europe. The incidence of syphilis was tremendous in all strata of society and the lesions it caused were of extreme severity, as described by Fracastoro and Ulrich von Hutten. Its virulence spontaneously decreased during the succeeding centuries while the disease continued to be widespread through all social classes. Observations reported by Bartolomé de Las Cases in his *General History of the Indies,* written in the sixteenth century, suggest that syphilis was then common among the American Indians but caused them little trouble. Even more convincing evidence of the spontaneous attenuation of syphilis was obtained some three decades ago. In a village of the Punjab nearly every living person was found to give a positive Wassermann reaction although no stigmata referable to the infection could be detected.

Remarkably enough, Fracastoro, the Italian physician who coined the word syphilis in the sixteenth century and gave the first classical description of the disease in his famous poem "Syphilis sive Morbus Gallicus," was also the first epidemiologist to postulate that contagious diseases exhibit ebbs and flows. The following quotations taken from his book *De Contagione* still constitute today a timely warning:

> I use the past tense in describing these symptoms, because, though the contagion is still flourishing today, it seems to have changed its character since those earliest periods of its appearance. I mean that, within the last twenty years or so, fewer pustules began to appear, but more gummata, whereas the contrary had been the case in the earlier years. . . . There will come yet other new and unusual ailments, as time brings them in its course. . . . And this disease of which I speak, this syphilis too will pass away and die out, but later it will be born again and be seen again by our descendants,

just as in bygone ages we must believe it was observed by our ancestors.

Another highly contagious malady that has changed its manifestations is the English sweating sickness. Between 1485 and 1551 several outbreaks of this mysterious disorder struck England and to a lesser degree the Continent. Beside the "sweat" even the dreaded plague paled into insignificance, as can be seen in the diaries and accounts of the Tudor era. Then, for unknown reasons, the English sweat vanished as swiftly as it had come and it has not returned to this day, at least not in a recognizable form. In contrast, bubonic plague remained rampant in Europe until the middle of the nineteenth century. It is possible that the sweat was an exotic disease such as dengue, but more probably its agent is still present in our communities masquerading perhaps as an adeno virus or as an attenuated form of viral influenza.

In general, the changes are less sudden than was the case for the sweating sickness. The more common situation is that presented by syphilis, which progressively decreased in severity, persisting in a milder form. Scarlet fever, measles, and mumps serve as convenient examples to illustrate this phenomenon, because of the relative ease with which they can be diagnosed.

In a book printed in 1701 scarlet fever was referred to as "this name of a disease, for it is scarce anything more." However, the celebrated Irish physician Graves took a different view of the subject a century and a half later:

> In the year 1801, scarlet fever committed great ravages in Dublin, and continued its destructive progress during the spring of 1802. It ceased in summer, but returned at intervals during the years 1803–4, when the disease changed its character . . . either so mild as to require little care, or so purely inflammatory as to yield readily to the judicious employment of an antiphlogistic treatment. . . . The experience derived from the present epidemic [1834–35] . . . has proved that, in spite of our boasted improvements, we have not been more successful in 1834–35 than were our predecessors in 1801–2.

In 1840 the mortality due to scarlet fever suddenly doubled in England and Wales. From then until 1880 it remained at the top of the infectious maladies of childhood and accounted for 4 to 6 per cent of deaths at all ages. The highest mortality from the disease in England was probably reached during the ten-year period 1861–70. In 1863 the death rate from scarlet fever was 1,500 per million; a large percentage of children died of it before attaining the age of five. After this devastating spell the severity of scarlet fever declined, and by 1900 first measles and then diphtheria surpassed it as a cause of death, but it retained some of its malignancy until the beginning of the present century. In the words of the English epidemiologist Major Greenwood, it was long known as "*the* fever" and required no qualifying adjective any more than the plague does. More recently, scarlet fever has been on the whole a relatively mild disease. Indeed, it is probable that much of the credit for its control, which is commonly given to drugs and to improved methods of treatment (as was done at the beginning of the ninteeenth century), should in reality go to the unknown factors that brought about the "spontaneous" decrease in its virulence.

A similar story can be told of the fluctuation in the severity of measles. William Heberden asserted in 1785 that measles "are usually attended with very little danger; it is not often that a physician is employed in this distemper." But the position changed about 1800. In 1804 measles caused as many deaths, chiefly among adults, as did smallpox, and actually surpassed the latter in 1808. The illness appears to have resembled what Sydenham in 1674 called "anomalous" or "malignant" measles, and it is probable that its mortality was even greater than indicated by statistics, since it must have caused many fatal pulmonary ailments that remained undiagnosed. Measles constituted the main cause of death in children until around 1840; then it began to ebb and scarlet fever displaced it for over forty years as the chief infectious disease. But measles again resumed some of its importance toward the end of the century, and almost every year until 1915 the deaths from it outnumbered those from smallpox, scarlet fever, and diphtheria combined. From then on its mortality fell and has continued to fall ever since.

b. Social factors and microbial disease

History provides many examples of the fact that social misery and its usual consequence, physiologic misery, are associated with increases in the prevalence and severity of microbial diseases. It has long been known that war, famine, and pestilence commonly ride together.

In many cases, of course, the infectious diseases associated with famine, crowding, or war are the direct result of increased contact with particular types of microbial agents. Armies in the field have been plagued by typhoid and dysentery whenever the breakdown of sanitary practices has permitted massive infection with salmonellae and shigellae. Napoleon's troops contracted typhus when they were heavily exposed to lice and rickettsia in Eastern Europe during the 1812 Russian campaign. Similarly, the Western allies suffered greatly from scrub typhus and malaria in the Asiatic theater during World War II, even though the adult native populations with whom they were in contact suffered much less from these diseases, to which they had been exposed throughout life.

Of equal importance and greater interest, perhaps, is the fact that famine, crowding, wars, and social upheavals can increase the prevalence of microbial disease through other mechanisms, namely by creating conditions which decrease general resistance.

The tuberculosis epidemic which prevailed throughout the industrialized countries of the Western world during the nineteenth century owed part at least of its severity to the long working hours, the poor nutrition, and the low living standards prevailing among the labor classes during the Industrial Revolution. As living standards improved, tuberculosis mortality immediately began to decrease. The improvement was already noticeable at the end of the century, long before any specific measure of prophylaxis or therapy had been introduced. Tuberculosis mortality again exhibited a sharp and sudden rise in parts of Europe during World Wars I and II, and once more it resumed its downward course as soon as the hostilities were over.

During the 1920s, inflation in Germany provided a spectacular illustration of the bearing of social factors on resistance to disease. Following an abrupt fall at the end of the war, the tuberculosis mortality increased sharply during the years of inflation and fell again as soon as stabilization of the currency allowed a return to more normal living conditions. The rapid and reversible manner in which European countries responded to changes in social and economic conditions by changes in tuberculosis mortality rates cannot be explained in terms of infection rates. The more likely explanation is that large numbers of persons in our communities are in a state of unsteady equilibrium with tubercle bacilli, as

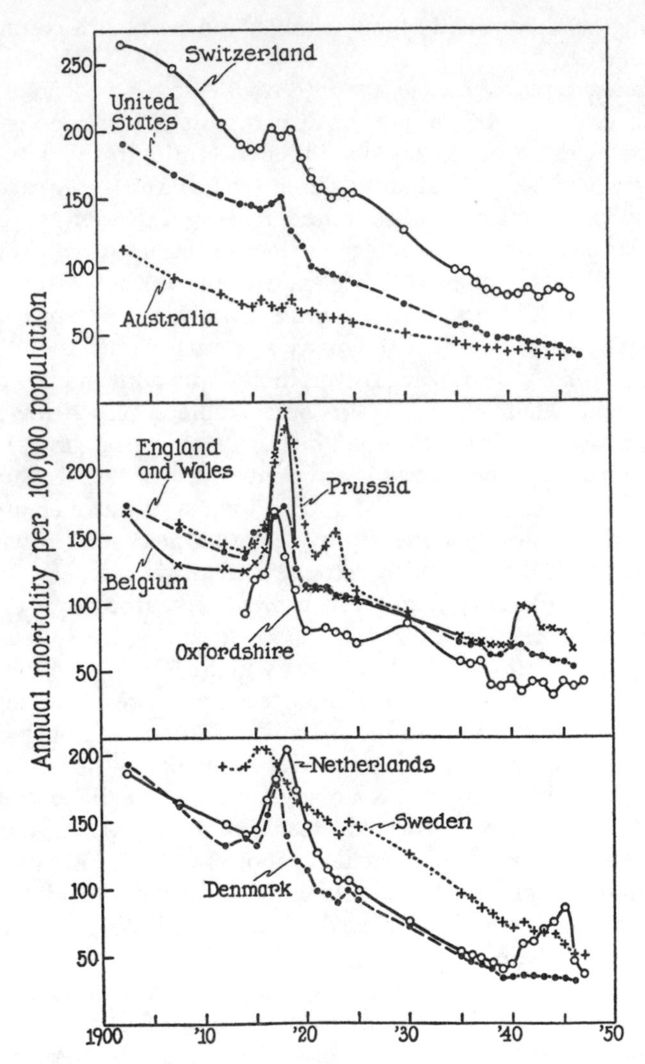

FIGURE 11. The effect of war on tuberculosis. The figures at the bottom indicate the years; those at the left, the number of deaths per year per 100,000 population. (Reproduced from R. Dubos, and J. Dubos, 1952.)

well as with many other pathogens, and that this equilibrium is upset whenever disturbances impair the general state of resistance.

This relationship has been brought out in an analysis of the disease problems in German concentration camps during World War II. Among infectious diseases, it was not the exotic, unusual epidemics like typhus, cholera, or even bacillary dysentery which proved to be the most troublesome in the camps, but rather ordinary skin ailments, colds, bronchopneumonias, staphylococcal infections, pulmonary tuberculosis—in other words, the type of diseases, minor or severe, caused by microorganisms *endemic* in the normal European communities. Here again, increase in contact infections could hardly account for the aggravation of these endemic diseases. Far more important certainly was the loss of natural resistance caused by malnutrition and other forms of physiologic misery. Indeed, most of the internees overcame their microbial maladies shortly after their return to a normal environment, usually without the help of specific therapy. Even in the case of tuberculosis, rapid recovery was the rule, though antituberculous drugs did not become available until the late 1940s.

The recent spectacular health improvements in Puerto Rico illustrate well that the evaluation of the role played by control programs is complicated by a number of unresolved epidemiological puzzles. As late as 1950, diarrhea and enteritis were the leading causes of death in Puerto Rico, with a mortality rate of 138 for 100,000 population. This rate had fallen to 33 in 1962! The reasons for this startling change are obscure, since the fundamental cause of most intestinal disorders is still today a matter for speculation, and since neither antimicrobial vaccines nor drugs could have been used for prophylaxis or treatment. During the same period, the mortality from measles, whooping cough, and scarlet fever also fell dramatically, again without the benefit of specific methods of prophylaxis or treatment. These epidemiological puzzles are not peculiar to Puerto Rico. As already mentioned, the mortality from various bacterial and viral infections began to fall precipitously in Europe and in the United States long before the development of specific methods of control (Figures 11 and 12).

c. *History and resistance to infection*

Granted that the levels of hereditary resistance and susceptibility may differ from one national group to another and change with time (Chapter V.2), the evolution of the microbial diseases of man strongly suggests that these characteristics are not necessarily of racial origin. Usually, they are rather the consequence of the extent and duration of the contact that a particular group of people has had with a particular microbial agent. The history of disease

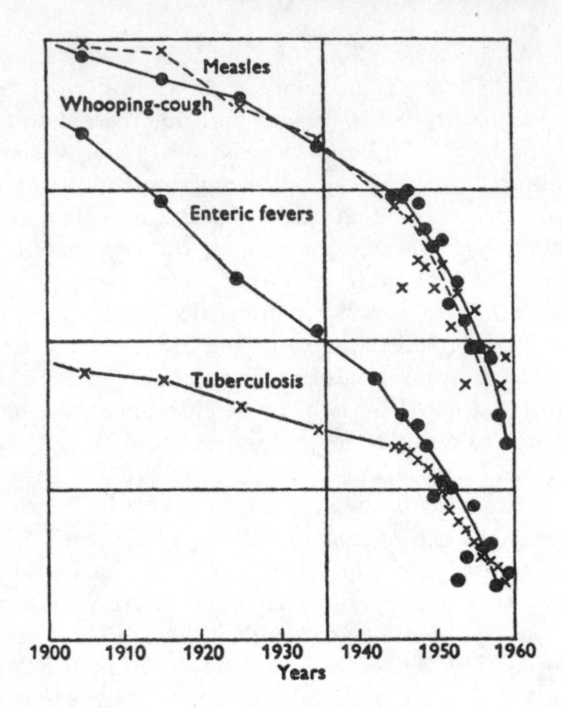

FIGURE 12. The trend of mortality from some infectious diseases in England and Wales during the twentieth century. Relative death rates are shown for measles, whooping cough, enteric fevers, and tuberculosis. The absolute scales differ, but all are shown logarithmically, the horizontal lines indicating a 10-fold difference. Between 1936 and 1960, there has been almost a 100-fold diminution in the deaths from whooping cough and measles. (From F. M. Burnet, 1962.)

among the various Indian and Polynesian populations is instructive in this regard.

Smallpox is said to have been introduced into the American continent early during the Spanish Conquest, perhaps by a Negro in Cortez' band. The Indians proved to be highly susceptible to the disease, which almost wiped out some of their settlements, and there is reason to believe that this disaster contributed to their rapid defeat by the Spaniards. The conquest of North America a century later provided further dramatic evidence of the susceptibility of the Amerindians to smallpox. Repeated outbreaks decimated village after village and at times whole tribes. Early in the seventeenth century, for example, the Massachusetts and Narragansett Indians were reduced from pre-smallpox populations of 30,000 and 9,000 respectively to less than a thousand in a few years.

In the epidemic of 1837, the Mandan population fell from 1,600 to 31, the Assiniboins lost whole villages, the Crows lost one-third of their population, while the total deaths among the Plains tribes amounted to 10,000 in a few weeks. Similar outbreaks occurred as late as 1870–71 among the Blackfeet.

Other microbial diseases brought in by the European invaders contributed still further to the rapid decline of the Indian populations during the nineteenth century. There were more than 700,000 Indians in North America before the arrival of the Europeans, but only 250,000 in 1850. Tuberculosis played a large part in this holocaust, as illustrated by the epidemics that decimated the Plains Indians of Western Canada. In the 1890s, the death rate from tuberculosis in the Qu'Appelle Valley reservation of Saskatchewan reached the fantasic figure of 9,000 per 100,000 population per year! More than half the families of this tribe were eliminated in the first three generations of the epidemic. Some 20 per cent of the deaths in the surviving families were caused by tuberculosis!

Microbial diseases among the Polynesians during the past century presents a picture similar to that seen among the Amerindians. The South Pacific was explored during the second half of the eighteenth century. Granted that the charm of the Pacific Islands and the amorous welcome of their women may have warped the judgment of the European visitors, there is no reason to doubt the validity of their unanimous opinion concerning the health and the vigor of the Polynesian people at that time. The Europeans saw throughout the islands robust and happy men and women, obviously well adapted to their physical environment as well as to their local pathogens. Yet, microbial diseases became rampant among the South Sea Islanders within a short time after these early explorations, and the Polynesian population soon began to fall. From approximately 300,000 at the time of Captain Cook's first visit in 1778, the population of native Hawaiians had fallen to 37,000 in 1860. During the same period, the population of the New Hebrides was reduced to one-tenth its original size.

There is no doubt that this catastrophy was brought about largely by venereal diseases, tuberculosis, scarlet fever, measles, and other infectious disorders that the Polynesians contracted during their short contacts with the Europeans. When measles was introduced into Hawaii, practically the whole population went down with the disease, and many thousands died. Epidemics of measles, whooping cough, and influenza occurred again in 1848, and every child born that year died. When smallpox struck in 1853, there were over 9,000 cases with 6,000 deaths out of a population of 70,000. During his second and third visits to the South Pacific, Captain Cook himself was much disturbed at the thought that his sailors were responsible for having introduced venereal diseases among the Polynesians in Tahiti. However, he found solace in the belief that the initial guilt was to be placed on Bougainville's French crew, who had preceded them.

The fantastic mortality caused in Europe by plague during the Justinian era and again during the Renaissance constitutes a classical example of the extreme virulence that a disease commonly exhibits when newly introduced into a population. This phenomenon was once more illustrated in a dramatic manner by the outbreak of measles in 1952 among the Eskimos of the Canadian arctic. The attack rate in this instance was over 99 per cent, including all ages, and the mortality rate reached up to 7 per cent at Ungava Bay. An outbreak of poliomyelitis among a group of Eskimos of the Hudson Bay in 1949 exhibited the same pattern. Fourteen per cent of the population died, and over 40 per cent developed paralytic disease. In this case again, all age groups were involved, the lowest clinical attack rate being in the infants.

Still another example, made familiar by the writings of Albert Schweitzer, is provided by African trypanosomiasis, also known as sleeping sickness. Trypanosomiasis was introduced in the Ogowe region of Equatorial Africa some 40 years ago by carriers who came with the Europeans from Loango, where it apparently has existed from time immemorial. The disease proved to be terribly destructive in its new territory, carrying off one-third of the population in the course of a few years. In Uganda it killed 200,000 persons out of 300,000 in six years. Of the 2,000 inhabitants in a village of the upper Ogowe, only 500 survived after two years of the epidemic.

The examples cited above, and these are only a few among many others that would tell the same story, leave no doubt that whole populations can be literally decimated by pathogens with which they have had little contact in the past. It is apparent, on the other hand, that the diseases introduced by the white man among the Indians, the Polynesians, and other primitive people during the eighteenth and nineteenth centuries no longer affect them as acutely and with as rapid a fatal outcome as was uniformly the case in the past. These people have developed, or are developing, biologic responses to infection similar to those seen in the Western world under normal conditions.

The high mortality rates caused by the plague bacillus or by the yellow fever virus among Europeans probably can serve as an analogy for the degree of virulence that the tubercle bacillus or the measles virus had for the Polynesians or the Indians two centuries ago. Tuberculosis and measles are still important causes of disease among these people today, but, as already mentioned, they rarely give rise to catastrophic epidemics in the areas where they have been established for several generations. Similarly, it has been observed that, after a while, African trypanosomiasis loses some of the destructive character it exhibited when first introduced into new districts of Equatorial Africa. The disease lingers on but it carries off small numbers of victims instead of killing two-thirds of the exposed populations as it once did.

Precise observations are available concerning the changes in the clinical manifestations of tuberculosis among some Indian tribes of North America. In the first and the second generations to suffer from the tuberculosis epidemic in the Qu'Appelle Valley reservation, extensive grandular involvement was the rule in school-age children. Meningitis, generalized miliary disease, and bone and joint disease were extremely frequent—evidence of inability of the host to localize infection. In 1921, at a time when the generalized epidemic was in the third generation, the disease showed a greater tendency to localize in the lungs and to exhibit a chronic course; the mortality was falling, and glandular involvement had dropped to 7 per cent among school children. This latter manifestation of high susceptibility to tuberculosis has continued to decline steadily and was seen in less than one per cent of children in the fourth generation. In other words, while tuberculosis among the Amerindians exhibited at first a very acute course, different in character from that observed in people who have had contact with the tubercle bacillus for several generations, it is now undergoing a change that makes it resemble the more chronic type of disease commonly seen among Western people under normal conditions. Several recent reports have emphasized the differences in susceptibility to tuberculosis among the various ethnic groups of Jews in the state of Israel. On the whole, the Ashkenazim and the Sephardim who originate from the cities of Central Europe and the Mediterranean basin, where they had been in contact with the disease for many generations, are far more resistant to it than are the Yemenite Jews, who originate from agricultural regions where tuberculosis was very rare.

3. Exogenous vs. endogenous microbial disease

a. The prevalence of endogenous infections

Irrespective of the specific nature of their etiological agents and of their pathological manifestations, microbial diseases can have two different kinds of origin. They can be either exogenous or endogenous.

Many pathological processes are of course the direct outcome of exposure to a virulent pathogen. Smallpox, yellow fever, typhus, syphilis, diphtheria, tuberculosis, typhoid fever, malaria, etc. illustrate situations in which the disease can usually be traced to contact exposure to the responsible microbial agent; the pathological phenomena then usually develop within a fairly predictable period of time after exposure. This truly *infectious* kind of microbial disease, with an *exogenous* origin, is still of immense importance throughout the world, but it has become less of a menace in our own communities, thanks to public health practices, vaccination, prompt diagnosis, and antimicrobial therapy.

In contrast, the microbial diseases now gaining prominence, especially in the prosperous countries of the Western world, are often caused by microbes formerly regarded as essentially harmless, for example, the coliform and other Gram-negative bacilli, the nonhemolytic streptococci, various kinds of yeasts as well as fungi, and probably many viruses still unidentified. Even though their virulence is low, these organisms can give rise nevertheless to serious pathological states under special physiological circumstances.

The expression *endogenous* microbial disease refers to any pathological state caused by microorganisms acquired at some prior time which have persisted in the body as part of its indigenous microbiota (Chapter V.1). It is important to emphasize, however, that highly virulent pathogens, such as tubercle bacilli, can act as etiological agents of endogenous disease. Most types of virulent organisms, if not all, can become established in the body, and persist in it, without manifesting their presence in the form of overt disease until the general resistance of the host is lowered. An extensive and critical review of this problem has recently been published in a book entitled *Attenuated Infections* (Simon, 1960).

Endogenous microbial diseases have always been responsible for a very large amount of human misery, but two developments have made them more obvious and perhaps increased their relative importance during recent years. One is that as acute exogenous infectious processes are being brought under control, it becomes easier to recognize the microbial diseases of endogenous origin. The other development is that, paradoxically enough, many therapeutic procedures introduced by modern scientific medicine favor the proliferation of certain components of the indigenous flora and thus allow them to cause disease. For example, enteritis caused by staphylococci, yeasts, or fungi frequently occurs when the normal gastrointestinal ecology is disturbed by the administration of antimicrobial drugs. Similarly, the microorganisms of the indigenous flora often proliferate after treatment with cortisone or antileukemic drugs, as well as following certain surgical procedures such as subtotal gastrectomy.

The increasing prominence of endogenous microbial diseases has thus brought to the fore an aspect of the problem of infection that has long been known but grossly neglected, namely that various microbial agents including the classical pathogens can exist and persist in the tissues without manifesting their presence by pathologic disorders.

As early as the 1840s investigations of the potato blight in Ireland revealed that most healthy potato plants normally carried the fungal agent of the disease, *Phytophthora infestans*. The parasite caused the blight only under unusual weather conditions. Pasteur observed that guinea pigs can harbor the causative agent of fowl cholera for many months or years in abscesses without suffering from the infection; he clearly realized and emphasized the theoretical importance of the carrier state for epidemiology. Koch proved that healthy men can excrete cholera vibrios in their stools, and this provided the first practical illustration of the fact that healthy carriers play an important role in the transmission of human disease. Thus, the existence of infection without disease was recognized early, long before the designation "carrier state" was coined for the concept, and before Typhoid Mary became its personification. But it is only during the past two decades that the generality of the phenomenon has been appreciated.

Awareness of the fact that most pathogens are widely distributed and yet do not cause clinical disease except in a very small percentage of persons and animals harboring them has come chiefly from the study of latent viral and rickettsial infections. Laboratory animals carry a large variety of viruses, which can be evoked into activity, as is the expression, by all sorts of nonspecific stresses. Most colonies of albino mice, for example, are infected with the pneumonia virus of mice (PVM), which is potentially capable of causing a fatal pulmonary disease in them but usually remains dormant. Many flocks of birds, of parakeets in particular, are normally infected with the psittacosis virus, but they do not suffer from the infection until the virus is caused to multiply by the stresses resulting from poor husbandry.

The isolation of adenoviruses from the tonsils and the adenoids of healthy persons constitutes evidence that latent viral infections occur also in man. In fact, as mentioned in Chapter V, some 150 different viruses have been recovered from man, and for many of them there is no evidence that they are associated with disease states. The recognition that the Brill-Zinsser disease is the manifestation of the reawakening of an old dormant infection with the epidemic strain of typhus provided the first well-documented evidence of latent rickettsial infections in man.

Latent infections of man with virulent bacteria, fungi, and higher parasites are at least as common as are those with viruses and rickettsia. Surveys made in different parts of the world show, for example, that 30 to 50 per cent of normal persons harbor coagulase-positive staphylococci in their nasopharynx. Until a few years ago, practically all human beings in the Western world became infected with tubercle bacilli early in life, and this situation still exists today in most underprivileged countries; yet only a small percentage of tuberculin-positive persons develop clinical tuberculosis.

In the United States, *Amoeba histolytica* can be isolated from the stools of a surprisingly large number of persons who have never suffered from amebiasis. Similarly, some 30 per cent of the population give a positive skin test for toxoplasmosis, and encysted trichina are often found in muscles in the absence of any clinical sign of trichinosis. As we shall mention later, the causative agents of syphilis and gonorrhea can also persist in apparently healthy persons. Indeed, the list of latent infections with all classes of pathogens continues to grow as the search for them is expanded.

b. Microbial persisters

The fact that most pathogens—viruses, rickettsia, fungi, bacteria, protoza, and even helminths—can persist in the body in a nonactive state for prolonged periods of time, and thus can be isolated from the tissues of normal individuals, is at the origin of some of the most important and most perplexing problems of epidemiology and clinical medicine. Furthermore, it has given rise to an extensive and rather confusing terminology. The expressions "carrier state," "latent, inapparent, or dormant infection," "masking and unmasking of viruses," etc. have never been clearly defined, and it is often difficult to differentiate the shades of meaning they are supposed to convey.

The recognition that even unicellular organisms normally carry particles capable of being reproduced intracellularly under special circumstances (for example, bacteriophages for bacteria, kappa factors for Paramecia), has contributed still further to the confusion (Chapter IV.1). The phrase "attenuated infection" is probably broad enough in meaning to encompass all the associations between host and potential parasite that do not express themselves in overt disease states (reviewed in Simon, 1960).

It will suffice to point out here that pathogens and other microbes can persist in the body in a number of different states. At one extreme are those situations in which the pathogen exists in the tissues in a form that can be demonstrated readily by standard technical procedures. The recovery of typhoid bacilli from typhoid carriers or of the virus of herpes simplex from the tissues of normal persons are examples in point. At the other extreme are the situations in which recovery of the pathogen is all but impossible as long as the latent infection has not evolved into an active pathological process. For example, rickettsia of the epidemic strain of typhus (*Rickettsia prowazeki*) can persist in vivo for many decades and remain undetected until they cause Brill-Zinsser disease; or again murine corynebacteria can persist in rodents in an atypical form difficult to detect, yet multiply explosively as a result of stress, nutritional deficiencies, or treatment with massive doses of cortisone.

Technical difficulties often account for the failure to isolate the pathogen. Indeed, it is only in specialized cases that the methods in use at present permit detection of small numbers of infective units. The report that a pathogen is "inapparent" or "masked" must always be judged, therefore, in terms of the sensitiveness of the procedure used to test for its presence. In many cases, infected tissues contain growth-inhibitory substances which, although not capable of destroying the pathogen that is being looked for in vivo, can nevertheless inhibit its multiplication in vitro. For example, neutralizing antibody transferred from infected tissue along with a virus is likely to prevent the latter from becoming established in a recipient animal.

Tissue inhibitors other than antibodies can also cause difficulties, as when tissue constituents prevent the growth of tubercle bacilli in otherwise favorable culture media. Elimination of neutralizing antibodies and of other inhibitory agents is therefore a prerequisite for the successful demonstration of the presence of small numbers of infective units in the tissues, but few if any are the cases in which this can be done adequately.

Even more subtle mechanisms may be involved in the failure to recover pathogens from infected tissues. Viruses can persist in vivo in the form of naked nucleic acid structures and bacteria in the form of protoplasts, as long as they are sheltered in the well-balanced environment of their hosts. However, such delicate infective structures are readily destroyed or inactivated when removed from their protective environment, and as a result they usually escape detection by the customary laboratory procedures. The role of such "naked" infective particles in disease has not yet been fully documented, but their very existence illustrates nevertheless that many different mechanisms can account for the well-established fact that microbes often persist in the tissues even when their presence cannot be demonstrated by ordinary microscopic or cultural techniques.

Finally, microbes can persist in a form so atypical as to prevent recognition. For example, the agent of murine corynebacterial pseudotuberculosis persists in vivo as an avirulent variant producing an atypical colony, which reacquires virulence and multiplies extensively when the rodent is treated with cortisone.

Oddly enough, the phenomenon of persistence has become more obvious, and perhaps of greater practical importance, as a result of the widespread use of chemotherapeutic agents. Whereas antimicrobial drugs are often highly successful in arresting the progress of disease, they rarely succeed in eradicating the causative agent from the tissues. For example, even after successful treatment of scrub fever with tetracycline, of gonorrhea with penicillin, or of tuberculosis with isoniazid and other antimycobacterial drugs, some of the causative rickettsia, gonococci, or tubercle bacilli persist in the tissues. And so it is with many other types of infectious agents even following treatment with highly effective drugs.

The spectacular increase in the incidence of venereal diseases all over the world gives special significance to the recent findings that *Treponema pallidum* can also survive in vivo despite apparently successful treatment with large doses of penicillin. It appears that in syphilitic patients a modus vivendi can become established between host and parasite, resulting in an attenuated infection. Experiments in rabbits have revealed furthermore that administration of large doses of cortisone can tip the balance in favor of the *T. pallidum.*

According to the report of an expert committee of the World Health Organization published in 1963, at least 65 million new cases of gonorrhea occur in the world each year, and the infection is becoming once more prevalent in the very young. Improved methods of treatment have made little impact on the epidemiological problem. All methods of control such as case finding, treatment of contacts, "cluster testing," prophylactic treatment, mass treatment of special groups, etc. have proved of limited value. One of the fundamental difficulties of the epidemiological problem in gonorrhea is that the gonococcus can persist after therapy, especially in females, and that consequently there exists a large reservoir of undiagnosed infection.

The word "persister" has recently been introduced to designate more specially the parasites that survive in tissues despite successful and prolonged chemotherapy. In most cases probably, such persisters are intracellular and exist in a metabolically inactive phase. It is important to emphasize that, contrary to common belief, persisters are not necessarily forms of the organisms that have become genetically resistant to the drug used for therapy. More often than not they are still genetically drug-susceptible, but either they exist in a physiologically resting form on which the drug is not active, or they are located in parts of the body or in tissue cells that provide a protective shelter for them.

The production of persisters is not a phenomenon peculiar to chemotherapy. It occurs also in the presence of protective antibodies; immunological processes are rarely if ever more successful than antimicrobial drugs in fulfilling Paul Ehrlich's ideal of absolute sterilization. In other words, viruses, bacteria, and other pathogens can persist in the tissues even when the level of specific immunity is high enough to prevent the infection from progressing.

c. Manifestations of endogenous infection

A few examples taken from the fields of animal and human pathology will illustrate some of the ways in which endogenous diseases manifest themselves under natural conditions.

Shipping fever, also known as transit fever, is a severe condition that strikes farm animals when they are moved from one area to another under conditions of stress; the disease is associated with a great variety of microbial agents. Calves, for example, commonly develop pneumonia or enteritis during transit; pigs become susceptible to the pathological effects of *Erysipelothrix,* which they carry in a latent state. Interestingly enough, most of the manifestations of shipping fever can be prevented by treating the animals with tranquilizers at the time of shipment. In some unexplained manner, these drugs minimize the physiological disturbances that activate the indigenous microbial agents involved in shipping fever.

Two surveys recently made in large American hospitals highlight the importance of endogenous microbial diseases in man. At the New York Hospital, 54 per cent of the fatal infections observed in 1957–58 occurred while the patients were under care on the medical wards; most of the infections were caused by microbes of "low virulence" and were brought about by therapeutic procedures directed at other disease conditions. At the Mayo Clinic, only half of the 294 cases of meningitis studied during the 1950s could be traced to the classical etiological agents of this condition, namely meningococci, pneumococci, or influenza bacilli; the other cases were caused by common organisms such as *E. coli, Pseudomonas,* coagulase-negative staphylococci, etc., not usually regarded as virulent.

Among a group of women working in a telephone exchange, carefully followed over a period of several years, the largest bulk of disease affecting either the respiratory or digestive tracts occurred in relation to various stressful episodes. Similarly, the initial stresses of life in army camps probably play a part in evoking into activity some of the respiratory disorders so common among new recruits. In this regard, it is of interest that the detailed study of ten young adults working in a sophisticated and well-equipped New York virology laboratory failed to give a clear etiological picture for their ordinary medical problems; only a very small percentage of the febrile illnesses among them could be identified by the best current diagnostic methods. In brief, it appears that many varied latent microorganisms can play a part in the etiology of disease when the conditions are favorable for them.

d. Infection and herd immunity

It has long been recognized that the persistence in the body of a given microbial agent is accompanied by a high level of resistance to superinfection with this same agent. Such a state of resistance was early recognized and designated as infection immunity or premunition. Although emphasized chiefly for its relevance to malaria, tuberculosis, syphilis, and relapsing fever, infection immunity is certainly of very general occurrence, but its study has been grossly neglected. Observations made during recent years should renew and stimulate scientific interest in this unorthodox yet highly interesting field of theoretical and practical immunology.

An illuminating example of infection immunity has been provided by the study of the comparative resistance of several strains of mice to corynebacterial pseudotuberculosis. Mice of the C57 Black strain are highly resistant to the causative agent of this disease, *Corynebacterium kutscheri,* whereas mice of the albino SS strain are highly susceptible. Cross-mating between the two mouse strains gives results suggesting that their difference in susceptibility has a genetic basis. However, when "normal" C57 Black mice are treated with large doses of cortisone, they succumb from spontaneous infection with *C. kutscheri,* a fact which proves that they normally harbor this organism. In contrast, albino SS mice subjected to the same treatment do not develop the endogenous disease and therefore must not be carriers. This correlation between carrier state and resistance to superinfection has been established for many other strains of mice. In other words, resistance to superinfection with this particular corynebacterium is in some manner correlated with the presence of bacterial persisters in the infected animal.

These findings bring to mind the response of experimental

animals to certain viral infections. For example, the virus of lymphocytic choriomeningitis produced a severe disease with fatal outcome when first introduced among albino mice, but it has now become established in several mouse colonies. Within a few years after its introduction, the virus was found to be present in all animals of these colonies, producing in them a latent infection. The infection was contracted in utero but did not manifest itself by any detectable sign of disease in adult animals, even though their organs continued to harbor active virus throughout life. In this case, therefore, a few years sufficed to change the type of relationship between mice and choriomeningitis virus from that of a virulent epidemic to a state of silent commensalism or symbiosis. In many colonies, the virus has become an unobtrusive guest, the presence of which is barely noticed by the host, but it renders the carrier mice resistant to superinfection.

The response of animals to the polyoma virus provides another striking illustration of the resistance afforded by contact early in life with an infectious agent. Injection of the virus into a variety of species of experimental animals is likely to result in the production of many different kinds of tumors. It seems, however, that this happens only if the animals are from a colony free of polyoma virus. In contrast, there is no evidence of tumor production in wild or laboratory populations of animals that normally carry the virus. As in the case of corynebacteria, therefore, the acquisition of a viral agent either in utero or very early in life may fail to produce any sign of disease, yet profoundly alter the response of the host to subsequent infection with this virus.

The findings just outlined point to the paradoxical conclusion that animals, and probably human beings, may behave as if they were genetically resistant to a given agent, yet owe this behavior in reality to some modification of their tissue response brought about by the fact that they carry persister forms of this very agent. Such infection immunity or premunition may also play some part in the lifelong immunity to viral infections that follows early exposure to the pathogen or vaccination with attenuated living vaccines.

It is probable that the conditioning of response brought about by infection in utero or very early in life operates through several unrelated processes, ranging from immune tolerance, through cellular and humoral immunity, to stimulation of nonspecific defense processes. Whatever the theoretical basis of their individual mechanisms, however, the various forms of infection immunity have in common certain practical consequences that suggest possible ap-roaches to the control of some problems of infection.

It has been observed, for example, that human beings who carry *E. coli* do not readily become superinfected with other strains of this bacterial species. Furthermore, recent clinical studies reveal that the colonization of human infants with virulent staphylococci can be prevented or retarded by early contamination, nasal or umbilical, with very small doses of a coagulase-positive strain of staphylococcus (selected because of its great susceptibility to penicillin and low virulence). Not only does the attenuated strain establish itself in the artificially contaminated infants, it also spreads to other human contacts. The overall effect is to limit the dissemination of more virulent staphylococci.

Thus it is apparent that the ubiquitous presence of a potential pathogen in a community can confer a high level of resistance to this particular agent. Infection immunity probably does not constitute an immunological entity, but through the various mechanisms that permit microbial persistence, it may play a very important role in the phenomenon of herd immunity.

4. Genetic factors in microbial diseases

The changes that have occurred during historical times in the relation between man and the various microbial agents of disease suggest that either the latter undergo fluctuations in their infectivity and pathogenicity, or that the resistance of man to them can fluctuate. It is probable in fact that both mechanisms can operate, in certain cases singly and in others jointly.

Viruses, bacteria, and other parasitic agents can of course undergo mutations affecting most of their characteristics, including their immunologic specificity and their virulence. While it is legitimate to assume that such mutations occur in nature, it has not been possible so far to demonstrate convincingly that they play an important role in modifying the course of human epidemics. However, observations made with the virus of rabbit myxomatosis in Australia and Europe have established that genetic changes in virulence can occur under natural epidemic conditions.

The European rabbit was introduced in Australia in 1859. As it had no natural enemies in its new habitat, it rapidly spread over much of the continent and multiplied enormously. The rabbit population eventually reached several billions, thus creating a grave threat for the agricultural economy of Australia. In 1950, the virus of myxomatosis was introduced in an attempt to control the rabbit population. Myxomatosis occurs naturally in the wild rabbits of Brazil but merely in the form of a benign tumor, associated with only a mild and transient viremia. The relationship between the Brazilian rabbit and the myxoma virus clearly denotes an ancient association resulting in ecologic equilibrium. In contrast, European rabbits, which had never had any contact with the virus, uniformly develop an acute, almost always fatal disease when inoculated even with very small doses.

The initial outbreaks of myxomatosis immediately after introduction of the virus in Australia were characterized by a case mortality rate higher than 99 per cent. Within a short time, however, the case mortality had fallen to 90 per cent in areas where a second spontaneous outbreak had occurred. Interestingly enough, it turned out that this lower mortality was due in part to a decrease in the virulence of the virus. Some details concerning this phenomenon are of interest here because the decrease in virulence was unexpected and probably has general significance for epidemiologic theory.

In Australia, the virus is transmitted from rabbit to rabbit almost entirely through mosquitoes, which act mechanically as "flying needles." Because the highly virulent strains of virus kill rabbits within a very few days, the chance for their transmission through the mosquito vector is rather limited. However, when attenuated mutant strains of the virus appeared spontaneously, they produced a less rapidly fatal disease, and the rabbit developed skin lesions of longer duration. Thus, the less virulent mutant strains had a better chance of being transmitted through mosquito bites, and they progressively displaced the original highly virulent strain in the field. A few years after the introduction of myxomatosis in Australia, viral strains recovered from the field caused a mortality of 90 per cent in laboratory rabbits instead of 99 per cent as had been the case with the strain originally introduced (reviewed in Fenner, 1959; Fenner and Ratcliffe, 1965).

While the evolution of rabbit myxomatosis in Australia can be explained in part by hereditary changes in the virus, there is no doubt that it has been influenced also by hereditary changes in the host. As already mentioned, the European type of rabbit introduced in Australia is immensely susceptible to myxomatosis, a disease with which it had no racial experience until a few years ago. Now, however, rabbits trapped in areas where the infection has been prevalent for several years exhibit a degree of resistance to the virulent forms of the virus much higher than that exhibited by rabbits before the beginning of the epizootic, or by rabbits carefully and constantly maintained in isolation. This increase in resistance has a genetic basis and therefore must result from the selection of mutant animals which had survived the initial outbreak. Genetic changes in resistance to infection are of course common among animals. In fact, it is possible almost at will to select and breed laboratory animals having a high resistance or high susceptibility to one or another type of bacterial or viral infections.

It seems reasonable to believe, but difficult to prove, that genetic changes also occur in the resistance of man to his pathogens. Since epidemics with a great killing power among young people tend to eliminate a large percentage of those having a high degree of susceptibility, the likely outcome is the selective survival of that segment of the population endowed with a higher than average innate resistance. Thus, genetic resistance progressively accumulates in the affected population. The history of family groups strongly suggests that such a selective process resulting in higher herd resistance operated in Europe and America during the nineteenth-century epidemic of tuberculosis, as it certainly did when tuberculosis eliminated half the families among the Indians of the Qu'Appelle Valley reservation. It is probable indeed that the decrease in the severity of tuberculosis in the countries of Western civilization during the past few decades has been brought about not only by improvements in standards of living and medical care but in no small part also by genetic changes.

There are indications that a similar genetic selection has taken place in the Orient with regard to leprosy. Among people living today in the Hawaiian Islands, the Chinese, the Japanese, and other Oriental people originating from areas where *Mycobacterium leprae* has long been ubiquitous seem to suffer from this disease far less commonly than do the Polynesians who became exposed to it for the first time during recent centuries.

A slight loss of virulence by the parasite and some increase in the genetic resistance of the host provide the right conditions for the development of acquired immunity in a large percentage of infected individuals. Here again, rabbit myxomatosis in Australia constitutes an enlightening model. Protective antibodies have been found in the surviving animals trapped in infected areas. It can be surmised, therefore, that the transmission of maternal antibodies helps the young to withstand infection during the first days of life and thereby allows them to develop their own active immunity.

The many interrelated aspects of this problem cannot be discussed here, but it seems worth pointing out that the genetic and immunologic changes favoring resistance to infection, which occur during a generalized epidemic, are the almost inevitable outcome of prolonged contact between any parasite and any host. For this reason, it seems unlikely that living species can ever be completely wiped out by epidemics, however virulent the parasite. The outcome of evolutionary forces is of necessity a modus vivendi, according to which the parasite and the host reach some sort of equilibrium which permits the survival of both. The concept of successful parasitism, so ably formulated by Theobald Smith (1934) and N. H. Swellengrebel (1940) a generation ago, is another expression of this evolutionary equilibrium resulting from prolonged racial contact between host and parasite.

Evolutionary adaptation between hosts and their parasites helps to explain why the ability or failure of a particular microorganism to cause disease cannot be fully explained in terms of the concepts encompassed by the word "virulence." The need to redefine virulence in terms of host-parasite relationships will now be illustrated by consideration of malignant malaria in primitive populations.

Plasmodium falciparum possesses attributes that make it highly virulent for most members of the human race. Yet this parasite produces only a rather mild disease in the Bush Negroes who live in the area included in former Dutch Guinea. While over 90 per cent of these Negroes become infected with the malignant form of plasmodium shortly after birth, the infection contributes little if anything to infant mortality among them. Most adults, if not all, harbor the parasite in their blood, but the intensity of parasitemia decreases with advancing age.

The extent of splenomegaly constitutes further evidence of the mildness of malaria in Bush Negroes. Among them, the spleen rate declines from a high of over 80 per cent in infants to less than 15 per cent in adults. The fact that such a large percentage of adults show parasitemia without splenomegaly illustrates the two complementary determinants of the infectious process. On the one hand, the parasite is virulent since it can become established, multiply, and persist in the tissues. On the other hand, the Bush Negro is resistant to the parasite since he can keep its proliferation under control and since his response to its presence is so mild that the infection usually fails to express itself in overt clinical disease.

The probable explanation of these findings is that the Bush Negroes of New Guinea have evolved in constant contact with plasmodia. Widespread infection constitutes a selective force favoring the kind of genetic endowment that makes it possible for human hosts and virulent parasites to coexist in a state of biological equilibrium. In the case of African Negroes, as is well known, possession of the sickle-cell gene is one of the attributes increasing resistance to malaria. Needless to say, immune responses reinforce the effectiveness of genetic mechanisms when the parasite is ubiquitous in the environment.

Similar phenomena of adaptive host-parasite relationships certainly occur in all other infectious diseases of mankind. On a few occasions during historical times, the social and economic circumstances proved favorable for the spread of *Pasteurella pestis* among the population of Western Europe, and every time the death rate was enormous. During the Justinian era in the Roman world, and again during the Renaissance in Italy, France, England, and other parts of Europe, whole areas were decimated by the bubonic and pneumonic forms of plague. *P. pestis* is still widely distributed today in certain parts of Asia; but granted that it constitutes an important cause of disease in these populations, it does not behave toward them as the terrible scourge that almost depopulated Europe on two occasions. Similarly, cholera can run a fulminating fatal course in people exposed for the first time to its causative vibrio, yet the disease tends to manifest itself as a mild annoyance in many bazaars of Asia.

Like protozoa and bacteria, viruses also tend to approach a state of biological equilibrium with man. Thus, a great variety of respiratory infections now extremely common and rather mild in our communities are caused by viruses that probably were at one time responsible for more severe pathological processes. Evolutionary adaptation may have taken place even in the case of paralytic poliomyelitis. Until recent times, most human beings in our communities became infected with polio viruses, but only a very small percentage of the infected persons developed paralytic disease; in general, the infection expressed itself in the form of symptoms so mild as to be overlooked. True enough, maternal antibodies did account in large part for the failure to develop disease following infection, but it is probable that genetic mechanisms of resistance were also involved.

Herpes simplex infection is of special interest in this regard because the virus is potentially capable of causing acute and fatal encephalitis. More generally, however, large numbers of children and young adults become infected without displaying severe pathological signs and they continue to carry the virus throughout life. In our communities, herpes infection is known chiefly in the form of fever blisters, a self-limiting disease that occurs only when the carrier of the virus is under some form of physiological stress. Man has come close to achieving a state of biological adaptation to the herpes virus, as well as to many other viruses that are ubiquitous in his communities.

5. Determinants of microbial diseases

In the late stages of evolutionary adaptation between a given parasite and its host in a given community, infection is extremely prevalent, but it rarely evolves into overt and fatal disease. Given enough time, a state of peaceful coexistence eventually becomes established between any host and any parasite. Microbial persistence, without clinical manifestations or destructive pathological processes, is therefore not a laboratory freak, not a rare phenomenon engendered by tricky manipulations of antimicrobial drugs or immunological responses. Throughout nature, infection without disease is the rule rather than the exception.

Latent infections, however, can become activated by many different kinds of changes either in the host or in the environment. For this reason, a large and most important aspect of the epidemiology of disease (as contrasted with the epidemiology of infection) has to do with the factors that upset the equilibrium between host and parasite and thereby convert dormant infection into overt disease.

Maintenance of health despite persisting infection means, of course, that the mechanisms of physiological resistance and the humoral or cellular processes involved in acquired immunity are capable of inhibiting the multiplication of the parasites, but do not succeed in eradicating them from the tissues. This does not imply, however, that acquired immunity necessarily results in freedom from disease, for there are many types of physiological disturbances that allow the parasite to multiply even if the host is specifically immune to it. For example, persons who carry the virus of *herpes simplex* commonly have a high level of neutralizing antibodies for this virus in their serum; nevertheless, they can experience transient episodes of virus multiplication under the influence of nonspecific stimuli—as varied in nature as certain types of fever, excessive exposure to the sun, fatigue, menstruation, or section of the trigeminal nerve. The result is then the production of herpes blisters even in the presence of humoral immunity to the virus.

On the basis of clinical experience, as well as of common observation, it is usually considered self-evident that susceptibility to all sorts of infectious agents is much increased by the various stresses and strains of life. Patients, as well as their physicians, tend to incriminate poor food, bad weather, lack of sleep, or emotional disturbances in the causation of disease, even when it is certain that the signs and symptoms result from the activities of viruses or bacteria. The life situations assumed to be responsible for the activation of endogenous microbial disease range all the way from the deprivations and physiological misery of internees in concentration camps to the emotional upsets resulting from business failure or an unhappy love affair.

Unfortunately, few are the cases in which a convincing correlation has been established between the physiologic state of the host and his susceptibility to infection. Diabetes constitutes one of the best-documented examples of this type of relationship, since there is little doubt that the poorly controlled diabetic patient is easy prey to staphylococci, tubercle bacilli, and many other kinds of bacteria. It is also true, on the other hand, that his resistance to infection becomes almost normal once the diabetes is controlled by adequate insulin therapy. In this case, therefore, the response of the host to infection is under the influence of events, perhaps biochemical in nature, that can be altered reversibly by the physiologic control mediated through insulin. Certain therapeutic procedures also can increase susceptibility to infection. As already mentioned, the very use of antibacterial drugs not uncommonly favors the multiplication in vivo of microorganisms that are not susceptible to these drugs, probably by first eliminating the indigenous microbiota (Chapter IV.3). Extensive surgery (and any form of trauma, especially if it results in a state of shock) is another procedure that can render partially ineffective the defense mechanisms of the body, at least for a while.

It is certain, therefore, that many nonspecific stresses increase the vulnerability of the host, but there is little, if any, understanding of the mechanisms through which these effects are exerted. Indeed, many of the relationships believed to be so obvious as to need no demonstration in reality have proved to be almost impossible to reproduce in the laboratory. For example, it is often difficult to increase the susceptibility of animals to infection by nutritional deficiencies, temperature changes, or behavioral upsets. More surprisingly, experiments in human volunteers fail in many cases to provide clear evidence that exposure to cold and other forms of bad weather increases susceptibility to upper respiratory infections. Because of these difficulties, it has not yet been feasible to identify the physiologic and biochemical mechanisms through which environmental stresses alter the body response to microbial agents.

A few forms of activation of endogenous disease have been consistently reproduced and studied in laboratory models. Thus, as already mentioned, the susceptibility of experimental animals to infections caused by the microbial agents they normally carry in their tissues can be increased by large doses of antibacterial drugs, by production of traumatic shock, or by extensive body irradiation. Administration of ACTH also can enhance infection in certain experimental models, and this aggravating effect can be neutralized by the proper dose of somatotrophic hormone. Injection of thyroid hormone into rabbits markedly increases their resistance to experimental tuberculosis, but in contrast hypothyroidism seems to be associated with increased susceptibility to other infections. Treatment with large doses of cortisone is probably the most uniformly successful laboratory technique of activation, a fact of great practical importance since administration of this hormone has been a frequent cause of grave infections in man. The infection-enhancing effect of cortisone may provide a clue for the analysis of a large and ill-understood problem, namely the manner in which environmental stimuli, including emotional disturbances, affect endocrine activity and thereby modify indirectly resistance to disease.

While it is certain that the course of infectious processes can be profoundly altered by a variety of hormonal influences, the mechanisms involved are far from clear. The fact that cortisone interferes with the production of antibodies under certain conditions seems at first sight to provide the clue to the problem, but this immunologic inhibition cannot account entirely for the infection-enhancing effect. In addition to interfering with antibody production, cortisone influences many other physiologic processes, some of which may affect directly or indirectly the response of the body to infection. Interference with the inflammatory response and with the activity of the reticuloendothelial system, disturbance of intermediary metabolism, and activation of proteolytic enzymes are but a few of the effects of cortisone that might be of importance in this regard.

Nutrition is one of the environmental factors for which it is easy to postulate hypothetical mechanisms explaining a role in infectious processes (Chapter VI). Once pathogens have become established in the body, there come into play a variety of responses that tend to limit the spread of infection beyond the site of the initial lesion. As we have seen (Chapter VI.1), the rate of wound healing is markedly influenced by certain dietary factors.

There is no doubt that nutritional deficiencies play a decisive role in the prevalence and severity of microbial diseases among underprivileged people. For example, the very high susceptibility of patients suffering from kwashiorkor to all sorts of viral, bacterial, and parasitic diseases is a direct consequence of protein malnutrition. In our own communities, however, it is far more difficult to relate microbial diseases to the nutritional state. The problem is further complicated by the fact that, paradoxically enough, certain nutritional deficiencies seem to increase rather than decrease the resistance to certain viral infections. (Reviewed in Dubos and Schaedler, 1959; Scrimshaw, Taylor, and Gordon, 1959, 1965.)

Thus, many are the ways in which the total environment can affect the interplay between man and the countless microbial agents which he normally carries, or with which he comes into contact. This interplay can result in disease or be compatible with the maintenance of health, depending upon the environmental circumstances under which the encounter between man and microbe takes place. In other words, the ecology of microbial disease is under the influence of factors, both general and local, independent of those which control the frequency of contact with infectious agents. To a large extent, endogenous microbial diseases are therefore indirectly the expressions of environmental forces.

The relative importance of the factors that determine the chance that infection will take place, and of those responsible for converting latent infection into overt disease, is conditioned naturally by the characteristics of each particular microbial agent, but also varies from one population and one community to another. When a group of people, whatever the race, first comes into contact with a pathogen, the chance is great that the general mechanisms of resistance will be of little avail and that infection will become manifest in the form of severe disease in a very large percentage of the persons infected. This type of situation occurred when smallpox, measles, tuberculosis, etc. were first introduced among the Amerindians and other primitive people. It was produced experimentally by introducing the myxoma virus among the rabbits in Australia and Europe. It probably would happen again if yellow fever, plague, or any type of infection with which Western man has not had contact recently were introduced for the purpose of biologic warfare. The possibility that mutants of common pathogens can behave as new agents of disease against which the general mechanisms of resistance is of little avail has been suggested to account for the virulence of widespread epidemics such as the influenza of 1918–19.

At the other extreme are the epidemiologic situations in which a particular microorganism is ubiquitous in a given community and becomes established in a latent form as a persister in most normal individuals. Since the event of infection is not the variable in this type of epidemiologic system, the factors that upset the ecologic equilibrium between host and parasite then become the effective determinants of disease. Needless to say, there are many intermediate situations between these two extremes, and the evolution of microbial diseases leaves no doubt, furthermore, that changes can occur in the parasite, in the host, and in the nature of their interrelationships. That the changes can occur rapidly has been shown with myxomatosis in rabbits and also with lymphocytic choriomeningitis in mice.

It is clear, in conclusion, that the type of relationship existing at any given time between hosts and their parasites is the outcome of many different factors, including past racial experience, evolutionary adaptation through genetic changes and immunologic processes, and transient disturbances in the internal and external environments. In the classical infections of exogenous origin, the determining etiological event of the disease is exposure to the infective microorganism. In endogenous microbial disease, the immediate cause is the environmental factor that upsets the biological equilibrium normally existing between the host and the microbial agents (persisters).

This profound difference in etiological mechanisms suggests that the methods used in the control of microbial diseases, both prophylactically and therapeutically, must differ from place to place and vary from time to time. The methods of sanitation and vaccination designed to cope with the great epidemics of the past will not prove effective in the control of the disease states caused by microbial agents that are ubiquitous in our communities in the form of dormant infections. It is also very doubtful that the usual antimicrobial drugs can be effective against them.

So far, the main goal of medical microbiology has been to prevent infection from taking place or, if it occurs, to treat disease once it has become established. The techniques designed to this end aim at attacking the microbial agents. It might be worth considering now whether useful practices of disease control can be derived from the fact that peaceful coexistence with pathogens often occurs in nature. This approach will require that the determinants of infection be separated conceptually from the determinants of disease; its objective will be to understand and control the processes responsible for converting infection into overt disease.

VIII. Environmental Pollution

1. Chemical pollution in modern life

Almost daily and in every part of the world, new health hazards arise from modern technology. Some of these hazards make an immediate public impact, for example radiation fallout, the acute effects of smog, or the congenital malformations produced by thalidomide. Others attract less attention because they lack drama and are not obvious in their effects; they behave as a new kind of "pestilence that stealeth through the darkness." Such is the case for the dangers posed by certain pollutants of air, water, and food, which remain almost unnoticed despite their potential importance for public health.

Two semipopular and extensively documented books on environmental pollution appeared almost simultaneously in 1962 under the evocative titles *Our Synthetic Environment,* by Lewis Herber, and *Silent Spring,* by Rachel Carson. The purpose of the authors was to alert the public to the dangers inherent in the thousands of new chemicals that technological civilization brings into our daily life. These two books were soon followed by a flood of articles and other books on environmental pollution. In response to public alarm, a report on the "Use of Pesticides" was issued in 1963 by an ad hoc panel of the President's Science Advisory Committee. Well formulated and useful as it is, this report nevertheless creates a limited and distorted picture of environmental pollution because it deals exclusively with the dangers arising from the use of insecticides and herbicides. In reality, these substances constitute only one aspect of the pollution problem, probably the least important.

Environmental pollutants affect every professional occupation and reach into every aspect of modern life. In composition, they range over the whole field of chemistry, including elements such as lead and arsenic, simple compounds such as oxides of sulfur and nitrogen, complex molecules such as the estrogens fed to cattle and chickens in order to bring them into condition, and sleep-inducing drugs, on which so many human beings have come to depend in ordinary life. The effects of the potentially dangerous products with which modern man now comes in daily contact encompass the whole gamut of toxicology—from acute poisoning to carcinogenesis, from chronic respiratory impairments to mental disorders.

One important aspect of environmental pollution has its origin in the enormous variety of risks that modern techniques introduce into industrial and office employment. A survey recently made by the newly created Occupational Health Research and Training Facility, of the Public Health Service, revealed that 400,000 cases of occupational diseases still occur annually in the United States. A few years ago, for example, workers in the garment industry began to develop serious irritation of ear, nose, and throat. The toxic agent turned out to be a formaldehyde component in certain resins used in wash-and-wear clothing that tended to volatilize in storage warehouses. So varied and unpredictable are the dangers of modern technology!

Surprisingly little attention has been paid to the potential health dangers of the ever-increasing exposure to lead in the industrial and urban environment. There seems to be no doubt that the concentration of lead is increasing rapidly in air, water, and food, and that the chief source of contamination is alkyl lead. Blood lead levels are the highest in auto mechanics and next in traffic policemen; they are higher in urban dwellers than in rural dwellers. The blood levels are sufficiently high in many cases to be associated with a variety of subacute toxic effects in persons chronically exposed.

Chemical hazards also exist in the home. Lead poisoning is commonly found among children in the poorer and older areas of all American cities; it results from their ingestion of paint applied on the interior woodwork of houses and on furniture. Because the characteristics of lead intoxication in children differ markedly from those found in adults, the diagnosis is often missed in areas without adequate facilities for case finding. In 1963 a bale of blue jeans became contaminated from a leaky drum of phosdrin concentrate during shipment. As this highly toxic phosphate ester pesticide is readily absorbed through the skin, six boys who wore unwashed jeans from the contaminated bale were poisoned, two of them seriously.

Self-medication is another source of environmental hazard in the home. For that matter, medication administered in hospitals by physicians themselves is also becoming a considerable factor in iatrogenic diseases. The dangers of medication will probably become still more widespread as time goes on, even though the thalidomide tragedy has had some influence, salutary but probably transient, by alerting the public and physicians to the hazards involved in the use of drugs. Indeed, drug-induced diseases are in the process of becoming some of the most common and dangerous forms of intoxication. They include conditions as varied as the gastrointestinal side effects of aspirin, the phenacetin-induced hemolytic anemia, the enterocolitis that follows the use of antibacterial drugs, and many forms of allergic sensitization. (For reviews of this topic see Chapter VIII.3.)

The pollution of air and water occupies a place of special importance in the field of environmental pollution because it affects everybody almost constantly and because it is rapidly increasing wherever life is governed by modern technology. In the following pages we shall consider some of the most conspicuous air and water pollutants, and emphasize the theoretical and practical difficulties posed by the study of their effects and by the development of methods for their control.

2. Aquatic and Atmospheric Pollution

a. The pollution of water resources

Throughout most of human history, domestic and industrial sewage was spread over the land. This practice took advantage of the fact that soil possesses certain built-in mechanisms—physical, chemical, and microbial—capable of destroying most substances and microorganisms, including those which might be dangerous for man. The countries of Western civilization began to abandon this practice approximately one century ago. Waste materials are now generally committed to water in the optimistic hope that purification will take place more or less spontaneously through the agency of natural processes in properly designed sewage plants.

The techniques of water purification that worked well at the beginning of this century are becoming inadequate wherever water pollution is intense, which means practically everywhere. An appalling amount of untreated or inadequately treated urban sewage is discharged into river water along with the wastes from slaughter houses, chemical factories, and metallurgical plants. In many cases, the river water is then used for public consumption by communities downstream after a so-called purifying process, which often does not go beyond filtration and chlorination. While such procedures are usually sufficient to rid drinking water of coliform bacteria, they do not render it free of viruses or nematodes, and they leave untouched most chemical pollutants.

The pollutants that enter water courses as a result of man's domestic, industrial, and agricultural activities can be divided arbitrarily into two groups, depending upon whether they persist and accumulate or are rapidly destroyed by chemical and biological processes. Until recent years, the first group consisted almost exclusively of inorganic chemicals, such as chlorides and the metallic salts from industrial wastes. But increasingly now, water is contaminated by new synthetic organic compounds that are partially or completely resistant to attack by the bacteria and other microorganisms normally present in streams and sewage plants.

Certain kinds of detergents widely used domestically and industrially are at present the most troublesome of these "persistent organics," but other synthetic substances also persist in water under special conditions. Thus, DDT, 2-4D, chlordane, cyanides, and other agricultural and pharmaceutical chemicals are intermittently found in streams. Water contamination with chemicals used in agriculture is often particularly intense because these products are commonly delivered to streams in storm run-off and are thus introduced so rapidly that there is not time for complete decomposition.

Drinking water increasingly contains a variety of detergents, bleaches, dyes, chemical and metallurgical wastes, herbicides and insecticides, etc., which have resisted the purification process. Even when biological oxidation of organic matter has proceeded effectively, the end products can be toxic if they are too concentrated. Evidence that some of the water pollutants are present in high concentration is provided by the frequency of the poisoning of aquatic life. In certain circumstances, oxidation of nitrogenous compounds can result in the accumulation of amounts of nitrates sufficient to cause methemoglobinemia and goiter.

Acute toxic reactions have not yet been observed to follow intake of water from city supplies, but this does not rule out chronic manifestations resulting from prolonged usage. As in the case of air pollution, the study of chronic toxicity of drinking water is rendered extremely difficult by the fact that the pollutants vary in composition from one place to another and from time to time. Toxicological studies and methods of control valid for one place at a given time are likely to be meaningless at some other place or at a later time as a result of the introduction of new technologies.

Furthermore, conventional methods of water analysis are rarely sensitive enough to detect the objectionable products, and even if detection were possible, it would serve little purpose because no practical method is available at present to rid water of its chemical pollutants. Clearly what is needed is an entirely new set of policies for the disposal of sewage and industrial wastes and for the chemical purification of drinking water.

The problems arising from the pollution of water by the chemical substances released by modern technology are naturally attracting the greatest scientific attention at the present time. It is important to emphasize, however, that the orthodox methods used for the control of microbial contamination still constitute the underpinnings of public health practice and remain as important as ever. Yet these methods are in danger of being neglected, through overconfidence and loss of interest. In New York City, some 60 to 65 per cent of Richmond Borough and 40 per cent of Queens Borough had not yet been provided with sewers in 1960! And similar situations are not uncommon in other urban areas of the United States. Overflowing septic tanks are becoming a menace throughout the country.

One hundred million persons in the United States depend on overburdened surface water. The production of synthetic organic chemicals has more than doubled in the period 1947–58, reaching over 40 billion pounds in the latter year, and the waste materials of their production and use are discharged into water and air. Yet the techniques of water purification and sewage treatment and disposal have remained essentially the same as they were in 1920. These techniques are ineffective against gross chemical pollution, and it can be anticipated that they will soon become inadequate also against the transmission of pathogenic microbial agents.

b. Air pollution in history

Mountaintops or skyscrapers disappearing in the clouds symbolized until recently a lofty mood of detachment from the trivialities of life. In contrast, the same pictures now evoke the threats to health posed by industrial and urban smog. Although news reporters caught the scent of air pollution only a few years ago, the problem is in reality very old. It began when primitive man first used fire, and it has occurred throughout human existence whenever and wherever fuel or refuse has been burned. Most industrial processes, however primitive, release a variety of toxic fumes; the air in large cities has long been accused of being insalubrious on this account.

In the "Anatomical Examination of the Body of Thomas Parr" (Chapter IX.1), William Harvey himself has left a written testimony concerning air pollution in the London of his time. After performing an autopsy on the centenarian "Old Parr," who had died of a "peripneumony" after a visit to London, Harvey stated that "the chief mischief [was] connected with the change of air, which through the whole course of [Parr's] life had been inhaled of perfect clarity," whereas the air to which he was exposed in London was polluted by smoke.

In the *Fumigufium or the Smoake of London,* John Evelyn also painted in 1661 a gloomy picture of the "hellish and dismal cloud of Sea-coale" over London, which was

> so universally mixed with the otherwise wholesome and excellent Aer, that her Inhabitants breathe nothing but an impure and thick Mist, accompanied with a fuliginous and filthy vapour, which renders them obnoxious to a thousand inconveniences, corrupting the Lungs, and disordering the entire habit of their Bodies; so that Catharrs, Phthisicks, Coughs, and Consumptions, rage more in this one City, than in the whole Earth besides.

Objective evidence of the fact that the air in certain urban areas has long been polluted is provided by observations on the prevalence of lichens. Whereas lichens of many varieties are ubiquitous in areas that are not industrialized and cover all kinds of rocks and the bark of trees, they have become very scarce in London, Paris, New York, or almost any large city. Only a very few species persist in areas where the air is polluted, notable among them being *Lecanora conizaeoides.* While this species is tolerant of pollution, it seems to be unable to withstand competition from other lichen species and is in fact displaced by them in areas that are not polluted. The descriptions left by botanists during the past two centuries indicate that the disappearance of most lichens in large cities began more than one century ago. The differences, both qualitative and quantitative, in the lichen flora in any geographical region between the farming districts and the industrialized areas are so striking that a kind of "lichen count" might provide a useful index of air pollution.

Historians of California have an interesting story to tell concerning the origin of air pollution in their state. When Juan Rodriguez Cabrillo surveyed the California coast and dropped anchor in San Pedro Bay in October of 1542, he could see the mountain peaks in the distance, but not their bases. He described how the smoke from Indian fires rose for a few hundred feet and spread out over the valley. This spectacle illustrated in a primitive, rustic form not only man-made pollution but also the phenomenon now designated as thermal inversion, or the trapping of warm air under cold air. Because of his experience, the Spanish explorer called the place he discovered "Bay of the Smokes," little realizing that the thermal inversion he had witnessed was a warning of far worse things to come.

Thus, air pollution is a very old phenomenon, but its importance as a health problem has been recently magnified by the introduction of many new kinds of air pollutants on a large scale. Chemical plants, factories and mills, power generators, heating equipment, automobiles, and refuse incinerators now belch out into the air huge amounts of pollutants, many of which had never occurred on this planet before. Pollutants exist as minute particles of dust, smoke, or fumes, as droplets or vapors of liquid, or as gaseous products. Some are common materials and are stable, others are rare and transitory. The thermal inversion witnessed by Juan Rodriguez Cabrillo remained but a picturesque circumstance characteristic of some topographical conditions until September 1943, when Los Angeles had its first daylight dim-out. Severe irritation of the eyes and the respiratory mucosa were the symptoms which revealed that smog had become a hazard to public health. From then on, California, followed by the rest of the world, became increasingly air pollution-conscious.

c. Air pollutants and thermal inversion

Soot and dust, the sulfur oxides, free hydrochloric and sulfuric acid, and pollen and other kinds of allergens used to be regarded as the most common and most important components in air pollution. These materials still constitute a very large percentage of the air pollutants all over the world. But the present-day pollution problem is further complicated by the presence in the air of many other groups of substances released on an enormous scale by the new industrial processes and by automobile exhausts.

It would be useless to try to list all the potentially dangerous products that now pollute the air in industrialized and urbanized areas because this list, however large, would certainly be incomplete. Even more important, it would be different from one place to another, and different next year from what it is today. In the words of Senator Maurine Neuberger, "We are dumping our garbage into the sky." This "garbage" includes solid particles such as ashes and bits of stone and carbon; microscopic droplets of oily and tarry matter; and especially many kinds of gases, some of which are obvious because they are pigmented, smelly, or irritating to the mucous membranes, while others of equal importance remain unnoticed. The range of air pollutants is determined by the nature of the local economy, and it changes with the economic structure. Air pollution is now identified with the word *smog*. Originally this word referred to the simultaneous occurrence of smoke and fog, as is frequently encountered in London, where coal is widely used both for industrial purposes and for domestic heating. However, the Los Angeles type of smog involves neither smoke nor fog. It arises from chemical reactions that occur in the air under the catalytic influence of solar energy; for this reason, it is referred to as a photochemical smog.

The London type of smog occurs on cold foggy days or nights; it traps pollutants consisting chiefly of soot and sulfur derivatives, and it holds them suspended in the air. In contrast, the photochemical Los Angeles type of smog occurs in the middle of hot sunny days. The thin haze that obscures the blue sky of Southern California is an outcome of the brightness of sunlight acting on olefins and nitrogen oxides.

At the present time in the United States, approximately half the air pollution is referable to exhausts from internal combustion engines. Carbon monoxide makes up some two-thirds of this outpouring of volatile poisons; its concentration can exceed 50 parts per million at breathing height in congested automobile traffic. Hydrocarbons (especially olefins), nitrous oxides, sulfur derivatives, benzpyrene, and many other products are also released in huge amounts and may have more important pathological effects than carbon monoxide. The highly pulverized rubber and the asphalt liberated by the abrasion of tires upon streets also add to the pollution load.

When the insolation is intense, the olefin hydrocarbons and the nitrogen compounds given off by the combustion of petroleum products react in the air. Their catalytic oxidation under the influence of light gives rise to nitro-olefins and to oxidized types of pollutants, such as ozone and other peroxidelike substances. Although these products were first recognized in the Los Angeles smog and are still characteristic of it, they attain toxic concentrations in many other areas, indeed, under any conditions where both automobile traffic and insolation are sufficiently intense. The photochemical oxidant type of pollution has been found not only in Los Angeles, San Diego, and San Francisco but also in Arizona, New Mexico, Detroit, Philadelphia, and New York.

As research progresses, other types of air pollutants are being recognized. For example, tetraethyl lead is being found in the vegetation growing along our highways! A sample of washed grass collected in the Denver area contained up to 3,000 ppm. of lead (in ash) near major intersections; the concentration was still above 50 ppm. 500 feet downwind. Similar observations have been made in New York State, Maryland, Canada, and England.

Whatever the chemical nature of the air pollutants, their toxicity is often increased by certain solid particles, which act as adsorbents, and by moisture droplets, which act as solvents. These agents condense the irritant gases that might be present in otherwise innocuous amounts and thus give rise to localized concentrations high enough to irritate or damage the tissues. It has been shown experimentally that the effect of sulfur dioxide is markedly enhanced by the presence of a solid or liquid aerosol not irritating by itself; likewise, vaporized phenols attached to carbon particles are much more toxic when introduced into the lung than are similar amounts of phenolic vapors suspended in distilled water. On the other hand, there is some evidence that particulate matter can in certain cases decrease the toxicity of pollutants.

While dust, soot, and smoke are the most obvious and most annoying elements of air pollution, they constitute but one limited aspect of the problem. Their removal goes far toward eliminating some of the most obviously annoying and unpleasant effects of smogs, and may decrease toxicity by reducing the number of particles on which the irritants become concentrated. But effective control of pollution demands much more than the removal of soot, smoke, and dust. In fact, as mentioned above, it is even possible that certain types of colloidal particulate matter, instead of being deleterious, exert some protective effect against the toxicity of volatile products. The one essential requirement for air pollution control is to avoid venting into the air the great variety of gases that are potentially dangerous even though invisible when they come out of automobile exhausts or smoke stacks. On a sunny day, the oxidant type of smog (whether in Los Angeles, New York, or elsewhere) may create a haze so thin as to be almost invisible to local residents, yet concentrated enough to obscure vision at a distance, as from an airplane for example. Polluted air free of soot is less annoying than smoke but is just as dangerous and may even be more so.

Excluding dust, pollen, and fog, about 100 air pollutants potentially dangerous to human health have been identified so far. They may be grouped as listed in Table 14. In 1956, the total release into the air of some of the pollutants then most common within the United States was estimated as follows:

> On an annual basis, emissions of organic compounds in combustion products of coal, oil, gasoline and refuse, etc., are estimated at 21 million tons; evaporation products from losses of hydrocarbons, natural gas, solvent vapours and other materials prior to combustion yield 11 million tons of additional

TABLE 14. Typical Concentration Values for Gases and Vapors in Urban Air *

		Maximum average during September 1962 (in parts per million by volume)			
Gas or vapor	*City*	*Monthly*	*Daily 24 hr.*	*Hourly 1 hr.*	*Single measurement 5 min.*
Hydrocarbon	Chicago	3.0	5.0	7.0	12.0
Oxidant	New Orleans	0.02	0.05	0.09	0.11
Ozone (O_3)	Philadelphia	<0.01	0.01	0.03	0.04
Nitric oxide (NO)	San Francisco	0.06	0.10	0.30	0.37
Nitrogen dioxide (NO_2)	Cincinnati	0.03	0.05	0.11	0.14
Sulfur dioxide (SO_2)	Washington	0.04	0.07	0.17	0.33

* Adapted from A. C. Stern, 1964.

> organics. Nitrogen compounds expressed as nitrogen dioxide amount to 8 million tons and halogen compounds to about 2 million tons [Rupp, 1956].

Needless to say, the figures would be much larger today in most cases.

Some effort has been made during recent years to prevent the release of particulate matter and of obviously toxic products from industrial plants, but the other sources of pollution are much more difficult to control. In particular, the materials released into the air by automobiles and trucks and by domestic fires (soft coal, garbage, brush, etc.) constitute a source of pollution that cannot be as readily regulated as industrial fumes. In most cities, pollutants of such origin exceed in amounts those derived from industrial operations. During the 1952 episode of London smog, for example, some 60 per cent of the smoke and sulfur dioxide came from domestic fires.

TABLE 15. Air Contaminants *

Group	*Examples*
Solids	Carbon fly ash, ZnO, $PbCl_2$
Sulfur compounds	SO_2, SO_3, H_2S, mercaptans
Organic compounds	Aldehydes, hydrocarbons, tars
Nitrogen compounds	NO, NO_2, NH_3
Oxygen compounds	O_3, CO, CO_2
Halogen compounds	HF, HCl
Radioactive compounds	Radioactive gases, aerosols, etc.

* Taken from L. C. McCabe, 1961.

Of special importance is the ever-increasing problem posed by automobile traffic. It has been estimated that the daily output of every 1,000 operating automobiles in an urban community burdens the air with 3.2 tons of carbon monoxide, 400 to 800 pounds of organic vapors consisting predominantly of hydrocarbons, and 100 to 300 pounds of nitrous oxides, plus smaller amounts of surfur and other chemicals. In the Los Angeles area, automobiles put out more than 6,000 tons of pollutants every day, of which 1,500 tons consist of olefin hydrocarbons, as against a mere 100 tons from the oil refineries!

Air pollutants accumulate to form smogs when thermal inversion interferes with normal vertical air movements. This phenomenon must be emphasized again because it plays such an important

TABLE 16. Total Emissions of Air Contaminants to the Atmosphere in Metric Tons per Day, Los Angeles County, July 1958 *

Hydrocarbons	1320	Oxides of nitrogen	635
Aldehydes and ketones	60	Sulfur dioxide	495
Other organic gases	125	Carbon monoxide	5070
Total	1505	Other	5
		Total	6205

* Adapted from M. Katz, 1961.

role in air pollution. Ordinarily, the air is warmer at the ground and colder above; indeed, the updrafts so essential for air cleansing arise from this temperature gradient. During thermal inversion, in contrast, a layer of warm air forms at higher altitude and traps a layer of cold air at the ground. In the classic type of London smog, the inversion forms at a very low level, 300 to 400 feet above the ground, and usually lasts several days. The smog then traps a large amount of coal smoke, abolishes visibility, and does not dissolve when it penetrates the houses, as a true fog would. In Los Angeles the inversion forms at a much higher level, 2,500 to 4,000 feet, but it is persistent and may last day and night during much of the summer.

Whenever and wherever an inversion roofs over the atmosphere of a heavily industrialized or populated region, the local air pollutants can no longer escape and consequently accumulate under it. This can happen at almost any season of the year in most cities. It is the chronic situation in several parts of Southern California because of the special topography of the region. A report published by the Stanford Research Institute (1948–50) states the problem in the following words:

> The inversion layer is like a canopy over the Los Angeles basin, preventing both vertical and lateral dispersion of contaminants. The natural haze, composed of salt from ocean spray and dust particles from soil and vegetation, is augmented by smoke, fumes, and gases from industrial and domestic activities, including vehicular traffic. The larger aerosol particles are removed by settling, but the fine aerosols, and gaseous pollutants, tend to build up at the top of the atmospheric layer which lies just below the base of the inversion stratum. The worst conditions occur when the man-made pollution accumulates over several days in calm weather and the base of the inversion layer is forced to its lowest level, thus bringing the polluted air to the ground.

TABLE 17. Mean and Maximum Concentrations of Selected Particulate Contaminants, American Cities, 1957–61 (in milligrams per cubic meter STP)

Pollutant	*Category*	*Arithmetic mean*	*Maximum*
Suspended particulate matter	Urban	0.118	1.706
	Nonurban	0.036	0.461
Benzene soluble organics	Urban	0.0099	0.1239
	Nonurban	0.002	0.0235
Nitrates	Urban	0.0024	0.0248
Sulfates	"	0.0118	0.094
Antimony	"		0.00023
Bismuth	"		0.000032
Cadmium	"		0.00017
Chromium	"	0.000047	0.000998
Cobalt	"		0.000003
Copper	"	0.00007	0.0037
Iron	"	0.0025	0.074
Lead	"	0.0005	0.0065
Manganese	"	0.00010	0.0078
Molybdenum	"		0.00034
Nickel	"	0.00004	0.00083
Tin	"	0.00005	0.001
Titanium	"	0.00006	0.0015
Vanadium	"		0.0012
Zinc	"	0.00048	0.014

* Adapted from A. C. Stern, 1964.

3. BIOLOGICAL EFFECTS OF AIR POLLUTANTS

The nuisance aspects of air pollution are obvious to all as well as are the irritating effects on the eyes and mucous membranes. But convincing evidence that pollution is a serious threat to health has been obtained only in the case of persons who were already suffering from some form of cardiopulmonary disease at the time of exposure to smog conditions.

Several dramatic episodes of air pollution due to the London type of smog have caused many fatalities during the past few decades: in the Meuse valley, Belgium, in 1930; in Donora, Pennsylvania, in 1948; in Poza Rica, Mexico, in 1950; and in London in 1952. In all these cases, most deaths could be traced to industrial or domestic fumes held in suspension by fog; the victims were chiefly the aged, especially persons with a history of cardiac or pulmonary disorder. But although these lethal episodes were spec-

TABLE 18. Maximum Allowable Concentrations of Air Contaminants in Los Angeles *

Pollutant	*Maximum allowable concentration (ppm.)*		
	Warning	*Alert*	*Danger*
Oxides of nitrogen	3.0	5.0	10.0
Ozone	0.5	1.0	1.5
Sulphur dioxide	3.0	5.0	10.0
Carbon monoxide	100.0	200.0	300.0

*Taken from L. C. McCabe, 1961.

tacular and have been much talked about, they have remained isolated events. Furthermore, there is no evidence that the oxidant, so-called Los Angeles type of smog has caused anywhere a sudden and large increase in mortality. At a recent seminar on respiratory diseases, it was indeed stated that "studies of air pollution in California have failed to show any effect on mortality." What is not known, however, is the extent to which constant, or repeated, exposure to the mixture of substances present in polluted air can become a cause of disease in the course of years for the general population.

Prophets of gloom do not lack examples to dramatize the damaging effects of air pollutants. The Los Angeles smog attacks rubber and makes it brittle. Nylon stockings have been known to pop and run during warm humid days in Richmond, Virginia, in Jacksonville, Florida, and in many other places, apparently as a result of high concentration of sulfur oxides in the air. Damage to growing plants is reported everywhere, and there are many lawsuits for livestock poisoning, especially by fluorides in the air. Even buildings are damaged and property values often go down in areas blighted by certain kinds of smogs. The situation is essentially as bad in other industrialized countries as in the United States or England. It is said that some of the picturesque and celebrated pines of Rome are dying from an oily air pollutant that coats their needles!

Although studies of the effects of pollution on experimental animals are still rather limited, the few facts available demonstrate that pathological states can be caused by exposure to concentrations of pollutants of the order of those which exist in the urban atmosphere. On the basis of these results, it can be surmised that pollution can also have deleterious and lasting effects on human beings. In no case, unfortunately, is there adequate information concerning the minimum toxic doses of the various air pollutants. The maximum allowable levels formulated by the Los Angeles Board of Health can serve only as guide posts in this regard (Table 18).

TABLE 19. U.S.A. Air Quality Standards and Threshold Limit Values for Solids or Liquids *
(in milligrams per cubic meter STP)

	Workroom air TLV [a]
Beryllium (Atomic Energy Commission)	0.002
Arsenic and its inorganic compounds (as As)	0.5
Chromium and its inorganic compounds (as Cr_2O_3)	—
Dust (nontoxic)	—
Lead and its inorganic compounds (as Pb)	0.2
Lead sulfide	—
Manganese and its inorganic compounds (as Mn)	5.0
Mercury and its inorganic compounds (as Hg)	0.1
Soot	—
Sulfuric acid	1.0

* Adapted from A. C. Stern, 1964.
a. Threshold Limit Values—for 8 hours a day, 5 days a week exposure of healthy adult worker.

Nitro-olefins seem to be particularly toxic. They are rapidly absorbed through all routes of administration and produce marked irritation of the eye, skin, and entire respiratory tract. Mice exposed to simulated Los Angeles smog containing nitro-olefins slowed down on their night prowling; their fertility and the survival rate of their young were reduced.

Levels of ozone as low as one ppm. proved lethal in 6 hours for animals made to exercise in a rotating cage for a few minutes of each hour. Human volunteers exposed to ozone concentration of 3 to 10 ppm. by weight also exhibited symptoms. In addition to headaches, with a tickling sensation of the nose and throat, they developed chest pains in 30 minutes. After one hour exposure, their vital capacity was markedly reduced, and the residual lung volume was increased. In animals and human volunteers, inhalation of concentrations of sulfur dioxide similar to those frequently found in the air produces spasms of the smooth muscles that usually go unrecognized but nevertheless result in tightening of the bronchioles. Many other tests have revealed that exposure to air pollutants always brings about disturbances in respiratory rates and an increase in total pulmonary flow resistance. Such exposure also slows down or stops completely the beating motion of cilia in the respiratory passages, and thus interferes with the sweeping up of mucus and particles out of the airways.

Histological evidence of lung damage has been observed in animals exposed 6 hours a day 5 days a week for various periods of time to concentrations of ozone only two or three times greater than those occurring in urban areas. Inhalation of sublethal levels of this gas resulted in pulmonary fibrosis; some of the exposed animals eventually died from pulmonary edema. Daily exposure of rats to sulfur dioxide for a period of 6 weeks produced likewise a thickening of the mucous layer lining the trachea.

While it is naturally impossible to carry out on man experiments designed to test the pathological potentialities of air pollutants, the statistical analysis of certain actual social situations bids fair to provide useful evidence on this point. For example, a comparative study of pulmonary function was recently carried out among persons 30 years of age and over, in two neighboring communities characterized by marked differences in air pollution. The communities were Seward and New Florence—two neighboring small towns in west central Pennsylvania. The average sulfur dioxide values were 6.2 times greater, dustfall was 3.2 times greater, and soiling was 1.5 times greater in Seward than in New Florence. A vegetation survey in June 1960 disclosed sulfur dioxide injury to plants in Seward and vicinity with no similar damage at New Florence. Airway resistance, measured by two different techniques, revealed statistically significant differences between the two communities for each sex, even after differences in height and age were taken into account.

Although most of the toxicologic studies of air pollutants have been focused on the respiratory tract, other toxic effects have also been recognized. It is worth mentioning in this regard that many particles inhaled with the air reach the digestive tract, a fact that is probably relevant to the gastrointestinal disturbances observed during the Donora smog of 1948, and also perhaps to the causation of stomach cancer.

Air pollutants exert toxic effects not only on man and animals but also on plants and microbes. The deleterious effects of pollutants on plant growth seem, generally, to be most severe in win-

TABLE 20. California Air Quality Standards, Specifications and Threshold Limit Values for Gases and Vapors *
(in parts per million by volume)

	Ambient air			*Workroom air*
	Single measurement	*Average*	*Time over which average is taken*	*TLV* [a]
	CARBON MONOXIDE			
Serious level	—	30.0	8 hr.	100.0
	—	120.0	1 hr.	100.0
First alert	—	100.0	30 min.	100.0
	—	200.0	10 min.	100.0
	—	300.0	2 hr.	100.0
Second alert	—	100.0	1 hr.	100.0
	—	200.0	20 min.	100.0
	—	300.0	2 hr.	100.0
Third alert	—	200.0	1 hr.	100.0
	—	300.0	24 hr.	100.0
	ETHYLENE			
Adverse level	—	0.5	1 hr.	—
	—	0.1	8 hr.	—
	HYDROGEN SULFIDE			
Adverse level	—	0.1	1 hr.	20.0
	OXIDANT (BY POTASSIUM IODIDE)			
Adverse level	—	0.15	1 hr.	—
	OXIDES OF NITROGEN			
First alert	3.0	—	—	5.0
Second alert	5.0	—	—	5.0
Third alert	10.0	—	—	5.0
	OZONE			
First alert	0.5	—	—	0.1
Second alert	1.0	—	—	0.1
Third alert	1.5	—	—	0.1
	SULFUR DIOXIDE			
Adverse level	—	0.3	8 hr.	5.0
	—	1.0	1 hr.	5.0
Serious level	—	5.0	1 hr.	5.0
Emergency level	—	10.0	1 hr.	5.0
First alert	3.0	—	—	5.0
Second alert	5.0	—	—	5.0
Third alert	10.0	—	—	5.0

* Adapted from A. C. Stern, 1964.

a. Threshold Limit Values—for 8 hours a day, 5 days a week exposure of healthy adult worker.

TABLE 21. Data on Which German (VDI) Air Quality Specifications Are Based *

Concentration, in parts per million by volume	*Effect*	*Species*
	CHLORINE (*September 1960*)	
600	Death	Animals
10–100	Sickness	Animals
1	Irritation	Man
0.3–3.0	Damage	Plants
0.05	Olfactory threshold	Man
	HYDROCHLORIC ACID (*February 1963*)	
100–1000	Leaf damage	Plants
10–50	No damage	Plants
10	Irritation	Man
5	No organic damage	Man
	HYDROGEN SULFIDE (*April 1960*)	
700	Quickly fatal	Man
400–700	Dangerous (30 min. or less)	Man
170–300	Intense local irritation	Man
100–300	Spontaneous injury	Animals
70–150	Slight symptoms	Man
20–30	Strongly perceptible odor	Man
1.5	Toxic limit	Plants
0.3	Definitely perceptible odor	Man
0.1	Slightly perceptible odor	Man
0.025	Olfactory threshold	Man
	NITRIC ACID (*February 1963*)	
25	Damage	Plants
	NITROUS OXIDES (*May 1960*)	
40–80	Pulmonary edema	Animals
10–20	Methemoglobin formation	Man
1–2	Minor irritation	Man
0.1	Olfactory threshold	Man
	SULFUR DIOXIDE (*November 1961*)	
100	Strong irritation	Man
10–50	Irritation	Man
1–10	Contradictory findings	Man
1–2	Damage	Plants
0.3–1.0	Perception threshold	Man
0.15	Tolerance limit	Plants

* Adapted from A. C. Stern, 1964.

ter annuals, which grow at a time when air pollution is at its worst. Exposure to aerosols obtained from natural smog, or from irradiated automobile exhaust, exerts a bactericidal effect on *E. coli;* this occurs with 20 parts per billion of oxidized type of pollutants, a concentration also sufficient to produce a mild degree of eye irritation. The toxicity of air pollutants for microbes may account for the fact that plants growing in polluted air are often relatively free of fungus infections.

One of the biological effects of air pollutants is to increase the susceptibility of animals to experimental infections. Thus, the susceptibility of mice and hamsters to *Klebsiella pneumoniae* can be increased by exposing these animals to pollutants during, be-

TABLE 22. Registered Deaths in London Administrative County by Age; Comparison of 7-Day Period before the 1952 Episode with the 7-Day Period that Included the Episode *

Age	*7-day period preceding the episode*	*7-day period including the episode*
Under 4 weeks	16	28
4 weeks to 1 year	12	26
1–14 years	10	13
15–44 years	61	99
45–64 years	237	652
65–74 years	254	717
75 years and over	355	949

*Adapted from H. Heimann, 1961.

fore, and after challenge with an aerosol infection. Definite increase in susceptibility was achieved by exposure to as little as 1 ppm. of ozone for 4 to 15 days, or 1 to 4 ppm. for 3 hours. The acute toxicity levels of this compound are 12.6 by volume for mice and 35.5 ppm. for hamsters, expressed as LD50 for 3 hours' exposure. Hemorrhage and pulmonary edema occurred in all species of animals exposed to lethal doses of this gas. Similar results have been obtained by exposing mice to nitrogen dioxide. In other words, exposure to pollutants in concentrations of much less than one-tenth of a lethal dose is sufficient to increase susceptibility to infection.

Probably of greater potential importance is the fact that exposure to 1 ppm. of nitrogen dioxide is sufficient to activate subclinical respiratory infections in animals and to produce thereby a full-blown and often fatal disease. Thus, mice exposed to an artificial atmosphere of ozonized gasoline spontaneously developed large lung abscesses and pneumonia; many died and *Actinobacillus lignieresi* was recovered from them. As no death occurred among the control animals, it can be concluded that the artificial smog had activated a latent infection. These experimental findings acquire special significance in view of the frequency of broncho-respiratory infections in populations exposed to toxic dusts and fumes.

Pollen and other plant allergens are not the only materials present in air that can elicit an allergic state. Certain highly reactive pollutants not allergenic as such can combine and form complexes with body proteins and thereby act as antigens and allergens. The so-called Tokyo-Yokohama asthma may be a case in point. This disease, which is common not only among Japanese but also among American servicemen and their families residing in Japan, is a form of bronchial asthma often referred to as "air pollution asthma." In general, its victims are relieved of the asthmatic attacks when they are evacuated to other areas, whereas all those who stay continue to suffer. This kind of asthmatic response appears to result from a chronic effect of air pollution, and there is some evidence that it involves an allergic component.

Suggestive correlations between asthmatic episodes and the presence in the air of certain pollutants have been reported also from Los Angeles, Nashville, New Orleans, and other places. In New Orleans a peculiar irregularity is observed in the numbers of visits by non-white asthmatics to emergency clinics; the outbreaks of asthma seem to be associated with certain wind movements. Recent studies suggest that these outbreaks are related to particles of a silicon-containing compound that is emitted into the atmosphere by poor combustion of garbage or refuse in the New Orleans city dumps. It has been suggested that the air pollutant acts as an allergen in certain susceptible individuals, but evidence on this point is not yet convincing. Wherever they occur, asthmatic responses to low concentrations of air pollutants provide examples of the difficulty in establishing cause-effect relationships in the pathological states associated with air pollution. As we shall see, this problem is of particular importance with regard to the etiology of chronic respiratory disease.

It is commonly stated that mortality from chronic bronchitis is very much larger in England and Wales than in the United States. However, the significance of this statement is difficult to evaluate because there is much confusion in the use of the words bronchitis, asthma, and emphysema. Most authors in Great Britain emphasize chronic bronchitis as the initial disease and regard emphysema as a complication; whereas most authors in the United States consider emphysema as a disease *sui generis*. In practice, chronic diffuse obstructive pulmonary disease is defined largely on clinical criteria such as diffuse airway obstruction, dyspnea on exertion, and less generally a chronic productive cough with repeated bouts of bronchopulmonary infection. Needless to say, every feature of this syndrome occurs at some time in almost every person. It is therefore difficult to measure the frequency of chronic respiratory disease by such ill-defined clinical criteria. Granted these limitations, a few epidemiological facts appear nevertheless clear.

In Northern Europe, chronic respiratory disease is much more common in industrialized than in rural areas. In Manchester, for example, it accounts for more absenteeism and actual unemployment than any other cause. A remarkable gradient appears when the low rates of the disease in Norway and Denmark are compared with the slightly higher rates in Sweden, France, and Switzerland, and the still higher rates in Western Germany, the Netherlands, and Belgium. The peak rates are reported from the British Isles in general and from England and Wales in particular. The black counties of England are the districts most affected by industrial and domestic smoke pollution, and they are also those where chronic respiratory disease is most prevalent; the smog of 1952 was most intense there. In these areas, furthermore, large numbers of semiskilled artisans and laborers are exposed by their jobs to industrial dusts and fumes under conditions demanding great physical effort and exposure to unpleasant weather throughout much of the year.

During the past two decades, the crude death rates from emphysema, bronchitis, and asthma have also increased sharply in the United States. More generally, epidemiological surveys reveal that death rates from cardiorespiratory diseases are higher in urban than in rural areas, and increase with city size, as does air pollution. The death rates from emphysema and primary carcinoma of the lung are lowest in rural areas, intermediate in areas surrounding large cities, and highest in the central portions of metropolitan areas.

In 1957, emphysema ranked second among the diseases in men for whom disability was allowed under the Social Security Act. In a survey carried out by the California State Board of Health in 1954–55, bronchitis and asthma proved to be among the ten most frequently disabling diseases, accounting for 6 per cent of the total days of disability. The California death rate from emphysema has risen from 1.5 per 100,000 in 1950 to 5.8 per 100,000 in 1957. Admittedly, this increase must reflect in part better diagnosis and greater interest on the part of the physicians, but it also certainly indicates a rising incidence of pulmonary disease.

Quite aside from any possible role of smog in initiating respiratory disease, the number of people with chronic bronchitis-emphysema is bound to increase in our communities. As a result of successful chemotherapeutic treatment, our population contains a steadily expanding pool of persons who have survived acute bronchopulmonary illnesses, but who still have defective lungs nevertheless. Large numbers of people are alive today in all age groups who would not have been alive in the days before antimicrobial therapy. They may show no signs of ill health until something untoward happens. But a prolonged period during which the air remains polluted is likely to precipitate a serious or fatal illness in such predisposed persons.

The aggravation of chronic bronchitis-emphysema by air pollution was dramatically illustrated by the smog disasters of 1948 in Donora and of 1952 in London. Donora, Pennsylvania, is on a bend of the Monongahela River and surrounded by high hills on all sides; its air is constantly polluted by the smoke and fumes of blast furnaces, steel mills, and other industrial plants. In October 1948, thermal inversion occurred and the usual smog persisted continuously for three days, instead of lifting at noon as it usually does. By the third day of constant smog, 5,910 persons were reported ill; more than 60 per cent of those 65 and older were seriously affected; and there were 17 deaths. Then a heavy rain fell, the smog disappeared, and the outbreak of disease stopped immediately.

During the 1952 smog in London, there was an "excess" mortality of 4,000 to 5,000 persons, and as in Donora, the deaths occurred almost exclusively among those with previous bronchopulmonary disease. Indeed, the veteran bronchitis patients in the London clinics "served almost as the canaries that miners once carried to detect noxious gases; they noted discomfort six to twelve hours before it was evident to others that an episode of smog was at hand" (McDermott, 1961). A less dramatic but nevertheless similar situation occurred in New York City a few years ago.

One of the striking features of both the Donora and London disasters was that no single smog component was present in an unusually high concentration. At no time during the London fog did the SO_2 concentration approach the conventionally accepted toxic limit of 10 ppm., and it did not exceed 0.5 ppm. in some areas where many deaths occurred. Some other pollutants may have approached or reached values at which physiological effects are said to occur. These findings suggest that the same smog breathed by everyone a day or two at a time without discernable ill effect can become highly injurious to substantial numbers of people when it is breathed continuously for a few days longer.

Still more disquieting is the subsequent experience of the persons involved in the Donora disaster. Before the episode, residents of Donora appeared to have the same health status as people in the rest of the United States. In the nine years that followed the smog, however, those who had become ill and recovered showed a higher mortality and incidence of illness than those who were present but were unaffected at the time of the smog. To some extent, this difference can be taken as reflecting the adverse effect of the smog on those with damage to their lungs and hearts anteceding the disaster. However, even Donora residents who had no history of heart disease prior to the 1948 episode, but became ill during the period of smog, have since had a higher illness rate. The deferred consequences of the Donora episode are among the scanty pieces of epidemiological evidence suggesting that contaminated air may actually initiate disease in man.

The fumes arising from the burning of coke and coal in open braziers are said to induce neoplasms in 20 to 30 per cent of commercially grown mushrooms. This finding is of small practical importance per se, but it acquires interest from the fact that several kinds of air pollutants are suspected of playing a role in the production or evolution of neoplasms in man.

Extracts of exhausts from diesel engines inefficiently operated release polycyclic hydrocarbons and can produce cutaneous papillomas and epidermoid carcinomas when painted three times weekly on the skin of mice. In fact, combusion of gasoline and diesel oil under any condition releases tar compounds that can produce experimental cancers in animals and are probably carcinogenetic also in man.

When measurements were made of the quantity of benzo(a) pyrene inhaled by human beings in a year, the values were found to range from 0.1 mg. in a state forest, to 150 mg. in a large city! By comparison, it is estimated that a person smoking one pack of cigarettes daily for a year might inhale at the most 60 mg. of this carcinogen. In other words, a person living in a large city might breathe as much benzo(a)pyrene over a year's time as he would get from smoking two packs of cigarettes daily. As far as is known, benzo(a)pyrene is almost exclusively a product of human activity and, in any case, its concentration is many times higher in metropolitan than in rural areas. On the other hand, the death rates from cancers of the respiratory tract are directly related to the population density, and such density in turn determines in large part the concentration of air pollutants.

Taken together, these facts, as well as others not to be considered here, strongly suggest that air pollutants play an important role in carcinogenesis. It has even been suggested that, in the causation of bronchial carcinoma, cigarettes may act merely as a co-carcinogen for some other true carcinogens originating from the environment, particularly from polluted city air. In this regard, it is of interest that the death rate from lung cancer among people in New Zealand who have immigrated from Great Britain is 75 per cent higher than among native New Zealanders if immigration occurred after the age of 30; yet the amount of cigarette smoking seems to be essentially the same in both groups. In other words, British-born immigrants carry with them a susceptibility greater than that of the native New Zealanders, and it would seem possible that this susceptibility has its origin in some carcinogenetic stimulus acquired during life in Great Britain.

Given enough time, living things eventually develop genotypic and phenotypic adaptations to adverse circumstances. It would be surprising, therefore, if biological mechanisms for adaptation to air pollutants did not exist. In fact, specimens of the plant species *Lupinus* collected in Los Angeles proved more resistant to smog than did other specimens of the same species collected from other parts of California. The likely explanation of this phenomenon is that smog can act as a selective agent in plant populations locally exposed to it for an extended period of time. In England, a peculiar genetic process of adaptation to air pollutants has been recognized in moths. In industrial areas of England, or in the districts where prevailing winds cover the trees with black soot, dark forms of certain species of moths tend to predominate; the pale-colored forms, in contrast, gain the upper hand in unpolluted districts. It seems that birds act in each case as one of the selective forces, detecting more readily and thus destroying more effectively the phenotypes of moth with a low camouflage efficacy—the pale insects against the smoked bark or the dark mutants against the lichen of the tree trunks.

Animals can develop some degree of resistance to ozone and to sulfur dioxide if they are exposed continuously to low concentrations of these pollutants. In man also a moderate degree of tolerance to sulfur dioxide has been observed to result from continued exposure, especially among industrial workers who are in more or less constant contact with low concentrations of this gas in the course of their occupations. The tolerance is believed to be due to the outpouring of increased amounts of mucus on the walls of the tracheobronchial tree, thus affording some protection to the epithelium. As this form of resistance corresponds in reality to a mild pathological state, it is not unlikely that it has dangerous consequences in the long run (Chapter X.2).

4. Delayed and indirect effects of biologically active agents

Public concern with the dangers of environmental pollution has led many different branches of government to encourage new lines of scientific research and to enact various control programs. Despite these efforts, however, it is clear from the proceedings of the many national and international conferences held during recent years that knowledge concerning the medical significance of the pollution problem is still extremely scant. In most cases, knowledge hardly goes beyond the description of a few *immediate* pathological responses. Hardly anything is known of the *delayed* effects of pollutants on human life, even though they probably constitute the most important threats to health in the long run.

As we have seen (Chapter VIII.3), epidemiological surveys in human populations, and studies of laboratory animals under various conditions of experimental exposure, have provided strong circumstantial evidence linking air pollution to chronic pulmonary disease and probably to certain forms of cancer. However, it is extremely difficult to establish convincingly and to evaluate quantitatively these dangers, because most of the effects of environmental pollution are delayed and probably cumulative. The situation is very similar to the latent period so commonly observed not only following exposure to ionizing radiations but also in industrial diseases. For example, aplastic anemia has been known to occur among workers in industries using benzene up to 10 years after cessation of exposure; moreover, the manifestations of the disease often seem to be precipated by a number of unrelated forms of stress ranging from pregnancy to intercurrent infection or to administration of barbiturates. According to very recent observations, even laser light seems to have delayed and cumulative deleterious effects. Nine months of repeated exposure to small doses of laser energy, which appeared to be innocuous, produced pathological changes in the skin and in the underlying vascular bed; erythema-edema reaction was followed by localized petechial hemorrhages.

The possibility of delayed and cumulative effects is not limited to any particular class of agent. It exists for a large variety of unrelated substances, including many of those used in agriculture and even in medicine. The results of recent field experiments will be quoted here to illustrate the perplexing problems posed to public health officers by the accumulation in the tissues of substances that do not create immediate toxic reactions, yet may be dangerous in the long run.

The experiments in question were designed to determine the dangers involved in the use of insecticides during agricultural operations. The findings demonstrated that the usual practice of dusting potato crops with endrin involves an exposure of the worker to no more than 1.8 per cent of a toxic dose. In related observations it was noted that apples in orchards sprayed for insect control carry an amount of residue so small that a child would have to eat more than 300 apples to ingest a lethal dose of poison. Stated in these terms, the dangers posed by insecticides appear very small indeed. They appear more alarming, however, when it is realized that endrin, dieldrin and other chlorinated hydrocarbons accumulate in fatty tissue and therefore can have a cumulative toxicity not revealed by acute toxicity tests. It is true that a boy will not eat 300 windfall apples at one sitting, but who knows how many he might eat in a season!

Several insecticides, including DDT, have been found to be

present in readily detectable amounts in the fat of people throughout most of the countries of Western civilization. In the United States, the fat of persons without occupational exposure had an average concentration of 12 ppm. DDT for the past 10 years, i.e. approximately 100 to 200 mg. per adult; the corresponding average concentration in England and Germany was only 2 ppm. It is obvious that dangerous consequences might follow the sudden release into the body of such large amounts of DDT, as a result of any condition causing rapid fat metabolism.

Many forms of delayed toxic effects have now been shown to follow treatment with various drugs (see for example Chapters VIII.3 and XIV). Two particular examples have been selected for illustration here because they concern substances that were used for many years before their potential toxicity became apparent.

Many patients receiving synthetic antimalarial drugs develop serious ocular abnormalities. Corneal deposits and retinopathy appear to be directly related to chloroquine medication. Equally disturbing was the very recent discovery that tetracylines become deposited in bone, in deciduous teeth, and in the nails of the fingers and toes. Babies born from mothers who had received tetracyline during pregnancy showed deposition of the drug in their teeth, though none of the babies had been breast-fed. Deposition had therefore occurred in utero, and the drug had persisted for months thereafter. In view of the ease with which many kinds of substances are transferred from the mother to the fetus in utero and since human beings are more and more constantly exposed to new synthetic chemicals, it is to be feared that defects contracted during fetal life will increase in variety and frequency unless steps are taken to protect pregnant women from exposure (see Chapter I.2).

The dangers posed by environmental pollutants are rendered even more difficult to predict by the fact that most of these substances probably have indirect effects that manifest themselves through a complex chain of reactions in the organism. Whatever its apparent selectivity, any biologically active substance is likely to react with more than one structure and to affect more than one function. Above and beyond the direct toxic effects that occur rapidly and are readily detected, there are others that develop more slowly along indirect channels. Most knowledge of toxicity has been derived from techniques of observation and experimentation designed for the study of effects detectable within a short time and involving single variable systems. This approach is adequate for revealing the direct effects of biologically active substances, whether beneficial or toxic, but it rarely lends itself to the recognition of the delayed and indirect consequences that always result from exposure to almost any kind of environmental stimulus. A few examples will be quoted to illustrate the wide range of the indirect and delayed effects exerted by biologically active substances.

It is well known, of course, that radiations, nutritional deficiencies, viral infections, poisons, drugs, etc. can be responsible for congenital malformations. But indirect toxic effects are probably just as common in adult life, and many are completely unexpected. It has been recently observed, for example, that consumption of certain kinds of cheese can cause severe reactions and even death in persons under tranquilizer therapy. This toxicity is due to the presence in the cheese of large amounts of sympatheticomimetic amines (tyramine for example), which are produced from casein by bacterial decarboxylation. In normal persons, such products are destroyed by the body enzymes, but their detoxification is retarded in patients treated with monoamine oxidase-inhibitors. Experiments in rats have shown furthermore that commonly used drugs or hormones, such as sulfonamides, cortisone, and progesterone, can have pathological effects that appear long after the end of treatment and may last throughout the life span of the animal. Some of these effects are not obvious and have to do, for example, with the rat's spontaneous activities, its eating and drinking habits, or its reproductive functions; in many cases the disturbances take the form of abnormal physiological cycles.

In many cases, biologically active substances act indirectly through other members of the ecological system to which the organism belongs. Thus, one of the unexpected findings to emerge from the A.E.C. Chariot project that was lichens have an extraordinary power to assimilate Strontium 90 from fallout. As lichens constitute the main source of food of reindeer, and as this animal in turn constitutes an important food for Eskimos and other residents in the arctic regions, it is obvious that the selective uptake of Strontium 90 by lichens magnifies to a considerable extent the hazards of fallout for human beings.

The best documented example of ecological upset by biologically active substances is provided by the effect of antimicrobial drugs on the indigenous microbiota, especially of the digestive tract (Chapter V.1). As we have seen, interference with the microbiota can result in histological and physiological abnormalities, as well as in increased susceptibility to certain infections and stresses. In brief, the continued administration of certain antimicrobial drugs modifies the composition of the autochthonous microbiota, and this alteration in turn can modify the histological structure of the intestine, the efficiency of food utilization, the rate of growth, the metabolism of cholesterol, the resistance to certain toxic products or to total body radiation, and probably many other important characteristics of the organism.

The study of the indirect and long-range effects of biologically active substances is further complicated by the fact that living things can develop various degrees of tolerance to many of them. The word "mithridization" recalls that King Mithridates had made it a practice to ingest small amounts of several poisons in order to protect himself from poisoning by his enemies. Human beings can even become accustomed to articles of diet that appear at first unpalatable and may even be physiologically objectionable (Chapter III.4). Similarly, city dwellers commonly develop such a degree of apparent tolerance to certain air pollutants that they become unaware of the presence of these substances in the air they breathe (Chapters VIII.2 and X.2).

Thus, complex and especially prolonged observations will be required to bring to light the pathological effects on man of environmental pollutants and other biologically active substances. On the one hand, these observations must deal naturally with all the direct effects, including those far removed in time from the first contact with the substance under consideration. On the other hand, they must be directed also to the secondary effects different in mechanism from those corresponding to the initial impact of this substance, and involving structures and functions other than those primarily affected. Man constitutes a complex ecosystem and he responds as an integrated whole to each of the forces that make up his total environment.

The existence of delayed and indirect effects of environmental pollutants makes it certain that many of their pathological consequences can be revealed only by long-range epidemiological studies, both retrospective and prospective. The recognition, then demonstration, of the deleterious effects of cigarette smoking illustrates well the importance of the epidemiological approach, but also its difficulties.

The great drawback of epidemiological studies is that they are cumbersome and commonly yield the needed information too late for the development of corrective measures. In the case of environmental pollution, especially, the situation may well become unmanageable if the accumulation of convincing epidemiological evidence is made a prerequisite of social action. Fortunately several types of physiological tests in man and of experimental studies in animals can be carried out now, or could be soon devised, to demonstrate that many pollutants have indeed adverse effects on essential structures and functions. Such physiological tests and experimental studies cannot entirely substitute for epidemiological surveys because they are not likely to reveal *all* the potential dangers of environmental pollution. But they have the advantage of providing rapidly incontrovertible information which might therefore be more convincing and meaningful than that derived from statistical epidemiology, not only for the general public but also for scientists. The need for striking information, even though incomplete, is here of great importance because environmental pollution will not be controlled until physicians and scientists take an active part in its study and until the general public is alerted to the dangers of the "Pestilence that stealeth through the darkness."

IX. Changing Patterns of Disease

1. The life span of man

There are 11,000 persons over 100 years old now living in the United States. Surprisingly high as this figure may seem, the same percentage of centenarians has been found to exist in all populations for which data are available. The biography of Charles W. Thierry, who was born in 1850 and died in 1958, presents a picture which in its general outline seems to be fairly typical for the life of many centenarians. Thierry was a silversmith in Cambridge, Massachusetts, and exercised his trade until the age of 93; he was still vigorous when he retired, and his blood pressure remained normal to the time of his death. At the age of 103 he developed influenza and had a somewhat stormy convalescence; but he soon recovered his vigor when his consulting physician, Dr. Paul Dudley White, urged him to resume his long daily walks, irrespective of the inclement weather. He lived to be 107½ years old and died of pneumonia almost accidentally; his disease was diagnosed when he was already moribund, so late that antimicrobial treatment was to no avail. Autopsy revealed a strong and healthy heart and good arteries.

A few years ago, a man from the mountains of Colombia, in Central America, was brought to New York as a guest of medical scientists at an age certainly exceeding 100 years, and estimated from circumstantial evidence to be approximately 150; he died of an unknown cause shortly after returning to his country. The old man, who was of short and wiry structure, had spent his entire life under very primitive conditions; he had always been physically active and was still spry shortly before his death. Having been myself a patient on the very hospital ward where he was staying as a guest during his visit in New York City, I can testify to his liveliness, which I then greatly envied.

In one of his books, E. Metchnikoff wrote of the lives of many centenarians he had studied. He presented a cheerful picture of their physical and mental state to support his claim that the potential life span of man is much longer than usually assumed. History provides indeed several well-documented examples of men who enjoyed unusual longevity and were still healthy and vigorous at the end of their long lives. In 1635 an Englishman called Thomas Parr, reputed to be 152 years old on what was deemed undisputable evidence, was summoned up to London where he was presented to King Charles I. He was appropriately wined and dined, and thereupon promptly died. An autopsy was performed by none other than William Harvey, who pronounced all the organs of "Old Parr," as the old man was affectionately called, to be quite normal, "as healthy and sound as the day he was born"; there was not even calcification of the costal cartilages. Old Parr was declared by Harvey to have died of a surfeit and perhaps also of exposure to polluted air (Chapter VIII.2); he was buried in Westminster Abbey, where his tombstone can still be seen. Dragenberg, a Danish sailor and soldier, lived from 1626 to 1772, most of the time both actively and dangerously. At the age of 111 he married a woman of 60 and wanted to remarry when she died 22 years later.

There has been much speculation naturally as to the kind of life which makes for great longevity, but no formula has yet been discovered. Some of the centenarians never married, while others had several wives; some never smoked and were teetotalers, while others were smokers and had their bottle of wine or hard liquor every day. A few years ago a puritan Utah pioneer died at the age of 100 after 82 years of married life; and on the same day a cigar-smoking, unwed Chinese celebrated his 103rd birthday in New York City. Most of the centenarians one hears about were farmers or small tradesmen; but Bernard de Fontenelle, who was exactly 100 years old when he died in 1757, was very much a man of the world, one of the key figures of Paris society and of the French Academy. On watching his funeral, a wit remarked that "this was the first night M. de Fontenelle was not going out for dinner."

Nor does a balmy climate seem to be a critical factor for longevity if one judges from the fact that Shetland has had more than its share of very old people. A Shetlander who lived to be 140 had married at 100 years of age, and pulled his small skiff ashore, unaided in very severe weather, the very year he died. As to the possibility of reaching old age in non-Western cultures, it will suffice to mention that the American Indian who served as a model for the profile which adorns the 5-cent piece was 99 years old and healthy at the time of this writing. A profile photograph taken of him in 1962 (when he was 98) looks very much like the engraving made for the coin.

Clearly then, the potential duration of human existence greatly exceeds the Biblical three score and ten (Table 23). Furthermore, as already mentioned, great longevity can be achieved under very different conditions of life, and the percentage of old people seems to be much the same in all kinds of societies. The very existence of many centenarians in all parts of the world constitutes proof that great longevity can be achieved without the benefit of modern medical care, since most of the life span of people who are now very old was spent before the advent of scientific medicine.

TABLE 23. Comparative Life Span (in Years) of Man and Animals

	Age at puberty	*Average life span*	*Maximum life span*
Man	13 yrs.	70	150?
Lion	2 yrs.	23	29
Ox	10 mo.	23	30
Grizzly bear	30 mo.	20	32
Cat	10 mo.	15	21
Dog	7 mo.	15	34
Monkey (*Macaca mulatta*)	3 yrs.	15	29
Goat	8 mo.	9	18
Mouse	1.2 mo.	2	3+
Elephant (Indian)	12 yrs.	—	57

In sharp contrast to the existence of so many centenarians in many different contemporary communities is the fact that all skeletons recovered from ancient burial grounds, and also from Paleolithic and Neolithic sites, are of men and women who were under 50 years of age at the time of their death. Out of 173 Paleolithic and Mesolithic skeletons of men whose age could be determined, only three appeared to have been slightly older than 50 years.

There is every reason to believe that the potential life span of prehistoric man was much the same as our own. In the distant past, however, death was usually the result of accident or violence, and few human beings consequently lived out their potential life span. Franz Weidenreich, who was a physician as well as an anthropologist, came to the conclusion that the primary cause of death of fossil man in China was homicide. This was true for the Sinanthropus of the Choukoutien Cave. All the long bones found in this cave had been split open, suggesting that a fellow man had opened the bones to get at the marrow. The skulls also showed evidence of heavy blows. According to Weidenreich, one skull of a late Stone Age fossil showed "a long and wide split-like hole at the superior part of the left temporal region . . . giving the impression that it was caused by a spear-like implement piercing through the wall from above."

Weidenreich's views concerning the life span of ancient man have received support from recent studies of the Xavante Indian tribe in the Brazilian Mato Grosso. These Indians are completely isolated and their way of life corresponds to a Stone Age civilization. They all appear remarkably healthy, yet few of them live beyond the age of 50; deaths among them appear to be chiefly from violent causes. To a large extent, the same conclusions apply to the isolated community of Tristan da Cunha.

A medical survey of the Tristan Islanders was made by a Norwegian scientific commission in 1937. At that time, the conditions of life on the island were extremely primitive and involved great hardships. Yet, according to the report of the Norwegian commission, "diseases of the heart, kidney, liver and other internal organs seem to occur seldom. On the whole, the Tristanites seem to have a fair chance of getting old unless they are drowned or fall down from the mountain." These conclusions are the more startling in view of the fact that, when the Tristan Islanders were examined after a volcanic eruption had compelled their evacuation to England, they were found to be heavily infested with round worms, whipworms, and cysts of *Entamoeba hystolytica.*

It is naturally impossible to establish reliable vital statistics for times gone by, but the general outline of the trends in life expectancy can be guessed even for the distant historical past on the basis of the information derived from paramedical documents. Burial grounds, for example, tell a tale that is the same in all countries of Western civilization. Everywhere in old cemeteries, the graves of very young children and of young adults greatly outnumber those of old people. Many men and women of the past did live as long as and perhaps longer than is the case now, but until the end of the last century a very large percentage of the population died before the age of 20. Indeed most deaths occurred shortly after birth.

A progressive and almost uninterrupted decrease in mortality during childhood and early adulthood has taken place during modern times in the Western world. Although this change for the better began long ago, its rate became particularly rapid after the turn of the century. Its chief cause was probably the control of nutritional deficiencies and microbial diseases, but other factors, environmental and genetic, may also have played a part much more important than is usually assumed. Whatever its explanation, the consequence of the change in the mortality pattern has been that life expectancy *at birth* has been constantly increasing. In the United States it was 35 years in 1789; 40 in 1850; 47 in 1900; 55 in 1920; and it has now reached 71 years.

The general trend has been the same in most other countries of Western civilization. This achievement, which constitutes the most concrete evidence of disease control in our societies, is always quoted as a quantitative expression of improvement in the general state of health. In reality, however, the much vaunted increase in life expectancy at birth takes on a much less glowing aspect when interpreted in the light of other facts.

It must be emphasized again that the increase in life expectancy is almost exclusively the result of the virtual elimination of mortality in the young age groups. In contrast, mortality during late adulthood and old age has changed much less, if at all. The life expectancy for a person having attained the age of 65 has risen at most from 12 years in 1900 to 14 or 15 years in 1963. Furthermore,

TABLE 24. Average Remaining Lifetime (in Years) of White Persons in the United States at Specified Ages *

	Age (in years)			
Year of birth	*0*	*20*	*45*	*65*
		Male		
1840	38.7	41.3	24.1	11.6
1880	46.3	44.7	26.2	12.9
1920	63.1	52.3	30.8	15.5
1960	75.7	58.6	35.2	17.8
		Female		
1840	40.8	41.4	25.4	12.3
1880	50.1	48.6	29.8	15.4
1920	70.8	59.1	35.8	18.2
1960	80.1	62.3	38.0	19.5

* Adapted from Metropolitan Life Insurance Company Statistical Bulletin, August 1964, *45*, p. 3.

a large part of this increased life expectancy in the older age groups represents merely prolongation of survival time through complex and costly medical procedures, rather than healthy years gained from the onslaught of disease. It corresponds to what has been called medicated survival. In other words, aged people do not retain their health much longer now than they used to; the chief difference is that many more people survive infancy and live to reach adulthood. The increase in expectancy of life at birth provides therefore no evidence that adult men and women are now healthier. In fact, our contemporaries do not have a much greater chance than our ancestors to enjoy longevity and health.

2. Disease, old and new

Whatever the actual cause of death in the prehistoric past, it is certain that ancient man could fall victim then to many of the diseases which affect mankind today. For the prehistoric era bones are of course the most useful surviving witnesses of disease, and their testimony is clear. Rickets, always an accompaniment of deficient insolation in the past, has been identified in Neolithic bones of Denmark and Norway. Arthritic diseases were then as today among the most crippling afflictions. Evidence of them has been found in the remnants of Neanderthal man, as well as of Neolithic man, in the caves of France, in the mountains of Peru, and even in sunny Egypt. Early man does not occupy a special position in this regard; arthritic deformations are so common in remains of cave animals and especially of bears that the name "cave gout" has been devised for the disease, the "Hohlengicht" of Rudolf Virchow.

Although malignant tumors of the bone are rare in prehistoric man and animals, they have been found in Egyptian mummies as far back as 3400 B.C., in human remains of France, North America, and Peru, as well as in fossil horses and cave bears. Studies of human Neolithic skulls in France have revealed an incidence of some 3 to 4 per cent of dental caries; similar findings at a prehistoric site in Iran in a population dating from 4000 B.C. to 2000 B.C. make clear that, although dental diseases were then far less frequent than now, they afflicted man long before he became addicted to soft drinks, refined food, and candy (Chapter VI.2).

Additional information on disease in ancient man has been derived from X-ray photographs of mummies and from paintings and sculptures. If the interpretation of these documents is correct, King Siptah of the Nineteenth Egyptian Dynasty and some of his subjects suffered from paralytic poliomyelitis. The first evidence of arteriosclerosis of the aorta has been found in Merneptah, the Pharaoh of the Exodus, who lived about 1200 B.C.; hardened arteries also occurred in other Egyptians from the Eighteenth to the Twenty-seventh Dynasty, as well as in later Greek and Coptic mummies. In fact, the list of diseases found in mummies reads almost like the catalogue of a pathological museum and includes silicosis, pneumonia, pleurisy, kidney stones, sinusitis, gallstones, cirrhosis of the liver, mastoiditis, appendicitis, meningitis, smallpox, leprosy, malaria, tuberculosis, and congenital atrophy of the liver. Schistosomiasis, a parasitic disease of backward nations, was prevalent in Egypt 2,000 years ago. The Egyptians were also plagued by lice and the Peruvians by sand fleas, as judged from the lesions on the soles of their feet. The paleopathological records leave no doubt therefore that most of the known organic and microbial disorders of man and animals are extremely ancient, and are indeed probably coexistent with human life. This is almost certainly true for mental disorders.

Anatomically, man looks very much the same as he did during the Stone Age; the same 20,000-odd pairs of genes still control the phenotypic manifestations of his nature. The tools he fashioned then still fit the hands of the average city dweller today. Moreover, if we judge from the artifacts he has left all over the world in the course of his travels and adventures, his mind has not changed much either. Human societies have evolved, of course, but man himself has retained the same fundamental appetites, urges, and reactions. The unchanging aspects of man's nature, and of his diseases, account for the eternal and universal characters of medical science and practice. These are concerned with the innate properties of the body and the mind, with the intimate mechanisms of their derangements, and with the general responses man makes to the various forms of threat. Such knowledge is applicable everywhere and always.

While the fundamental aspects of man's nature and of his responses to the environment have not changed, the relative frequency and severity of particular diseases have been continuously changing with time and from one place to another. The historical fluctuations in the importance of the various infectious processes (discussed in Chapter VII) illustrate that each civilization is characterized by a peculiar pattern of diseases. This fact does not negate the statement made earlier that the fundamental nature of human diseases has remained essentially the same. On the one hand, balanced polymorphism in man makes possible rapid shifts in the resistance of human populations without requiring real changes in genetic structure. Such mechanisms may account in part for the ebbs and flows in infectious diseases that have been discussed in Chapter VII. On the other hand, and perhaps more important, man continuously modifies his environment and his ways of life, thus bringing about a concomitant change in the prevalence and severity of his various diseases, because these are always the expression of his responses to environmental insults and stimuli.

Unfortunately, the detailed medical history of these changes is never convincing, since the information must be derived from diagnoses at best uncertain and from vital statistics that are usually inaccurate and incomplete. Furthermore, words that appear precise have had very different meanings at different times. The famous plague of Athens described by Thucydides was not the disease known today under this name; it may have been measles or, more probably, typhus. The instructions to the lepers in Leviticus certainly referred to a variety of minor skin ailments in addition to leprosy caused by the Hansen bacillus. Not all pathological conditions reported as phthisical or consumptive in old writings can be attributed to tuberculosis. The word gout also has been associated with a host of diseases, including disorders of the stomach and heart; William Pitt the Elder had attacks of insanity which were euphemistically called gout in the brain! Even the meaning of the word gonorrhea has been questioned recently, and the view has been expressed that the disease referred to under this name by ancient authors was in reality spermatorrhea. "Gonorrhoea is an unwanted excretion of semen," Galen wrote, "which you might also call involuntary; or to be more precise you might say a persistent excretion of semen without erection of the penis." That Galen was not referring to an inflammation of the urethra is made plain by his statement, "Gonorrhoea is an affection of the organs of seed, not of the pudenda, which are organs for excretion of seed."

Granted all the difficulties in assessing the significance of old medical documents, some generalizations can be made nevertheless concerning the prevalence of various diseases at different periods. There is little doubt, for example, that the thousands of lazarettos that existed in Europe during medieval times had been founded to shelter and segregate lepers infected with the Hansen bacillus. In contrast, it is also certain that leprosy disappeared almost completely from Europe in the sixteenth century, and that it has never re-established itself in the countries of Western civilization, although it still constitutes a major disease problem today in certain other parts of the world.

For more recent times, other types of paramedical documents provide useful information concerning the attitude of the general public toward disease. Poets, novelists, and artists have always derived much inspiration from human suffering and therefore from the manifestations of disease they experienced or witnessed among the people of their communities. One need only recall how often malaria, tuberculosis, and the various "fevers" used to be depicted in the different forms of art, to realize what a large place these maladies occupied in the emotional life of our forebears. Gout and obesity must also have had social meaning in Europe during the eighteenth and nineteenth centuries if one judges from the frequency with which these disorders are represented in paintings and cartoons depicting the life of the bourgeois classes at that time. Nutritional disorders were also a widespread concern since so many persons were obviously poorly fed or grossly underfed, and a few much overfed. Rickets was extremely common in all social classes, even in the nobility; chlorosis was a frequent disorder in young genteel women, probably as a result of iron deficiency; indeed, these two conditions were regarded as an almost inescapable fate, much as we accept baldness or defective eyesight today.

The diseases most prevalent everywhere a century ago are still present among us, but their importance has decreased so much that they are progressively disappearing from the awareness of the layman. There is no need to illustrate the fact that words such as heart attack, stroke, cancer, leukemia, mental disease, etc. have now largely replaced the names of the various types of infectious disorders in the conversation of both the man in the street and his physician. Along with chronic and degenerative diseases, automobile accidents and other consequences of violent behavior are commonly referred to as the modern plagues, not only by statisticians but also by laymen. More and more, indeed, the conviction is growing among the general public that health is being threatened by new "diseases of civilization," which have replaced the epidemics of the past. Like the general public, laboratory scientists, medical students, and even physicians are progressively losing interest in many of the diseases that occupied the center of medical attention a few decades ago. This shift in interest becomes obvious when one compares the tables of contents of present-day textbooks of medicine and medical journals with those of earlier editions. The scientific knowledge of medicine has increased, of course, but the shift in the areas of emphasis is even more striking.

Vital statistics give substance to the general belief that there has been a change in the pattern of diseases. Wherever the living standards are high, the mortality from infectious diseases has been rapidly receding ever since the turn of the century. In contrast, heart diseases now constitute the leading cause of death in most prosperous countries, with cancer in second place, vascular lesions affecting the central nervous system in the third, and accidents in the fourth.

According to the report issued by the National Vital Statistics Division of the United States Public Health Service, the ten leading causes of death in the United States and their percentage of total deaths were as follows in 1962: diseases of the heart, 39.3 per cent; malignant neoplasms (including neoplasms of lymphatic and hematopoietic tissues), 15.9; vascular lesions affecting the central nervous system, 11.3; accidents, 5.5; certain diseases of early infancy, 3.9; influenza and pneumonia (except pneumonia of the newborn), 3.2; general arteriosclerosis, 2.0; diabetes mellitus, 1.7; congenital malformations, 1.2; and other diseases of the circulatory system, 1.2. All other causes of death accounted together for 14.8 per cent of the total. These figures stand in sharp contrast to those reported from the underprivileged parts of the world. In these countries, malaria, tuberculosis, diarrheal diseases, gastroenteritis, and colitis are still today the greatest killers, just as they used to be in the Western world half a century ago.

The availability and quality of medical care influence, of course, the toll taken by disease in the different countries. But it is certain that the general improvement in public health has been accelerated by factors that are poorly, if at all, understood. The tuberculosis mortality in New York City was 700 per 100,000 in 1812; it had fallen to 370 in 1882 (the year Koch discovered the tubercle bacillus); to 180 in 1910 (when bed rest was just beginning to be accepted as a method of treatment); and to fewer than 50 in 1945 (before the beginning of chemotherapy).

In most European countries, likewise, the tuberculosis mortality has been declining at an almost constant rate for more than a century, without the benefit of vaccination or antimycobacterial drugs. Equally striking is the fact that several viral infections also have been decreasing steadily. Figures 11 and 12 illustrate that, in the United States and Great Britain, the mortality from most infectious diseases began to decrease long before any method of prophylactic or therapeutic control had become available. In contrast to mortality rates, infection rates have not decreased significantly, except in a few cases, during very recent years. For example, a very large percentage of the population was still tuberculin-positive in all the large cities of Europe and North America during the 1940s, at a time when mortality rates had already been reduced to less than one-tenth what they were a century earlier. There is reason to believe, therefore, that the fall in mortality rates is due in part at least to an increase in the general state of resistance.

Similarly, a striking decline in the severity of rheumatic fever was well under way before the advent of the sulfonamides or penicillin. The nodules, pleuritis, pericarditis, congestive heart failure, and other features of the severe form of the disease are now much less frequent in prosperous countries than they used to be, even among untreated cases. This change is the more striking because streptococcus carrier rates are still very high everywhere, a fact which suggests that the ways of modern life have rendered man more resistant to some of the secondary effects of streptococcal infection. This interpretation is further supported by the continued prevalence and severity of rheumatic fever in certain countries where living standards are still low, as for example in Peru and Yugoslavia. Thus it is certain that environmental changes have played a dominant role both in the disappearance of certain maladies and, as we shall now see, in the emergence of others.

3. Social patterns of disease

a. Population changes and emergence of new diseases

Demographic changes have been so large and so rapid during recent decades that they have come to play an increasingly important role in determining the social patterns of disease. In particular, a host of difficult medical problems have been created by the mass population movements so characteristic of our times—whether involuntary, as in the case of refugees, or voluntary, as in the case of pilgrims and persons who migrate to cities and other places of opportunity. These problems include infections caused by microbial agents acquired through new contacts with men or animals; the introduction of disease vectors; nutritional disturbances resulting from shortages in essential nutrients and even more from disturbances in food habits; and last but not least, the ill-defined but nevertheless severe sociocultural and psychic stresses always associated with life under new conditions.

Even highly desirable technological and social innovations can create indirectly conditions dangerous for human health. There are good reasons to believe, for example, that malaria, and consequently the sickle-cell anemia associated with hemoglobin S, became prevalent in Africa as a result of the slash-and-burn agriculture that created stagnant pools favorable for the breeding of *Anopheles gambiae* larvae. More generally, the spread of primitive agriculture into potentially malarious areas brought man in contact with insects and built populations of sufficient density to maintain a reservoir of infection. At the present time, a somewhat analogous situation is being created by the networks of canals designed to carry water to arid territories. These canals have proved ideally suited for spreading bilharziasis and other water-borne diseases to the people they are intended to benefit. Since bilharziasis was unknown in many of these arid countries before the advent of perennial irrigation, the disease can truly be regarded as man-made. Irrigation of the San Joaquin valley in California has also provided favorable conditions for mosquitoes, and thus resulted in outbreaks of Japanese B encephalitis.

The problems resulting from social uprooting and rapid social changes are naturally at their worst among populations living in squalid squatter areas—the shanty towns, favellas, and *bidonvilles* that surround all rapidly growing cities. But, the worldwide urban sprawl is creating a disease pattern of its own even in prosperous settlements. Crowding is increasingly becoming an almost inescapable feature of human life, whether it takes the form of congestion in the urban centers, or of anonymous existence in the housing developments associated with new industries. One consequence of the explosive growth of large cities and of the urban sprawl is that the old problems of air, water, and food pollution are reappearing everywhere with new and intensified manifestations. As we have seen, more and more people must by necessity breathe polluted air and depend for their water supply on individual wells that are contaminated both biologically and chemically (Chapter VIII.2).

Other public health problems are created by the stresses and anxieties of life in unsettled societies. Many human beings break down when they are compelled to live in a constantly changing environment and when they have to face an unforeseeable future in a new social context. The effects seem to be much the same irrespective of country or race. Thus, identical patterns of psychiatric disorders have been found among the Yoruba tribesmen of Western Nigeria and among the white people in northern Canada, two regions very different in all regards, except that both are in an early stage of industrial development. The increase in the incidence of venereal diseases among teenagers also seems to be a universal manifestation of rapid social change, in the developing as well as the developed countries. Industrialization, urbanization, labor migration, and the ease of travel seem to have created a worldwide Common Market in syphilis.

The increase in the prevalence of certain chronic and degenerative diseases was first observed in the heavily industrialized, prosperous countries of the Western world. But the phenomenon is not peculiar to them. It has occurred and continues to occur among all people moving from a primitive civilization or from a simple pastoral or rural economy to a way of life more comparable to that of Western people, irrespective of nationality, race, or color. City life per se is not the cause of the change. Chronic and degenerative diseases are approximately as common in the farming and suburban districts of wealthy nations as they are in their large industrialized cities, while they are said to be rare in the teeming but poor conglomerations of Asia. By a cruel irony of fate, these diseases come to the forefront precisely at the time when abundance replaces poverty, and when acute infectious processes are brought under control. This is just as true in Asia and Africa as in Europe and America.

It must be emphasized, furthermore, that many environmental factors other than the prosperous ways of civilized life are involved in the causation of degenerative and chronic diseases. As mentioned in Chapter II.1, the composition of drinking water seems to affect the frequency of cancer and vascular diseases. We shall see also that the incidence of gastric cancer is steadily decreasing in the United States whereas it has remained high in certain less prosperous countries.

Examples of the varied effects of the environment and the ways of life on the prevalence and severity of degenerative and chronic diseases are so numerous, and most of them so well known, that it seems best to limit illustrations here to a few reports very recently published. One concerns the Xavante Indian tribe, already mentioned, in which extensive medical studies revealed a wonderful general state of health among men, with hardly any evidence of dental caries or cardiovascular afflictions. In contrast, all the American Indians who have become Westernized have the same pattern of diseases as the white people among whom they live.

A similar state of affairs has been observed among the Mebans, a rather primitive and isolated African population in the Sudan. The 20,000 members of this tribe live in the forbidding swamps of the White Nile "seared for six months of the year by the blistering sun, pelted by rainstorms the rest of the time." As long as they live under primitive conditions in their villages, the Mebans retain the ability to perceive high tones until old age. It was thought at first that this attribute was due to the fact that their village environment is almost noise-free, but other observations place the finding in a broader light. Tests recently carried out by a group of Western physicians revealed that the Mebans enjoy low blood pressure from childhood to old age; they are virtually free of obesity, coronary thrombosis, duodenal ulcer, ulcerative colitis, allergies, bronchial asthma, and dental caries. Many of them are said to live to be 90 years of age. The Mebans' relative freedom from degenerative diseases is not due to racial immunity, as shown by the fact that these afflictions are common among the members of the tribe who have settled in Khartoum and other African cities.

In Puerto Rico, the recorded death rates from cancer, heart disease, and vascular lesions of the central nervous system have rapidly and continuously increased during the past 15 years, precisely at the time when the island was beginning to emerge from poverty and was enjoying a spectacular decrease in mortality from infectious diseases. There are indications that a similar pattern will emerge from a systematic study of the Navajo Indians, the complete results of which are to be published in the near future. As far as can be judged at the present time, little or no evidence of hypertension or coronary heart disease could be detected in the tribe before industrialization had modified its ancestral ways of life (personal communication from Dr. W. McDermott). Bronchial asthma is another disease which apparently was almost unknown among American Indians before 1931, but is now increasing in all their tribes.

It was generally thought until recently that the increase in degenerative diseases wherever the Western ways of life are being adopted is merely an accidental byproduct of the control of infectious and nutritional diseases. More people now live long enough to fall victim to disorders which develop so slowly that they become manifest only during late adulthood or old age. This explanation, however, fails to account for the fact that the various chronic and degenerative disorders do not increase uniformly. Some of them become particularly widespread or severe in certain geographical areas or segments of the population and not in others. It seems almost certain, furthermore, that the increase in degenerative diseases is occurring faster than can be explained by the longer expectancy of life.

b. Social determinants of degenerative diseases

Granted that chronic and degenerative diseases are becoming more prevalent in urbanized and industrialized societies, there is little convincing knowledge of the forces responsible for this change. It may be worthwhile therefore to report here a few epidemiological facts which appear to indicate that environmental factors can affect the occurrence of vascular diseases and cancers.

Attempts to correlate the incidence of coronary heart disease with environmental factors have been spurred by the increasing awareness that the overall frequency of deaths from this disease differs from one country and one social group to another. In Johannesburg, South Africa, myocardial infarction is extremely rarc among the Bantus; in contrast, the death rate from coronary heart disease is much larger in the white population of the same city, and more so among its non-Jewish members than among the Jews. But while such differences have been observed in many other situations, it is now certain that they are not racially determined. They are instead correlated with the economic status; indeed, the differences are as pronounced among the various social strata of a given race in a given country as they are among national groups.

As is well known, many physicians and epidemiologists believe at the present time that there is a direct relation between the prevalence of coronary heart disease and the dietary intake of saturated fats and cholesterol. But other aspects of the ways of life are now gaining prominence as causative or contributory factors, in particular the lack of physical exercise and certain emotional states. A few recent findings will be selected from an enormous amount of relevant information, merely to illustrate how epidemiological surveys can help in the elucidation of these complex etiological problems.

Two independent studies, one of Swiss mountain villagers and the other of railroad employees in the United States, indicate that physical exertion may play a role in cardiovascular disease. The first study deals with farmers living in the southern Swiss Alps, a group of people whose physical activity is extremely vigorous and maintained for long periods of time. Their nutritional intake per capita is over 3,500 calories daily, of which 34 per cent comes from fats; indeed, animal fats make up 27 per cent of the total calories they consume. Despite this large fat intake, the cholesterol levels in the serum of the mountain villagers is remarkably low, ranging from 165 at age 20–39 to 200 at age 50–59; such levels are comparable to those recorded for Guatemalan Indians living on a diet much more restricted quantitatively and containing much less fat. In contrast, the serum cholesterol levels are considerably higher in men of comparable age living in the city of Basel. The Basel inhabitants are of the same ethnic origin as the mountain villagers, and the proportion of animal fat in their diet is approximately the same, but their cholesterol levels are similar to those commonly found in the United States.

The other study deals with mortality from arteriosclerotic heart disease among men employed in different kinds of jobs by the railroads in the United States. During the period covered by the study, the mortality was 5.7 per 1,000 for clerks, 3.9 for switchmen, and 2.8 for section hands. In other words, it appeared as if the incidence of deaths from heart disease was directly correlated with the amount of physical activity during working hours. These results are similar to those obtained in England showing that bus conductors, who are always on the move collecting fares, have a lower mortality from heart disease than have the bus drivers.

While many factors such as the self-selection of persons for certain jobs complicate the interpretation of these two studies, the findings nevertheless suggest that vigorous physical exertion may be as effective as low fat diets in keeping the cholesterol levels low and in minimizing the incidence of cardiovascular disease. (Reviewed in *Nutritional Reviews,* 1963.)

A study of Benedictine monks living in a cloistered Dutch monastery revealed a few years ago that their blood cholesterol levels were much higher in all age groups than the levels found in a comparable group of Trappist monks also cloistered in Holland. These different cholesterol levels probably reflected differences in the diet of the two monastic orders, since the Trappists have a stricter dietary discipline than the Benedictines and are almost completely vegetarian. Irrespective of cholesterol levels, however, the incidence of coronary occlusion and other cardiovascular diseases was the same among the Benedictines and the Trappists, being much lower than among ordinary male citizens of comparable age living at the same time in Holland.

From the results of these and similar epidemiological studies, it appears that acute occlusion of a major coronary artery, leading to a macroscopic myocardial infarct, may be less directly related to nutrition and blood cholesterol levels than has been claimed. As important as these biochemical factors, perhaps, are certain psychosocial stresses common in the Western ways of life, against which the members of both the Trappist and Benedictine orders are protected to some extent within the walls of their monasteries. The pressures and anxiety resulting from social and professional competition, the effects of crowding, and more generally the complexities of urban living might thus constitute factors that play a part in the causation of vascular diseases. While there are several physiological facts in favor of this hypothesis, it is very difficult to obtain convincing epidemiological evidence of its validity (Chapter I.4). Observations made in animals acquire a special interest for this very reason.

Autopsies carried out at the Philadelphia Zoo have revealed that arteriosclerosis has increased 10 to 20-fold in frequency during the period 1916 to 1956, both among mammals and birds. This increase seems to be independent of age and diet. During the past two decades, more and more animals have been added to the zoo's collection, resulting in greater population pressures among them, and probably affecting their adrenal function. In reality, of course, the immediate and delayed effects of so-called social tensions are just as difficult to identify and to measure in animals as in men. Nevertheless, the available observations, sketchy as they are, give some credence to the hypothesis that the competitive ways of modern life and a sudden increase in population density may contribute to hardening of the arteries through indirect physiologic and metabolic channels, with consequent circulatory disorders, heart attacks, and strokes.

The large geographical and social differences in the relative frequency of the various forms of cancer provide intriguing facts to illustrate the role of the total environment in the causation of this disease. So many relevant observations have been made during recent decades that only a few representative cases can be quoted here to illustrate the relationship between environmental factors and cancer. (For recent general review see World Health Organization Report, 1964.)

The national differences in cancer distribution suffice by themselves to emphasize the importance of such environmental determinants. Thus, pulmonary carcinoma is the most common cause of cancer deaths among men in the United States, England, Wales, and several other Western countries, but is much less frequent in Iceland. In contrast, stomach cancer accounts for 50 per cent of cancer among men in Iceland and Japan, but only for 10 per cent in the United States and for even less in Indonesia. Liver cancer causes half of all cancer deaths among the Bantus in Africa, but less than 4 per cent in Europe and North America. Breast cancer is over eight times more common among women in Israel than among women in Japan. Cancer of the cervix accounts for half of all cancer deaths among Hindu women.

Profound local differences also exist within a given country. Thus, skin and lip cancers are twenty times more common among white people living in the southern half of the United States than among those living in the northern part. In the USSR they are five to six times more common in the south than in the north, and are particularly frequent on the coast itself. One is reminded here of Hemingway's hero in *The Old Man and the Sea,* with precancerous lesions on his face.

Equally striking are the differences among social groups living in the same region. A study carried out during the 1950s in the Los Angeles area revealed that the death rate from lung cancer among Seventh Day Adventists was less than one-tenth of that expected among comparable groups of other American citizens.

Indeed, the only two cases noted were in converts who had joined the movement in middle age. The Seventh Day Adventists also showed a lower incidence of bladder cancer. Cancer of the cervix was abnormally low among their women, about as low as it is among Jewish women, who have long been known to be relatively free of this disease. Most other forms of cancer, which are responsible for the vast majority of cancer deaths, were found in this study to be as common among the Seventh Day Adventists as among the general population.

Knowledge concerning the nature of the specific factors in the environment that are responsible for increased carcinogenicity is still in a very primitive state, but encouraging progress has been made in a few cases. The carcinogenic activity of radiations, certain viruses, and coal tar products has long been recognized and therefore need not be illustrated here. It may be worthwhile, however, to mention once more that cancer of the stomach is extremely prevalent in Iceland, where it accounts for 35 to 45 per cent of malignant tumors in men in the northwestern part of the country. The hypothesis that nutritional habits play a role in this epidemiological peculiarity has received some support from the finding that the incidence of the disease is correlated with the consumption of large amounts of smoked mutton and smoked salmon. Smoked foods contain relatively high concentrations of various polycyclic hydrocarbons and regularly produce cancer in male rats when added to their diet. These findings may be of practical significance because they indicate that some forms of food may play a role in carcinogenesis. Smoked salmon, one of the great delicacies of the table, would seem to be one of the culprits in this case, but it is probable that the problem is more complex than this explanation would suggest.

Deaths from cancer of the stomach in the United States have decreased from 22,257 in 1955 to 19,368 in 1962. Age-adjusted rates, particularly for the sixth decade of life, reveal an even more pronounced fall. The situation is approximately the same in the prosperous countries of Western Europe. In contrast, the incidence of gastric cancer remains high not only in Finland but also in Yugoslavia, the Soviet Union, and Japan. In these cases, evidence seems to relate the frequency of the disease to large intake of starchy foods and perhaps to certain nutritional deficiencies.

It has been suggested that the great prevalence of primary cancer of the liver in the aboriginal races south of the Saraha Desert may also be of dietary origin. Of interest in this regard is a recent report that as many as three-quarters of the trout in certain hatcheries developed liver cancer when fed an artificial diet consisting of cottonseed meal, fish meal, with added vitamins and minerals. As the disease has not been observed either among trout under natural conditions or in hatcheries where other kinds of untreated food are fed to the trout, it is likely that the artificial diet contained a substance that acted as a carcinogenic agent.

In India, the very high incidence of cancer of the mouth is associated with the habit of slowly sucking a mixture of tobacco and lime. Also in India and Ceylon, cancer of the mucous lining of the cheek is extremely common in persons who chew tobacco mixed with betel leaf and betel nut. Lymphosarcoma in African children deserves special mention because its incidence suggests an environmental effect mediated through an infectious process. In some parts of Central Africa, extranodal lymphosarcoma accounts for nearly half of all neoplasms in children, and half of these cases present multifocal deposits in the jaw bones; these jaw lesions can therefore serve as a marker to study the distribution of the disease. The tumor exhibits a peculiar geographic localization. It occurs from coast to coast of the equatorial belt, but its distribution seems to be determined by altitude in Eastern Africa and by factors affecting humidity and vegetation in West Africa. Taken together, these epidemiological facts are compatible with the hypothesis that the condition is transmitted by an insect vector, and perhaps induced by a virus infection. (For a critical review of these findings see Dalldorf et al., 1964.) Needless to say, the latter hypothesis receives support from the recent findings that several kinds of animal tumors can be elicited by viral infections. However, convincing evidence that viruses are etiological agents in human tumors is still lacking.

As is well known, many metals and synthetic chemicals are carcinogenic in animals and probably in man, a fact that gives great importance to the problem of occupational hazards in cancer causation. Lung cancer, for example, is unusually frequent among workers engaged in the refining of nickel, in the manufacture of asbestos, and in the production of coal gas.

The World Health Organization Technical Report 220, published in 1961, listed the recognized and suspected occupational carcinogens:

(1) Based on well-documented evidence

(a) coal distillation and fractionation products, including

tar, pitch, creosote, anthracene oil, tar oils and soot—containing aromatic polycyclic hydrocarbons (e.g., skin, lung);

(b) distillation and fractionation products of shale oil, soft coal, petroleum and hydrogenated coal oils, including tar, asphalt cutting oils and crude waxes—containing aromatic polycyclic hydrocarbons and other carcinogenic constituents (e.g., skin, lung);

(c) aromatic amino-, nitro- and azo-compounds (e.g., 2-naphthylamine, 4-aminodiphenyl, some of their analogues, benzidine and 3,3′-dichlorobenzidine) (e.g., urinary bladder);

(d) some products of chromium- and nickel-ore processing (e.g., lung, nasal sinuses);

(e) ionizing radiation, both by X-rays and from radioactive compounds (e.g., skin, bone, lung, haemopoietic and lymphopoietic systems);

(f) ultraviolet radiation (e.g., skin);

(g) inorganic arsenicals (e.g., skin, lung);

(h) asbestos (e.g., lung);

(i) "isopropyl oil" (e.g., nasal sinuses).

(2) Suspected carcinogenic hazards

Beryllium, commercial benzol, mustard gas, etc.

It is never easy to prove that an agent, a substance, or a situation is truly carcinogenic. The task has been rendered even more difficult by the recent discovery that certain substances, harmless by themselves, can potentiate the cancer-producing action of other agents. Thus, citrus oils are said to give rise to stomach cancer in mice receiving doses of benzpyrene too small to be active by themselves. Of more immediate relevance to man is the fact that certain air pollutants seem to act synergistically with cigarette smoking to increase the incidence of pulmonary carcinoma among heavy smokers living in industrialized areas. It has been claimed, indeed, that the true carcinogen for lung cancer does not come from the cigarette itself, but is rather to be found in some environmental agent for which smoking acts as a potentiator (see Chapter VIII.2 and 3). Needless to say, etiological complexity is not peculiar to the causation of pulmonary carcinoma. In most pathological states, etiology involves a constellation of factors, rather than a single causative agent.

Epidemiological correlations of the kind illustrated in the preceding pages do not explain of course the intimate processes responsible for vascular diseases or by which body cells become malignant. These processes can be understood only through laboratory analysis at the cellular and biochemical level. But medical history richly demonstrates that essential knowledge of the etiology of a disease and effective methods for its control have often been derived from purely epidemiological studies, without the benefit of cellular or biochemical analysis.

Epidemiological studies can help first in ruling out false preconceived notions. For example, the fact that the frequency of vascular diseases or certain types of tumors among Oriental people or Negroes residing in the United States is much higher than that observed among people of these races who have remained in Asia or in Africa points to the importance of environmental as against racial determinants. Everyone now realizes that, whether they be white, yellow, or black skinned, all human beings who consume 3,500 calories a day and move only by motor cars and elevators are in greater risk of suffering from certain diseases than are wiry and half-naked people of the same color whose way of life is to plant rice knee-deep in mud, to cast heavy nets every day at the break of dawn, or to toil half-naked under the burning sun to bring in the harvest.

In many cases, furthermore, searching epidemiological studies should make it possible to identify the trigger factors that are responsible for the production of degenerative diseases. This knowledge will certainly suggest control policies based on manipulation of the environment and on changes in the ways of life, just as has been done for many infectious diseases.

The very fact that the etiology of most degenerative diseases probably involves a constellation of factors complicates the analytical work of the epidemiologist and the experimenter, but it increases the potential range of control measures by making it possible to break the chain of causation at different links.

c. Environment and genetic diseases

Changes in environment and in the ways of life can affect health and disease through the agency of genetic mechanisms. In fact, it is not possible to separate genetic from environmental factors when discussing the influence of the ways of life on human health. Favism, an anemia that occurs in some persons when they eat fava beans, can be regarded as a symbol of the manner in which environment and genetic endowment constantly interplay in all manifestations of life. As far as is known, the only persons who suffer from favism are those who genetically have erythrocytes susceptible to the constituent of the fava bean which brings out the anemia. In this case, the culprit is a recessive gene that causes a reduced activity of glucose-6-phosphate dehydrogenase. The affected persons are essentially normal unless they consume fava beans.

The relation of barbiturates to porphyria is another illustration of the interplay between environment and genetic constitution. The porphyria gene is present in about one per cent of the Afrikaner (Dutch) populations of South Africa. There is strong evidence that this gene was introduced in the country in 1686 by a woman who emigrated from Holland and that all the porphyrics today, approximately 8,000, are her descendants. Porphyrics suffer little discomfort, other than developing minor skin abrasions, as long as they live under conditions of rustic simplicity on their farms or in villages, but they are extremely susceptible to barbiturates and sulfonamides. Porphyric patients who receive these drugs become emotionally deranged and later paralyzed; the resulting mortality is between 50 to 90 per cent. Thus the porphyric gene, essentially harmless by itself, becomes lethal in the special environments that are created by modern medicine. "In genetic porphyria, it is usually the doctor who kills the patient."

It can be taken for granted that many other human ills, far more frequent in occurrence than favism or porphyria, occur only when genetic and environmental peculiarities happen to coincide. It is beginning to appear, for example, that rheumatoid arthritis has a familial distribution and is in some way related to lupus, which also runs in families. Neither of these conditions is a purely genetic disease, but both may be the result of some environmental insult happening at the right time to people with a certain genetic constitution.

Infectious diseases have probably acted as powerful selective agents in the past, but their role in this respect is obviously becoming less and less significant. As a result, one of the factors of human evolution is being eliminated, with consequences that cannot be foreseen. More generally, it is to be expected that medical advances will modify the hereditary structure of prosperous populations by permitting the survival of persons with genetic defects who in the past had little chance of reproducing their kind. For example, the number of potential diabetics is probably increasing in our communities as a consequence of the general availability of insulin.

Unquestionably, new environmental forces will act as agents of selection for our descendants; the increase in ionizing radiations might come to occupy in this respect the place that infectious and organic diseases occupied in the past, and it is to be feared that the effects will be dysgenic for the human race as a whole. Social forces also might act as agents of selection, since the ability to remain healthy and to function effectively in a crowded environment will increasingly become a condition of survival. Furthermore, the new ways of life may give greater social survival value to certain behavioral characteristics, such as aggressiveness or conformity, than to other more appealing human traits, such as altruism or love of independence.

It is pleasant to postulate, on the other hand, that some of the hygienic improvements will have eugenic effects. The decreased frequency of the sickle-cell anemia trait in regions where malaria has been brought under control is the now classical example of this possibility. Changes as advantageous as the disappearance of abnormal hemoglobins will probably eliminate other characters that persist in the population only because they minimize the damage caused by certain infective agents.

d. The seemingly healthy

Above and beyond their role in the causation of overt disease, the environment and the ways of life that prevail in our communities also have other nefarious effects which are more subtle but nevertheless important. Even when supplemented with morbidity rates, mortality rates fail to provide a picture of these subtle effects. The dead and the obviously sick are readily counted, whereas the people who are seemingly healthy but who have suffered from some kind of organic or psychic damage usually escape attention. Yet persons whose pathological experience is not manifest in readily detectable signs or symptoms may nevertheless be of great importance from the point of view of disease control. On the one hand, many of those who are seemingly healthy today will be sick tomorrow. Furthermore, they will contribute to the decrease in the quality of the population, even if they continue to carry out their daily routine as usual and to perform their social duties. In final analysis, many of them will eventually add to the burden of social and medical care.

The existence of a large amount of submerged potential illness has come to light during recent years through several accidental findings. Military screening revealed that a substantial percentage of young men were unfit to serve because of physical and mental deficiencies. In addition, pathological studies carried out on soldiers during the Korean War yielded the unexpected finding, since then extensively confirmed, that arteriosclerosis is common in a more or less advanced degree among vigorous and seemingly healthy young Americans.

In civil life, many persons who die of various diseases or accidents are found at autopsy to have vascular lesions in the nervous system. Approximately one-third of such persons have had actual hemorrhages, and many exhibit damage to the neural tissue resulting from an inadequate blood supply. It has been stated that of 33 per cent of autopsy cases with neural lesions, about 25 per cent can be accounted for by vascular disease, the remainder being due to tumors, trauma, malformations, and infections. In view of these facts, it is not surprising that strokes constitute the third principal cause of death in the United States. In 1959, one half million persons experienced strokes in this country, and one and one-fourth million hemiplegic patients were surviving in hospitals or at home.

Further evidence that potential disease commonly exists in an unrecognized form was provided by a recent study showing that a very large percentage of the Manhattan population presents some degree of psychiatric disability. In most cases, needless to say, this disability is not severe enough to demand hospitalization, but it is often sufficiently serious to create difficulties for the maintenance of health under the exacting conditions of modern city life.

Unfortunately, it will not be easy to work out techniques for recognizing among the population at large the persons who are in need of medical surveillance even though they are free of obvious symptoms and are seemingly healthy. In fact, the greatest difficulty will come not from deficiencies in the scientific knowledge required for medical diagnosis but rather from social attitudes. Any disease, or any kind of deficiency, that is very widespread in a given social group comes to be considered as the "normal" state and consequently is accepted as a matter of course within that group.

Primary and secondary yaws, for example, were so common among certain African tribes until a few years ago that the affected people did not regard the condition as a sickness. Similarly, dyschromic spirochetosis, also known as pinta, is so common in a certain South American tribe that the Indians regard as healthy those among them who have the disease and as ill those who do not. Such acceptance of disease is not peculiar to primitive people; it is a common attitude all over the world. The various forms of chronic pulmonary disease are almost accepted as a matter of course in many industrialized areas (Chapter X.2). In Japan, chronic bronchitis and asthma are so common that even the medical profession tends to regard these disorders as unavoidable complaints determined by the very nature of the country.

Surprising as it may seem to us, there was a tendency in Europe and America not so many decades ago to accept tuberculosis as an inescapable part of human fate. In the Report of the Sanitary Commission of Massachusetts, L. Shattuck wrote in 1850:

> The dreadful disease [tuberculosis] is a constant visitor to all parts of our Commonwealth, but creates little alarm because it is so constantly present, whereas the occasional visit of cholera or some other epidemic disease creates alarm and, therefore, precautionary measures are taken.

A similar concern was expressed by Bigelow in 1867:

> Should the cholera continue to prevail for three years throughout this continent, it would cease to interrupt either business or recreation. Mankind cannot always stand aghast; and the wheels of society at length would be no more impeded by its presence than they now are by the existence of consumption, of old age, or of drunkenness.

How relevant to several of our own problems these perceptive words are still today! Witness our lack of public concern with the slaughter on the highways.

Social acceptance of many forms of disease is in fact almost universal. In our own communities, all sorts of physiological, mental, and emotional defects are assumed to be of little consequence because substitutes or correctives for them are supplied by modern technology and medicine. Poor eyesight, deficient hearing, and diabetes are considered of relatively small importance because we have glasses, hearing aids, and insulin. We are less and less interested in physical stamina because we have automobiles, labor-saving devices, and air-conditioning equipment. And increasingly we come to believe that the time-honored mental disciplines are unnecessary, because we have drugs to stimulate the lethargics, to quiet the hyperactives, and to tranquilize the worriers and the compulsives.

The scientific justification usually given for such tolerance, or at least for this permissive attitude toward anatomic, physiological, and mental disabilities, is that adaptation must be defined not in terms of absolute values, of an ideal concept of man, but rather in terms of the modern environment with its technological and medical devices. While this concept of social adaptation has a sound biological basis, it is nevertheless fraught with great dangers in the long run (Chapter X.2). Many undesirable characteristics are obviously compatible with survival and even with success in our society. Overeating, obesity, and food fads; a low level of physical stamina and of general resistance; alcoholism and dependence on drugs; even juvenile delinquency and various forms of neurosis can in fact be regarded in many cases as the outcome of adaptive reactions to the modern world. But just the same, these characteristics are the expression of biological defects, and their ultimate consequences are often disastrous for the individual person concerned and even more for the community of which he is a part. As we shall see, adaptation to the present is often bought at the cost of dangers for the future (Chapter X.2). Furthermore, a time may soon come when social and economic necessities will limit the practical applications of medical knowledge, to the detriment of those who depend too extensively on its help (Chapters XV and XVI).

Environmental and social changes will in the future create new problems of disease, just as they have in the past. Man will have to learn to outwit his genetic endowment by modifying his ways of life, on the basis of knowledge of environmental risks entailed by special genes. But it is certain that physical comfort and economic abundance will not be sufficient to assure good health. The pathology of urban and suburban life, of antiphysiological leisure in a mechanized, automated, and crowded environment, may be the twentieth-century counterpart of the tenement pathology that has prevailed in industrial countries until recent years. The present epidemics of ischemic heart disease, lung cancer, duodenal ulcer, and mental diseases are for our communities what tuberculosis was for Europe during the first Industrial Revolution, and what typhoid and rickets were for Victorian England. Environmental pollution, chain smoking, dependence on drugs, overeating, underexercise, lack of social stability, and excessive mobility are but a few of the environmental factors of modern life that determine the pattern of disease wherever social success is identified with the present concept of high standards of living.

X. Adaptation and Its Dangers

1. Man's responses to his environment

a. Mankind adapting

Small bands of naked men with their women and children live without shelter in the deserts of South Africa and Australia, quenching their thirst from mudholes, feeding on wild plants and insects, occasionally starving, always deprived of the most elementary kind of hygienic and medical care and of the shelter and comfort regarded as absolutely essential by Western man. In the tropical rain forests of the Congo and Amazon valleys, other primitive tribes have long lived isolated from the rest of the world, eking out a precarious existence from the meager resources of an environment that appears to us overpoweringly hostile. Likewise, the Eskimos long managed to survive in the most desolate barrens of the arctic, equipped only with the simple tools of their Stone Age civilization.

It is remarkable enough that so many different kinds of populations can survive hardships and shortages which appear to us as almost incompatible with human life. But it is even more remarkable that, however poor their resources and hostile their environment, the most primitive people are human in the fullest sense of the word. All of them have developed complex social structures and original artistic skills; they have imagined poetical legends to account for creation and have composed romantic songs to celebrate springtime and romantic love. When at play with their children, the smile on their lips and in their eyes has the tenderness seen on the faces of Mary and Joseph in paintings of the Holy Family. Thus, primitive people demonstrate that mankind can survive, retain endearing traits, and create independent and viable cultures without technology, even under the harshest circumstances. Writing of the Caribs in his journal, Christopher Columbus expressed admiration not only for their physical stature, but also for their social manners: "I believe that in the world there is no better race. . . . They love their neighbours as themselves, and they have the softest and gentlest voices in the world, and they are always smiling."

Everyday experience in our cities provides endless illustrations of the wonderful ability of human beings to carry on a rich and diversified life even when completely estranged from their ancestral environments. Blond, blue-eyed Scandinavians, who trace their origin to some bleak land frozen most of the year, work side by side in most parts of America with men of black or brown skin, whose forebears lived in tropical rain forests. Irrespective of origin and color, Nordics, Southern Europeans, Latin Americans, Negroes, and Asiatics who have settled in New York, Chicago, and Los Angeles now eat the same food, wear the same kind of clothing, use the same drugs, depend on the same type of medical care, and as likely as not adjust their homes to the same temperature before sitting down in the same style of furniture, to watch the same television program at the same hour.

Nothing illustrates better the adaptability of man than the progressive improvement of the general state of health in the crowded cities of the Western world, even though life in these cities is so completely alienated from the ways of nature. It is true, of course, that urban life has often brought about a physical deterioration of mankind in the past. The huge problems of disease that prevail today in the congested cities of the underprivileged areas of the world provide further evidence for the very ancient belief that industrialization and urbanization can be inimical to human health. This belief is so widespread indeed that cartoonists are still wont to contrast the gray complexion and flabby muscles of the city dweller with the ruddy cheeks and muscular build of his country cousin.

In reality, however, and contrary to the conclusions that could be drawn from history, prolonged habitation in large and crowded cities is not necessarily deleterious to human health. The prosperous countries of Western civilization now enjoy much the same expectancy of life in urban as in rural areas. The largest, most congested, and most industrialized metropolis can provide conditions compatible with a long and healthy life, whatever its geographical location, topography, or climate. Holland, which is the most densely populated country in the world, with a climate that is often trying, nevertheless has at present one of the healthiest populations according to any of the generally accepted medical criteria.

b. Homeostasis and adaptation

The ability to adapt to a wide range of environments is of course not peculiar to man. Cats seem to enjoy the warm comfort of a bourgeois living room, but they also prosper in the back alleys of destitute neighborhoods. Adaptability is found throughout life and is perhaps the one attribute that distinguishes most clearly the world of life from the world of inanimate matter. Living organisms never submit passively to the impact of environmental forces; however primitive they may be, all of them attempt to respond adaptively to these forces, each in its own manner. The characteristics of this response express the individuality of the organism and determine whether it will experience health or disease in a given situation.

As first clearly stated by Claude Bernard, survival and health depend upon the ability of the organism to maintain its internal environment in an approximately constant state, despite the endless and often extreme variations of the external environment. Recognizing a fundamental and far-reaching truth long before it could be demonstrated, Bernard boldly asserted that "the fixity of the *milieu intérieur* is the essential condition of free life." He further expressed his conviction that the constancy of the internal environment is achieved through the orderly performance of the body's physiological and biochemical processes. Walter B. Cannon enlarged Bernard's concept by emphasizing the importance of the regulatory systems which enable the body to make useful responses to environmental stimuli. He acknowledged that the physiological adjustments required for such responses necessarily entail internal changes and that a certain degree of transient departure from the ideal state of internal constancy is therefore inevitable, but he particularly emphasized the homeostatic aspects of these physiological changes.

Homeostasis implies that the body can function well only to the extent that it can rapidly make the necessary adjustments so that its internal composition remains within limits precisely defined for each organism. On the other hand, the responses to environmental changes made by each individual organism must be such that ideally they help it to function adequately under the changed conditions. These two demands are equally essential for the success of life. Furthermore, they apply to populations as well as to individual organisms. The very continued existence of a biological system, whether primitive or complex, implies that it possesses mechanisms which enable it to maintain its identity, despite the endless pressure of external forces, and yet that it is able to respond adaptively to these forces. The complementary concepts of homeostasis and adaptation are therefore valid at all levels of biological organization; they apply to social groups as well as to unicellular or multicellular organisms.

Homeostasis however is only a concept of the ideal. Living things do not always return exactly to their original state after responding to a stimulus. The range of variations that is compatible with their survival and the threshold values beyond which corrective mechanisms must come into play constitute data just as pertinent to the definition of the organism as are those which describe its ideal state of internal constancy. It is true, of course, that the regulatory mechanisms built into the fabric of each and every living thing constitute the determinants of what Cannon termed the "wisdom of the body." But the most common expression of this wisdom is not so much the maintenance of an absolute constancy of the internal environment as the selection for each particular case of one kind of response out of several alternative possibilities available to the organism. Not only does the response often fail to return the internal environment to its original state; in many cases the response is not even appropriate for the welfare of the organism. It may be excessive or misguided and thereby elicit injurious or destructive reactions from even mild stimuli. Disease is the manifestation of such inadequate responses. Health corresponds to the situation in which the organism responds adaptively while retaining its individual integrity.

While so essential in every day discourse, the word adaptation is treacherous because it can mean so many different things to different persons. The layman, the biologist, the physician, and the sociologist use the word, each in his own way, to denote a multiplicity of genetic, physiologic, psychic, and social phenomena, completely unrelated in their fundamental mechanisms. These phenomena set in motion a great variety of totally different processes, the effects of which may be initially favorable to the individual organism or social group involved, and yet have ultimate consequences that are dangerous in the long run. Furthermore, an adaptive process may be successful biologically while undesirable socially (Chapter X.2). The concepts and the consequences encompassed by the word adaptation as it is used in practice are so diversified that it may be helpful to symbolize them by a few concrete illustrations.

An old Chinese fisherman, of the type so often depicted on scrolls of the Sung era, symbolizes the meaning of the word adaptation as applied to the life of primitive people. The fisherman appears fully at ease and relaxed in his primitive boat, floating on a misty lake, or even on a polluted and crowded harbor. He has probably experienced many tribulations in the course of his years of struggle and poverty, but he has survived by becoming almost completely identified with his environment. In fact, he is so well adapted to it that he will probably live for many more years, without modern comfort, sanitation, or medical care, just by letting his existence be ruled according to what he considers to be the inalterable laws of the seasons and of nature.

The state of adaptedness achieved by the old Chinese fisherman is the outcome of a great variety of unconscious processes. A very large percentage of his siblings and contemporaries have died in youth or early adulthood; he was one of the few who survived because he was lucky enough either to be so endowed genetically that he could overcome the assaults of his environment or to be protected from them by chance. In the course of his life, he developed many different protective mechanisms that increased his immunologic, physiologic, and psychic resistance to the physicochemical hardships, the parasites, and the social conflicts which threatened him every day. Finally, he has elected to spend the rest of his life in the environment in which he has evolved and to which he has become adapted. Robust as he appears and really is, he probably would soon become sick if he moved into an area where the parasites, physiologic stresses, and social customs differ from the ones among which he has spent his early life.

Unconscious adaptive processes are not peculiar to ancient civilizations. They constantly occur in our times around us and everywhere. Anyone traveling through Latin America or the American Southwest can see evidence of them in the Indian villages. Many Indian men and women reach old age and retain great vigor, eating food and drinking water that would be unacceptable to the people of another culture, especially our own. On the other hand, the same Indians would be struck by disease if compelled suddenly to adopt the ways of life that prevail along Broadway or even in prosperous and sanitary Suburbia.

The man in the space ship provides an image of adaptation which contrasts sharply with that provided by the life of the old Chinese fisherman or of any other people in prescientific cultures. Whether Russian or American, the astronaut is an adult who has willfully, consciously, and systematically trained himself to perform a particular task for a short time in an unnatural but well-defined environment. He can function effectively in the space ship for a few hours, a few days, or a few weeks, but he could never find in it material for the cultivation of all the potential wealth of his body and mind. The space ship does not provide the ambiance in which to develop the fullness of human life; it constitutes merely a limited environment to which man must adjust himself in order to perform a highly specialized function.

c. The mechanisms of adaptation

The ambiguous character of the word adaptation is not a product of carelessness in language but rather is inherent in the wide range of meanings attributed to it by the different groups of human beings, especially scientists. The difficulties of usage become painfully apparent when these meanings are applied to the human condition.

For the general biologist, the best and perhaps the only measure of adaptive fitness to a particular environment is the extent to which the organisms of the species under consideration can occupy this environment, make effective use of its resources, and therefore multiply abundantly in it. The biologically successful species can invade new territories by broadening its adaptive range. These criteria satisfy the tenets of Darwinian evolution, especially in the light of population genetics, but they are obviously inadequate when applied to man. Many of the human populations which are increasing in numbers most rapidly at the present time, and which are desperately trying to spread to other parts of the world, are precisely the ones whose conditions of life are considered the least desirable, even by their own standards. There is no objective evidence, true enough, that high population density is inimical to health or to a pleasurable way of life—witness the fact that most human beings are eager to move into the huge and congested cities of Western civilization. But the examples of large sections of South Asia and Latin America make it obvious that ability to increase in numbers is not a sufficient criterion of biological success for human beings. Social man has now reached a stage at which his biological problems can no longer be defined merely in terms of Darwinian fitness.

It is tempting to believe that the problem of adaptive fitness in the case of man can be restated simply by supplementing the classical biological criteria with new social and cultural criteria. This would be consonant with the fact that, even though man is still slowly evolving anatomically and physiologically, social and cultural forces have become of increasing importance in his evolution because they determine in a large measure the goals toward which he is moving. In other words, sociocultural forces are now more powerful than biological ones in orienting the evolution of his ways of life.

Unfortunately, it is at best very difficult, and in fact probably impossible, to define adaptation in terms as subjective as social strivings, ways of life, and cultural goals. A type of human being who would find comfort and happiness in the social environment provided by a casual, semi-anarchistic community would be ill suited to a society having as its ideal the submission of the individual to the state. Throughout history and all over the world, men who regard Sparta as their ideal have been at odds with the admirers of democratic Athens. Selection for brotherly love and otherworldliness would be an essential requirement for producing human beings bent on creating the City of God, but such a social attitude would certainly handicap the success and even the survival of the group in a competitive world. Even physical attractiveness would prove to be a poor characteristic for defining adaptive fitness in social terms. Esthetic criteria differ from one place to another, and furthermore they vary with time, a fact entertainingly illustrated by many thousand years of painting and sculpture of the female body in the nude.

Physiologists have their own meanings for the word adaptation, and these differ from one physiological school to another. Bernard's concept of the fixity of the *milieu intérieur* and Cannon's homeostasis are interpreted by modern physiologists as aspects of the mechanisms by which the organism maintains its state of adaptedness against the pressures of the ever-changing environment. Physiological and biochemical homeostasis, however, do not account for all the mechanisms through which living organisms respond adaptively to environmental stimuli. All adaptive changes have of course a genetic basis, but there are many phenotypic adaptations that are long lasting and yet do not involve modifications of the genoplasm. The various immune states, the tanning induced by prolonged or repeated exposure to sunlight, the increase in hemoglobin concentration and in the depth of respiration engendered by life at high altitude are common examples of adaptive changes that are phenotypic yet lasting, and different in nature from classical homeostatic processes.

At a higher level of integration, the organism responds adaptively to many kinds of stimuli by behavior patterns designed to abolish or neutralize the stressor stimulus, or to withdraw from it. Organisms with a highly developed nervous system have several alternative mechanisms of behavioral response and, furthermore, they possess the ability to ignore some of the stimuli that impinge upon them. The higher the organisms is in the evolutionary scale, the more numerous and varied are the types of responses at its disposal and the greater is its ability for selecting limited aspects of the environment to which it responds. The most evolved types of responses are the processes of social adaptations, through which the individual organism and the group modify either their environment or their habits, or both, in order to achieve a way of life better suited to their needs and tastes.

Whether they emphasize the biological, social, or cultural aspects of adaptation, theoretical biologists and sociologists agree in considering the problem from the point of view of the species as a whole. The practicing physician, in contrast, is not so much concerned with the species *Homo sapiens* as he is with individual human beings suffering from particular diseases or experiencing difficult life situations. He is impressed by the fact that many disease states result from inadequate responses made by the body or the mind to noxious influences, in a clumsy effort to adapt to the environment. The consequence of this attitude has been that medical scientists have come to think of adaptation in terms almost opposite to those commonly accepted by general biologists.

In much of medical literature, the word adaptation is used to denote phenomena associated with disease. Although this practice may appear objectionable to geneticists and evolutionists, the fact is that it has a long and distinguished history. Ever since Claude

Bernard, disease has been regarded by physicians and pathologists as the outcome of *attempts* at adaptation, without regard to the precise mechanisms involved, or to the conditions which elicit the response. The general acceptance of this theory of disease in medical circles, and of this use of the word adaptation, is illustrated by the following quotation from an essay "Adaptation in Pathological Processes," the presidential address given in 1897 by William Welch before the Congress of American Physicians and Surgeons.

> We have seen that in the sense in which adaptation was defined we can recognize in the results of morbid processes frequent and manifold evidences of adjustment to changed conditions. These adjustments present all degrees of fitness. Some are admirably complete; more are adequate, but far from perfect; many are associated with such disorder and failures that it becomes difficult to detect the element of adaptation. The teleological conception of a useful purpose in no case affords an explanation of the mechanism of an adaptive process. I have suggested that the adaptability of this mechanism to bring about useful adjustments has been in large part determined by the factors of organic evolution, but that in only relatively few cases can we suppose these evolutionary factors to have intervened in behalf of morbid states. For the most part, the agencies employed are such as exist primarily for physiological uses, and while these may be all that are required to secure a good pathological adjustment, often they have no special fitness for this purpose [Welch, 1897].

The use of the word adaptation to denote an essential mechanism of disease achieved its most descriptive and also most popular form a quarter of a century ago in the expression "General Adaptation Syndrome." This phrase refers to the complex sequence of pathological processes that result from nonspecific responses made by the body of animals and man to a large variety of unrelated stressful situations. While opinions differ as to the validity of the General Adaptation Syndrome concept, all students of disease take it for granted that pathological manifestations often arise from the fact that the attempts at adaptation, even though appropriate in kind, are either misguided or too intense. Thus, the patient may be gravely damaged through the very magnitude of his so-called "adaptive" reactions.

Needless to say, the judgment that the bodily or mental responses to a particular stimulus are inappropriate is often a value judgment. Responses which appear inappropriate in the light of conventional homeostasis may in certain cases serve useful functions and purposes not readily apparent. The destruction of a part of the body by a pathological process, for example, not uncommonly protects the other organs and thus helps the organism to survive. Some forms of neurosis also are said to have a protective value in man and perhaps in animals, and it has been claimed by certain psychiatrists that what is usually called disorganization of behavior may in reality be a form of defensive reorganization. Thus, the meaning of the word adaptation in the language of the physician extends all the way from classical homeostatic processes to damaging but nevertheless protective mechanisms, and to reactions that may be potentially dangerous and even fatal in the long run.

The layman is naturally unconcerned with the diverse concepts and the complex processes that biologists, sociologists, and physicians associate with the word adaptation. For him, to be well adapted simply means to be able to function effectively, happily, and as long as possible, in a particular environment. It is this ideal that the family, the school, and the other social institutions have in mind when they concern themselves with training children and citizens to make them "well adapted" to their community. Thus no group of scientists can claim priority or exclusivity in the use of the word adaptation or has a right to limit its usage. The word has been in the public domain for so long, and has been applied to so many different purposes, that it denotes a great variety of unrelated mechanisms and processes. In many respects a word is a living organism; its meaning changes with the conditions of its use. Like other manifestations of human behavior, the word adaptation has adapted itself to the changing needs of the human condition.

Darwinian fitness achieved through genetic mechanisms thus accounts for only a small part of the adaptive responses made by man and his societies. This does not mean, however, that purely genetic phenomena are no longer of importance in the responses that man makes to his environment.

Now as always, the human body undergoes spontaneous genetic changes which, although not adaptive by themselves, become so through the operation of selective pressures. Human adaptation is greatly facilitated by the fact that *Homo sapiens* is so eminently a polymorphic species. Balanced polymorphism consists in the simultaneous occurrence in the same habitat of two or more discontinuous forms of a character in such proportions that the genetic heterogeneity cannot be solely the consequence of mutation. In polymorphic systems, both heterozygote and homozygote must be regarded as normal; in other words, there is not one normal genetic type but several existing simultaneously in any given environment. Depending upon the local circumstances, selective pressures determine which type tends to predominate, the result being usually a nearer approach to Darwinian fitness.

Even a gene that has deleterious effects may be of selective advantage under certain circumstances, as illustrated by the prevalence of the gene for sickle-cell anemia among African Negroes. This gene persists in populations exposed to malaria because it contributes greater resistance to the parasite; but it will certainly become less and less frequent as the infection is increasingly brought under control. On the other hand, even when the deleterious genes are in the recessive state, their carriers are often slightly impaired as regards viability and fertility. It follows that such carrier states may be of greater importance than recognized heretofore in defining the characteristics and susceptibilities of a population, and therefore its response to environmental stresses.

Balanced polymorphism is therefore a costly method of adaptation, since it entails the constant production of some relatively unfit individuals. But it is beneficial in the long run because genetic systems based on it have great adaptational plasticity and possess characteristics that correspond to a kind of genetic homeostasis.

d. Responses to social stimuli

The responses man makes to environmental factors are profoundly conditioned, as already mentioned, by the symbolic interpretation he puts on the stimuli that impinge on him. The understanding of the role played by the cerebral cortex in the body responses progressively emerged from the writings of the Russian school, first under the influence of the physiologist I. M. Setchenov (1829–1905). Even though he had worked in Claude Bernard's laboratory and had been greatly influenced by him, Setchenov was less impressed by the constancy of the *milieu intérieur* than by the organism's ability to modify its composition in order to adapt to changes in the external environment; he postulated that this adaptation was under the control of the nervous system. In Russia also, S. P. Botkin (1832–89) taught that most body processes, if not all, are subject to regulation by neural integrative mechanisms. Thus, Setchenov and Botkin created the intellectual atmosphere for Pavlov's studies of the influence that the cerebral cortex exerts on visceral functions and metabolism.

Neural reactions integrated at the highest level always modify profoundly man's response to the environment, but the ultimate effect can be either favorable or deleterious. On the one hand, neural reactions stimulate important protective and adaptive devices of the body, for example by strengthening the inflammatory processes in the peripheral tissues. On the other hand, this protection is often bought at the cost of injury to a particular structure; in fact, reactions originating in the cerebral cortex can result in highly destructive processes.

It has been known ever since Pavlov's experiments with dogs that experimental neuroses can be produced by the manipulation of conditioned reflexes. However, body changes can also be elicited more directly by stimuli acting on the nervous system. For example, it is possible through electrodes implanted in the proper area of a rat's brain to induce the animal to stimulate itself as often as 8,000 times an hour, just for the pleasant sensation resulting from the stimulation. Such stimulation can have a great variety of physiological effects, and can even affect the rate of healing of experimental wounds.

In an even more dramatic experiment, young goats were placed in isolation shortly after birth and subjected for one hour to repeated sound signals at two-minute intervals. The consequence of these abnormal stimuli was that two out of ten goats died within a week with hemorrhagic adrenal cortices; seven others died in less than a year, leaving only one survivor. As mentioned earlier, experiments in rats have shown that the proper kind of stimulation of the pregnant female can exert lasting effects on the young in utero (Chapter I.2).

The preceding examples illustrate that environmental stimuli operating through the higher neural centers can affect physiological functions and even anatomical structures. In certain cases, the results are favorable for the organism; in others they express themselves in grave pathological states. Most of the emphasis in the study of the physiological mechanisms involved has been on the role played by the endocrine system. It is certain, however, that the highest neural centers can affect bodily processes through a variety of other mechanisms, which are not yet understood, as illustrated, for example, by hypnotic phenomena.

The medical applications of hypnosis are now well authenticated. They range from the relief of pain to easier child birth, and perhaps even, according to certain reports, to the treatment of warts. Tissue vulnerability can also be modified by hypnotic suggestion. Thus persons in deep hynosis were made to immerse one arm, then the other, in a tank of water at 24° C, and were told that one of the arms was burned and painful, whereas the other was intact and comfortable. When stimulation was subsequently induced in both arms by thermal radiation or by administration of trichloracetic acid, the skin damage was more serious on the arm in which injury had been suggested. Measurements of pain threshold and of finger pulse amplitude indicated local vasodilatation during the phase of increased vulnerability, thus pointing to a mechanism which might account for the alteration of tissue response induced by hypnosis.

The spectacular effects of placebos provide further proof of the influence exerted by suggestion and the cerebral cortex on body mechanisms. The word placebo (meaning etymologically "I shall please") was used in the fifteenth century as a synonym for flattery and a flatterer, then later for a courtesy designed to soothe or gratify; hence its present use to designate an influence which mimics that of a drug or procedure for the relief of pain or treatment of disease. The ability of placebos to provide relief has been established beyond doubt in a great variety of conditions, such as postoperative wound pain, angina pectoris, headache, seasickness, cough, and anxiety. Recent findings prove that even sham surgery can have a placebo effect. For example, it was believed for a while that ligation of the internal mammary arteries provided complete relief of angina pectoris, whereas in reality mere exposure of these arteries without ligation was just as effective.

The placebo effects are not limited to subjective changes; they include objective manifestations that can be measured in terms of various functions of the pulmonary, gastrointestinal, urogenital, and adrenocortical systems. Furthermore, it is a remarkable fact that the percentage of success is the same whatever use placebos are put to; approximately 35 per cent of all treated cases exhibit a placebo effect. Clearly then, some personality characteristics and habits of mind predispose to placebo response. These characteristics, however, probably exist to some degree in all human beings, perhaps to a greater degree among those with a stronger neurotic cast. In general, the greater the stress, the greater the effectiveness of placebos. As is the case with morphine and many other drugs, placebos are much more effective in relieving pain of pathological origin than experimentally induced pain, a fact which confirms that their operation requires the proper mental state.

The placebo effects are so many and varied that the meaning of the word must be enlarged to include not only those which are favorable but also those which are unpleasant and even dangerous. In any case, the widespread acceptance of the placebo concept in present-day scientific language constitutes a tacit recognition of the fact that man has an infinite capacity for self-deception, and more importantly perhaps that his mind is often as effective as the most powerful drug or stimulus in mobilizing or influencing organic processes, for good or for bad.

The voluntary modifications of pulse rate and blood pressure by the yogis, the well-documented cases of death by suggestion, and the unquestionable effects of the will to live or its converse constitute further examples of the complex interrelationships between the mind and body functions. The well-documented occurrence of voodoo deaths in their various forms and among different groups of people constitutes one of the most spectacular illustrations of the fact that the wisdom of the body is no match for the power of the mind. Deaths caused by the "bone-pointing" curse among Australian aborigines have been described by several authors. A bone or stick is pointed at the person to be killed, and the medicine man goes through certain routines accompanied by a curse repeated many times. According to the accounts of anthropologists, there is no question that in many cases the cursed person will die. Death does not necessarily happen immediately, but the will to live completely disappears in the cursed person, and the end comes in due course. A striking account of such voodoo deaths among the Arunta tribe is given in Baldwin Spencer's *Wanderings in Wild Australia* (1894).

Ignorance of the mechanisms through which the mind acts on the body does not decrease the importance of the role that psychic phenomena play in the responses of man to his total environment.

The neural reactions integrated at the highest level do condition most of the processes that relate man to the physicochemical world, and it is for this reason that the response to a given stimulus can be either protective or destructive, depending upon the person affected and the circumstances. For the purpose of discussion, experimentation, and therapy, it is often convenient to consider the purely physicochemical and physiological aspects of the response apart from those which are psychically determined. However, both aspects always coexist and in many cases they can hardly be differentiated one from the other. It is therefore unjustified conceptually to divide illnesses according to their presumed physical, psychosomatic, or psychic origin. A very large percentage of illnesses are the expressions of inadequate responses to the environment; they include in varying proportions effects that are directly caused by physicochemical forces and effects that are primarily determined by the highly individual interpretation the person puts on the events of the external world.

The complexity of the conditioned responses to external forces accounts for the importance of the physician-patient relationship in determining the therapeutic results of medical action. There is still much truth in Thoreau's remark, in *A Week on the Concord and Merrimack Rivers:* "In the most civilized countries the priest is still but a Powwow, and the physician is a Great Medicine." Thoreau would not have been surprised by the present emphasis on the fact that the physician's personality and attitude does influence profoundly the therapeutic effectiveness of any drug. He might not have expected, however, that this is true also for surgical procedures. When posterior gastroenterostomy was advocated for the treatment of peptic ulcer, the patients operated on by surgeons who were skeptical as to the validity of the method developed 20 times as many marginal ulcers as did those whose surgeons were believers in the effectiveness of the technique.

The highly individual peculiarities of human response to any situation are as yet unexplained, but the complexity of the problem does not justify the widespread tendency to ignore it or to place it in a mystical domain beyond the range of scientific study. As a working hypothesis, it might be useful to assume that one or a few common organic pathways account for healing agencies as different as psychoanalysis, ataractic drugs, Zen practices, yogi meditations, witchcraft, a good vacation, and of course the proper physician-patient relationship.

It is certain in any case that many environmental stimuli are dangerous for health only because of the symbolic meaning the person attaches to them. In ordinary life, as was pointed out by Sherrington, the best manner of response to stimuli and threats is motor activity—so to speak, acting it out; the next best response is symbolic motor activity, verbalization either in the form of language communication or psychotherapy; the least successful is internalized motor activity, fantasy work. To a large extent, these modes of response are conditioned by the social milieu. The ability of man to adapt to his total environment is thus markedly influenced by the manner in which he relates emotionally to his fellow men.

2. The Adaptive Potentialities of Man

a. Adaptedness vs. adaptability

Throughout prehistory and history, *Homo sapiens* and his societies have utilized many different genotypic, phenotypic, psychic, and social mechanisms in order to adapt to new environmental situations. This biological and social versatility accounts for the spectacular and continued success of the human species. Now that man can alter his physical environment so profoundly and modify it so rapidly to his ends, there is a tendency to believe that the biological mechanisms on which he has depended for adaptation in the past have become of negligible importance. It is commonly stated that the human species can without danger afford to lose the physical and mental qualities that used to be essential for its survival, because it can create an environment in which these attributes are no longer needed.

It is true indeed that Western man is able through his technology to shut out many of the environmental insults he could not avoid until a few decades ago and to which he had therefore to react adaptively. Air conditioning allows him to work in an office, at the North Pole or in Timbuktu, dressed as if he were in a South Seas paradise; he is almost unconcerned with seasonal shortages of food; he need never experience actual hunger and thirst and may have the diet of his choice anywhere in the world at any time of the year. He can in theory make his environment so free of pathogens and other microbes that his immune mechanisms come less and less into play. He can control illumination, both quantitatively and qualitatively, filter out some of the solar radiations in summer, and expose himself to the southern sun or to artificial radiation in winter. He can withdraw from noise in a cork-lined room, yet maintain the desired degree of auditory stimulus with background music if he so wishes. He is learning to control his biological and emotional responses by the use of drugs that selectively inhibit or stimulate various biochemical and physiological processes. In brief, modern man can almost orchestrate at will the nature and intensity of the stimuli he receives from the external world, and he can exercise some measure of control over his responses to them.

Because he can manipulate so many aspects of his environment, and also govern to some extent the operations of his body and his mind, modern man has entered a phase of his evolution in which many of his ancient biological attributes are no longer called into play and may therefore atrophy through disuse. As a whole, the human race has lost much of the physical strength it had in the days of the caveman, when survival required the ability to fight savage beasts with clubs, or even with bare fists.

For all his strength and primitive resourcefulness, on the other hand, the caveman would not get along well in a modern city. The present environment demands a different kind of man. Natural selection cannot possibly maintain the state of adaptiveness to an environment that no longer exists, any more than it can adapt human populations to environments that have not yet been created. Yet because of technological advances, such new environments will continue to appear at an accelerated rate. In order to keep pace with them, human evolution will therefore have to depend even more than in the past on cultural and social evolution.

Since man is able to eliminate or avoid many of the struggles and stresses which used to be his fate, it seems to follow that his biological mechanisms of adaptation have become useless, or at least obsolete. Paradoxically, however, the very avoidance of stresses may in itself constitute a new kind of threat to health if it is carried too far, because the body and the mind are geared for responding to challenges; they lose many of their essential qualities in an environment that is so bland as to make life effortless. As we have seen, man retains in his biological and emotional make-up the structures, functions, and needs that emerged during his evolutionary past, and it is certain that some form of challenge continues to be necessary for his normal development and performance. For example, the personality structure breaks down as a result of sensory deprivation (Chapter I.3), and the intestinal tract fails to develop normally in animals raised under germ-free conditions (Chapter V.3). The complete absence of challenge can thus be as deleterious as excessive intensity of environmental stimuli.

Human history shows, furthermore, that the same kind of knowledge that permits man to alter his environment for the purpose of minimizing effort, achieving comfort, and avoiding exposure to stress also gives him the power to change his environment and ways of life in a manner that often entails unpredictable dangers. The ability to adapt to the unforeseeable threats of the future therefore remains an indispensable condition of survival and biological success. Even if it were possible for man to achieve a perfect state of adaptedness to the environmental conditions which prevail now, such a state would not be suited to the conditions of the future. The state of adaptedness to the world of today may be incompatible with survival in the world of tomorrow.

Paleontological studies have established that narrow adaptation of a species is a frequent prelude to its extinction because it jeopardizes its ability to adapt itself to changing conditions. Similarly, it is dangerous to assume that man need be adapted only to the conditions of the present, created by today's form of urban living and industrial technology. This state of adaptedness gives a false sense of security because it does not have a lasting value and does not prepare for the future. It must be supplemented by the attribute of adaptability.

Balanced polymorphism provides a basis for a kind of adaptability based on genetic mechanisms (Chapter X.1). But adaptability involves, in addition, many phenotypic processes which have not been defined, let alone studied experimentally. The most that can be done at the present time perhaps is to approach the evaluation of adaptability through measurements of the tolerance or reserve of body processes for various inputs, loads, and resistances.

The various tolerance tests used in clinical medicine make it possible to acquire some quantitative knowledge of the range of activities of which a person is potentially capable. Mild physiological defects that otherwise remain unnoticed can be revealed by the application of increased loads or resistances on body processes. For example, administration of glucose, or evaluation of effect of cortisone on its utilization, can reveal levels of diabetes so low as to be inapparent under basal conditions. Similarly, an increased venous inflow load, or aortic outflow resistance, can bring out unsuspected hemodynamic evidence of cardiac failure. More generally, it can be said that when input load and opposing resistances exceed available capacity, the process fails to deal effectively with all of the input presented to it, and disease ensues. Conversely, when available capacity exceeds input load and resistances, the process exhibits a high degree of reserve or tolerance (reviewed in Frenster, 1962).

Tolerance tests, in their various forms, have been systematically developed so far only for a few physiological systems, and their use has been considered only with regard to the diagnosis of disease states. It would seem possible, however, that their extension could enlarge the knowledge of man's potentialities. In other words, judicious utilization of tolerance tests might make it possible to obtain some objective knowledge of what a particular person, or mankind in general, is capable of doing, overcoming, and learning, both physically and mentally. Such an approach might provide data for the evaluation of adaptability, i.e. of the potential ability for adaptation to new environmental circumstances and new challenges.

b. Adaptation to modern life

The social history of man leaves no doubt that he was in the past endowed with a wide range of adaptive potentialities and that he has retained most of this attribute at least until very recent times. Pollution of the air in the industrial cities of Northern Europe provides a striking illustration of the extent to which modern man can make some kind of successful adjustment to environmental conditions that appear at first sight extremely hostile to both his body and his mind.

Ever since the beginning of the nineteenth century, certain parts of Northern Europe have experienced a characteristic kind of smog determined by the cold humidity of the Atlantic weather and by the burning of large amounts of soft coal for domestic and industrial purposes (Chapter VIII.2 and 3). This smog, loaded with soot, sulfur oxides, and mineral acids, is extremely unpleasant to breathe and creates an atmosphere of gloom that is almost unbearable for those who are not accustomed to it. But the inhabitants of the industrial areas of Northern Europe have come to accept their dark, polluted skies as a matter of course. They grumble, of course, against soot, grit, and irritating fumes, but they manage nevertheless to create a stimulating and satisfying human atmosphere out of their dreary surroundings—witness the following description, written in 1963, of Huddersfield, the Yorkshire town where the technical revolution began two centuries ago:

> The long huddle of the textile mills, creeping away to the moor's edge; the marching file of tall brick chimneys, their vapours drifting into the dusk; the coveys of cramped terrace houses, jammed hugger mugger against the hillsides; the dingy red brick everywhere, the patina of dirt, labor and middle age. . . . Here in these grim moorlands of Northern England, the technical revolution began. . . .
>
> She is now a clubable town, rich in choral societies and brass bands—she has evolved a society that is, as human institutions go, strong, decent and kindly. . . .
>
> So she looks, two centuries after the event, and this is a comfort. In Cairo or Kiruna or the Indian steel towns—they can draw upon her for encouragement. The cruelty, she seems to say, need not be permanent. The degradation will not last [Morris, 1963].

A recent social survey of the Lancashire city of Leigh, near Manchester (England), provides a similar picture of human adaptation to unhealthy and dreary surroundings. The mill dominates the town, the air is gritty, obscured by smog, but human life is not defeated by the murky environment. The sociologists found in Leigh a vigorous, thriving culture, and a deep involvement in community affairs. Everywhere in Northern Europe the people of the industrialized areas have indeed managed to create a way of life that is spirited and emotionally warm despite soot, grit, and gloomy skies. They have remained highly active and productive physically, biologically, and intellectually. Their expectancy of life is not very different from that of people of the same stock and economic status living in uncontaminated areas, nor is their birth rate smaller. Two hundred years of history thus demonstrate that human beings can become adjusted to contaminated and darkened atmospheres. The raping of nature by technology does not necessarily make the environment incompatible with human life.

Much the same story can be told of other forms of threats created by technological innovations. There is little doubt concerning the toxicity of most of the pesticides, food additives, and other synthetic chemicals that contaminate air, food, water, and many of the objects with which the people of Western civilization now come into daily contact. But paradoxically, it is also true that the countries in which these potentially dangerous synthetic chemicals are most widely used are also the ones in which life expectancy is now the longest. We shall see later that there is another side to this picture. But for the time being, and in the Western world at least, mankind seems to take environmental pollution in stride.

The dangers posed by the agitation and tensions of modern life constitute another topic for which public fears are not based on valid evidence. Most city dwellers seem to fare well enough under these tensions; their mental health is on the whole as good as that of country people. Indeed, there is no proof whatever that mental diseases are more common or more serious among them now than they were in the past, or than they are among primitive people. They are not even different in kind.

The experience of urbanized and industrialized societies bears witness therefore to the fact that everywhere man can make some sort of adjustment to crowding, to environmental pollution, to emotional tensions, and certainly to many other kinds of organic and emotional stresses. It is not difficult to imagine how the people of the pre-industrial era would have reacted if they had been suddenly transported to our world of automobile traffic, polluted air, shrill noises, blinding artificial lights, and intense competition. All would have suffered, physically and mentally; many would have died; only very few would have been capable of functioning effectively in our "unnatural" environment. Yet, people of all origins, races, and colors, have now become adjusted to the artificial world created by modern industrial and urban civilization. In fact, city dwellers would find it difficult to return to the ways of life of not so long ago. The mechanized world of the modern city, which would have appeared so unnatural to our ancestors, has now become the natural environment, outside of which Western man can hardly survive.

c. The price of adaptation

If the ability to survive, multiply, and function were the only criteria by which to evaluate the dangers posed by modern technology, or by demographic and economic growth, the future of mankind would provide little cause for worry at the present time. On the one hand, there still exist in human nature large untapped reservoirs of biological adaptive potentialities, genotypic as well as phenotypic. In addition, there is no foreseeable limit to the variety and extent of the social adaptative mechanisms that man can bring to bear on the external world, in order to modify it according to his needs and wishes. Short of nuclear warfare, it would seem therefore that mankind will be able to take the stresses of the future in stride, just as it has survived destructive famines and epidemics in the past. One can almost take it for granted that mankind will adapt itself to the new ways of life created by the second industrial revolution, and even to the crowding, shortages, and other ordeals likely to result from the much dreaded increase in the world population.

Recent history confirms indeed that modern man can still make adjustments to an astonishingly wide range of threatening situations, even to some that appear at first sight almost incompatible with life. During the last war, many different kinds of persons managed to survive and to function, under circumstances so appalling as to be almost beyond the range of human imagination and belief—either the exhausting and frightening experience of combat, or starvation and torture in concentration camps.

However, the potential ability of mankind to survive crowding, emotional misery, environmental pollution, shortages of natural resources, and other kinds of threats constitutes but a limited aspect of the problem of adaptation. Other aspects have their origin in the fact that human life involves values which have little to do with biological needs and which transcend the survival of individual persons. All too often, the biological and social changes that enable mankind to overcome the threats posed by the modern world must be eventually paid for at a cruel price in terms of human values.

Sudden and profound changes in the ways of life, whatever their nature, always bring about a decrease in the resistance of the body and the mind to almost any kind of insult. But the social and medical problems associated with this early phase of change tend to be transitory. More important in the long run, even though less dramatic, are the distant consequences of some of the adaptive processes that make it possible for man to survive the deleterious consequences of the rapid changes brought about by technological innovations and by social revolutions.

One of these consequences is the accumulation of hereditary defects brought about by medical and technological procedures that allow many persons to survive and reproduce who would have left no progeny in the past. While it is likely that mutation rates are increasing, the modern ways of life are interfering with the elimination of undesirable genes; natural selection is more and more embarrassed by social and medical practices. True enough, the effects of genetic degeneration are so slow that they are usually overlooked or are accepted as a part of the natural order; but the most destructive operations of the living world are precisely those of a creeping, secular character.

It is misleading, of course, to speak of biological defectives without regard to the kind of environment in which man lives and functions. Medical techniques can make up for many forms of genetic and other deficiencies that would be lethal in the wilderness. While it is certain that the physically handicapped could not long survive under "natural" conditions, it is also true that medical and other social skills make it possible for men to live long and function effectively in the modern world even though they be tuberculous, diabetic, blind, crippled, or psychopathic. Fitness is not an absolute characteristic, but rather must be defined in terms of the total environment in which the individual has to spend his life.

On the other hand, it is often overlooked that fitness achieved through constant medical care has grave social and economic implications that will become manifest only in the future. We can expect that the cost of medical care will continue to soar, because each new discovery calls into use more specialized skills and more expensive equipment and products. There is certainly a limit to the percentage of society's resources that can be devoted to the prevention and treatment of disease (Chapters XV and XVI). Furthermore, it must not be taken for granted that the power of science is limitless. Only during the past few decades and in but a few situations has medical treatment enabled the victims of genetic disabilities to survive and to reproduce on a large scale. If the numbers of biologically defective individuals continue to increase, therapy may not be able to keep pace with the new problems that will inevitably arise.

The possibility of genetic deterioration is not the only threat that arises from the biotechnological advances through which mankind becomes adapted to the modern world. Many of the processes thereby set in motion have phenotypic effects that remain unnoticed at first, but eventually play havoc with the human values of life. Environmental pollution illustrates the fact that many of the adjustments that facilitate life in a hostile environment commonly express themselves at a later date in disease and human misery. As mentioned earlier, the inhabitants of the industrial areas of Northern Europe behave as if they had made a successful adjustment to massive air pollution (Chapter X.1). But while they function effectively despite the presence of irritating substances in the atmosphere they breathe, the linings of their respiratory tracts register the insult. Each exposure leaves its marks, and after several years the cumulative effects of the irritation result in irreversible pulmonary disease (Chapter VIII).

Chronic bronchitis in the industrial areas of Northern Europe provides a model for the kind of medical problems likely to arise in the future from the various forms of environmental pollution in our own communities. In the great majority of cases, control over chemical pollution of air, water, and food is sufficiently strict to prevent obvious toxic effects. But while neither the acute toxicity nor the nuisance value of environmental pollution is great enough to interfere seriously with social and economic life, a vastly more important danger comes from the likelihood that repeated exposure to low levels of toxic and irritating agents will eventually result in a great variety of delayed pathological manifestations (Chapter VIII.3).

As in the case of environmental pollution, the possibility exists that the apparently successful adjustments to the emotional stresses caused by competitive behavior and crowding can result in delayed organic and mental disease. Through the experience of social intercourse, man learns to control the outward manifestations of his emotional responses. He usually manages to conceal his impatience, irritations, and hostile feelings behind a mask of civil behavior. But inwardly he still responds to emotional stimuli by means of physiological mechanisms inherited from his Paleolithic ancestry and even from his animal past. The ancient fight and flight response still operates in him, calling into play the autonomic nervous system and various hormonal mechanisms that generate useless and potentially dangerous physiological reactions. It is probable, as we have seen, that these misguided responses leave scars which eventually threaten the body and the mind as they accumulate through the years.

The fight and flight response is only one of the many traits that man has retained from his evolutionary past. His bodily mechanisms are still linked to the daily rotation of the earth around its axis and to its annual rotation around the sun, as well as to other cosmic events. The natural rhythms that are built into his body fabric thus often conflict with those which govern his social life (Chapter II.2). Man is so adaptable, true enough, that he can carry his day into the night and move at supersonic speed from one latitude to another, unconcerned with the instructions of biological rhythms. But while he can usually make the necessary physiological adjustments, these always involve severe hormonal disturbances. It is not unlikely that the ultimate outcome of these disturbances will become more serious, as life becomes more mechanized and more acutely dissociated from the natural cycles. In fact, as we have seen, evidence of pathological effects has been obtained in experimental animals maintained for prolonged periods under physical conditions differing from those of their natural rhythms (Chapter II).

A last example will suffice to illustrate the wide range of threats to health arising from the adjustments that man makes to the new ways of life. Immense strides have been made toward the control of the acute infectious diseases which used to be responsible for so many deaths. However, many other kinds of diseases are still caused by microbes that are ubiquitous and become active when the resistance of the body is lowered (Chapter VII.5). Yet, physicians, as well as laymen, have come to accept as a matter of course a situation in which many days every year are sacrificed to so-called "minor" infections, despite the fact that these ailments have an importance that goes beyond their nuisance value. They erode the functional integrity of the organism, progressively damaging the respiratory, digestive, or urinary tracts, as well as the kidneys and perhaps also the blood vessels. They too probably play their part in the so-called diseases of civilization.

The aspect of the problem of adaptation that is probably the most disturbing is paradoxically the very fact that human beings are so adaptable. This very adaptability enables them to become adjusted to conditions and habits which will eventually destroy the values most characteristic of human life.

Millions upon millions of human beings are so well adjusted to the urban and industrial environment that they no longer mind the stench of automobile exhausts, or the ugliness generated by the urban sprawl; they regard it as normal to be trapped in automobile traffic, to spend much of a sunny afternoon on concrete highways among the dreariness of anonymous and amorphous streams of motor cars. Life in the modern city has become a symbol of the fact that man can become adapted to starless skies, treeless avenues, shapeless buildings, tasteless bread, joyless celebrations, spiritless pleasures—to a life without reverence for the past, love for the present, or hope for the future.

Man is so adaptable that he could survive and multiply in underground shelters, even though his regimented subterranean existence left him unaware of the robin's song in the spring, the whirl of dead leaves in the fall, and the moods of the wind—even though indeed all his ethical and esthetic values should wither. It is disheartening to learn that today in the United States schools are being built underground, with the justification that the rooms are easier to clean and the children's attention not distracted by the outdoors!

The frightful threat posed by adaptability when the concept is applied to human beings in a purely biological context is that it implies so often a passive acceptance of conditions which really are not desirable for mankind. The lowest common denominators of existence tend to become the accepted criteria, merely for the sake of a gray and anonymous peace or tranquility. The ideal environment tends to become one in which man is physically comfortable, but progressively forgets the values that constitute the unique qualities of human life.

The biological view of adaptation is inadequate for human life because neither survival of the body, nor of the species, nor fitness to the conditions of the present, suffice to encompass the richness of man's nature. The uniqueness of man comes from the fact that he does not live only in the present; he still carries the past in his body and in his mind, and he is concerned with the future. To be really relevant to the human condition, the concept of adaptability must incorporate not only the needs of the present, but also the limitations imposed by the past, and the anticipations of the future. Above and beyond all, man is still of the earth, earthy, notwithstanding all the technological and medical advances that superficially seem to dissociate him from his evolutionary past. As happened to Anteus of the Greek legend, his strength will probably wane if he loses contact with the biological ground from which he emerged and which still feeds him, physically and emotionally.

XI. The Population Avalanche

1. The continuous growth of the human population

a. Facts and assumptions concerning population growth

The most compelling evidence of the ability of the human race to become adapted to a wide range of environmental conditions is that mankind has progressively spread over the entire globe. The most obvious and immediate danger arising from this adaptability is that the world population is now growing so fast that it might soon reach a level incompatible, first, with the maintenance of the human qualities of life, and then with the very survival of the human species. Thus, the world population problem illustrates the indirect long-range dangers of adaptability (Chapter X.2).

It is not for the sake of originality that I shall use here the word "avalanche" instead of the more orthodox "explosion" to refer to the present increase in the world population. My reason is that the word avalanche conveys more accurately the important truth that this increase is not a sudden event, as the word explosion would suggest, but rather corresponds to a continuous process that has to reach a certain momentum before it becomes dangerous. Public alarm is justifiable because the increase is now occurring simultaneously all over the world, and because its rate seems to be accelerating.

The total human population was probably of the order of 10 million at the end of the Neolithic period; it had increased to approximately 300 million by the beginning of the Christian era, and had reached 500 million around 1650. Then the rate of growth began to accelerate, bringing the world population to 2,500 million in 1950 and to over 3,000 million in 1961. As far as can be judged, the yearly rate of increase, which was about 0.3 per cent between 1650 and 1780 and 0.9 per cent between 1900 and 1950, is now 1.8 per cent. At present the world population is increasing by more than 100,000 persons a day, close to 50 million a year! It would reach about 4 billion by 1975 and 6 to 7 billion by 2000 if the present rate of growth were to continue. In brief, it took hundreds of thousands of years for *Homo sapiens* to achieve a population of 3 billion, but his numbers might be doubled in the next 40 years!

The title of a recent book on the population problem, *Standing Room Only* (Sax, 1955), expresses vividly the obvious fact that growth cannot continue very much longer at an accelerated rate. The population will have to stop growing within the foreseeable future, simply for lack of space on earth, even assuming that science could find satisfactory substitutes to overcome the shortages of natural resources. The upper limit of the world population has been set at between 10 billion, a level that might be reached within a century, and 50 billion, according to estimates made by experts who have great faith in the power of technology to provide new sources of subsistence for man.

The really important question, however, is not so much the maximum population level that can be reached and maintained as the manner in which the rate of increase can be slowed down and ideally brought to zero. Will the break come from the operation of natural growth-limiting factors, such as those which maintain so many animal populations at stable levels in equilibrium with their environment? (Chapters IV.2 and XI.2); from a biological catastrophe similar in kind to the population crashes that occur repeatedly among lemmings and other rodents? (Chapters IV.2 and XI.2); from a technological accident, such as nuclear warfare or wholesale poisoning by environmental pollution? (Chapter VIII); or from concerted human action based on a reasonable assessment of what the upper population level *should* be in order that human life retain the qualities that give man his unique place in creation?

As is well known, several programs are presently under way in many parts of the world to bring the population avalanche under some form of rational control. But whereas there is general agreement on the dangers of unchecked population growth, there is great diversity of opinion on many other aspects of the problem. Surprisingly enough, however, both the lay public and many students of demographic problems have accepted as a matter of course a number of assumptions for which there is in reality no factual evidence.

It is usually taken for granted that (a) the most immediate and gravest consequence of continued population increase will be shortages of food; (b) that recent advances in medicine and public health have been responsible for the accelerated rate of population growth in our times; (c) that the one effective approach to population control consists in the development of new, convenient, and inexpensive contraceptive techniques, and in the education of the public in their use. All these statements are based on unproven assumptions and are certainly erroneous in part at least. It is therefore likely that control programs based on them will meet with failure. An attempt will be made in the following pages to identify the biological as well as technological and social forces that play a role in determining population size, both in animals and in man.

b. Food supplies, medical progress, and population growth

After being so long discredited, Reverend Malthus' famous essay is once more regaining prominence. True enough, Malthus' dire predictions concerning the hopelessness of the race between means of subsistence and population growth have been shown wrong by the events of the past century in the Western world, but their fundamental premise is nevertheless inevitably correct. Food production has so far grown faster than population size in the countries of Western civilization, and there is evidence that this trend may hold today, although to a smaller degree, even in the underprivileged parts of the world. But it is obvious that a critical state will be reached *eventually* unless the population stops growing. Malthus was not a poor prophet; he was an unlucky one. The immediate misfortunes that he foretold failed to materialize, not so much because his biological reasoning was wrong as because social events that he could not have foreseen occurred shortly after his time.

Throughout the nineteenth century, the Industrial Revolution in Europe created new sources of wealth and especially of power, thus making possible the rapid exploitation of many unsuspected resources. The lands of the New World, which were then almost empty, were rapidly brought under cultivation; railways, steamships, and other means of transportation facilitated the transfer of the enormous amounts of food and raw materials produced in the virgin lands of America, Asia, and Africa to the growing countries of the Old World. Thus, the period immediately after the writing of Malthus' essay was characterized by a new social and economic situation, which he had not foreseen. However, this situation cannot possibly recur. There are few empty habitable lands left in the world and the supply of natural resources is necessarily limited. More important perhaps, while science can invent new sources of power it cannot create new space. Unless population growth soon levels off, Malthus' predictions will come to pass, even in the wealthy countries of the Western world.

Granted the general validity of these statements, the complexity of social factors makes it difficult to formulate an optimum population level on the basis of natural resources. For a hunting and gathering economy, the maximum population density is one person per square mile. For a purely agricultural economy it can exceed 1,000 per square mile when all available land is exploited, as it is in Java or Egypt; but then the standard of living is low. For an industrial economy, the figure falls to a lower level, a few hundred or less, because so much of the land is used for purposes other than food production, especially for the manufacture and transportation of commodities identified with a high standard of living. At present, the population density is around 50 for the United States, as well as for the state of Minnesota in particular. This happens to be the density for the world as a whole.

Japan, China, the Philippines, India, Ceylon, Puerto Rico, etc. have only from 0.2 to 0.5 acres of arable land per capita and are considered much overpopulated. In the United States, by contrast, there are 3 acres per capita, and in Canada, 6 acres. These differences suggest a clear relation between population density and prosperity. However, their significance for the problem of overpopulation is questionable in view of the fact that Holland, Belgium, Switzerland, the United Kingdom, and Western Germany are just as land poor as are the countries of Southeast Asia; like the latter, these European countries possess only 0.2 to 0.5 acres of arable land per capita and should therefore be considered overpopulated, yet they are prosperous. Clearly then, population density per se is not the dominant factor in determining the level of prosperity and civilization that a country can achieve and maintain. In terms of average population density, Europe is more crowded than Asia (85 persons per square kilometer as against 60).

While it is unlikely that food shortages will constitute the most alarming consequences of population increase in the future, it must be acknowledged that population changes have been often linked to the food supply in the past. According to the Chinese census, the population of China, which was under 64 million in 1578, climbed to 108 million by 1661 and to 144 million by 1741. This extremely rapid growth seems to have resulted from the introduction into China of three kinds of crops easily grown and giving very large yields; corn around 1550, the sweet potato around 1590, and peanuts a little later.

Interestingly enough, the most spectacular population spurt in Europe occurred at about the same time as in China, and seems to have been due in part to the introduction of the white potato from the Andes. Following this event, the population of Ireland increased from 3,200,000 in 1754 to 8,175,000 in 1846, not counting some 1,750,000 who emigrated during this period, and despite great poverty. A similar situation, on a smaller scale, was created by the introduction of the bean among the Pueblo Indians in the Rio Grande valley.

Thus, increased availability of food has often brought a rapid increase in population. The change resulted in some cases from the generalized acceptance of new crops, such as corn, potatoes, or beans; in other cases from the importation of animal and plant foodstuffs produced elsewhere, as occurred in Europe shortly after the beginning of the Industrial Revolution. The population of England and Wales trebled between 1700 and 1850, even though mortality rates did not decrease significantly if at all. This phenomenon was certainly in large part an indirect result of industrialization; export of industrial goods financed the import of food from other lands.

The effects of food shortages on human populations are more varied and more complex. Hunger certainly reduces the desire or the ability to reproduce. Prisoners starved in concentration camps have reported that one of the earliest effects of undernutrition was a loss of sexual desire. The acute famine that prevailed in the Netherlands during the late phase of the Second World War was followed one year later by a marked decline in the number of births. These examples, however, correspond to extreme situations, in which people accustomed to a high level of nutrition and comfort were suddenly exposed to acute physiological misery. In contrast, food shortages do not, unfortunately, result in decreased birth rates in the underprivileged parts of the world. The poorest families are extremely ill fed, yet they commonly have the largest numbers of children.

The overall birth rate in South Asia and Latin America has not been reduced by either a chronic state of malnutrition or by the periodic famines of the past centuries. At least two very different factors can be invoked to account for the paradoxical fact that the populations least favored with regard to food supply are the ones with the highest birth rates. One factor is that the human body can achieve some sort of physiological adaptation to low food intake (Chapter III.4). It is not justified, therefore, to equate the physiological state of normally overfed Dutchmen suddenly placed on a famine regimen at the end of the war with that of Asiatics or Latin Americans for whom malnutrition is such a common and constant state that survival implies the ability to make physiological and emotional adjustments to food shortages.

The other factor, at least as important, is that human behavior is governed by mental states that usually take precedence over physiological and biochemical mechanisms. Despite lack of comfort and even of food and irrespective of sexual appetite, the desire for children is one of the most fundamental urges of human beings, perhaps even greater in the poor than in the prosperous countries. Under the social and economic circumstances that control human behavior in most of the underprivileged parts of the world today, and for that matter prevailed almost everywhere in the past, children constitute the easiest and most rewarding source of emotional satisfaction, as well as a kind of insurance against the future. Only in prosperous countries have other incentives weakened this fundamental human urge.

While malnutrition does not reduce birth rates in the underprivileged parts of the world, there is no doubt that it shortens the life span. Indeed, a very large percentage of malnourished children die in infancy, carried away by infectious diseases to which their susceptibility is increased by their poor nutritional state (Chapter VI). During the past two decades, however, the percentage of children who survive into adulthood has increased, thanks to modern techniques of public health and medicine. Insecticides, vaccines, antimicrobial drugs, nutritional supplementation, have all contributed to the lowering of death rates, especially in the early years of life.

From the point of view of population growth, the most important contribution of medical science is that several types of infectious diseases can be controlled in part, and infant mortality thereby much reduced, by techniques so simple and inexpensive that they could be applied almost anywhere. This does not mean, however, that modern medicine and public health have been the most important causes of the population avalanche. The world population would probably be just about as large as it is now, even if no medical or public health procedures had been brought into play.

It will be recalled that the growth of the European population began to accelerate markedly around 1750, long before modern medicine could have exerted any impact, indeed before the beginning of the sanitary revolution. In China also the population rose from approximately 150 million around 1750 to 300 million around 1850, and to 600 million around 1950. This spectacular increase, corresponding to a geometrical progression, occurred despite recurrent periods of famine, pestilence, war, and social disturbance. Yet there was so sanitary revolution or modern medicine in China during these two centuries!

All over the world and throughout historical times, the long-range population trends have always proved to be independent of epidemics, wars, famines, and other catastrophes. The widespread and fantastically destructive epidemics of plague that ravaged Western Europe during the Justinian era, and again during the Middle Ages, did of course sharply reduce population size for a while, but this effect was soon obliterated. The influence of the notorious London plague in 1665 was no longer perceptible 15 or 20 years later. In fact, the periods that followed the epidemics of plague from the fourteenth to the seventeenth centuries were among the most vigorous in European history! The famines and epidemics that have periodically laid waste in China, India, and Egypt during historical times have not made these countries less densely populated. The four years of the First World War, and the pandemic of influenza that followed it, caused at least 20 million deaths, but it took only a few months to make up this number again! Interestingly enough, concern with overpopulation became acute immediately after the end of the Second World War, before DDT and antimicrobial drugs could have exerted any significant impact on world health.

The paradoxical truth is that the phenomenal increase in world population during the past 50 years has coincided with great epidemics, two world wars, several minor ones, and deep disruptions of social and economic life everywhere. Furthermore, as is well known, the most destitute and disease-ridden populations of the world are precisely the ones that are increasing the fastest. This is particularly the case for many rural areas in which the state of health is hardly affected by physicians, drugs, or sanitation. In fact, the shape of the curves depicting the growth of the world population makes it clear that the acceleration far antedates the introduction of vaccines, insecticides, and drugs. In many countries of tropical Africa, for example, the mean increase of the population was about 1.5 per cent per annum for the period 1950–60, even though malaria eradication had then barely begun in these areas.

It is therefore unjustified and dangerous to oversimplify and falsify the world population problem by picturing the malariologists and other medical groups as irresponsible "sorcerer's apprentices" who have set in motion a force they do not know how to stop. Much more attention should be directed to the influence exerted on the growth of underpriviliged populations by the social and economic infrastructure derived from other Western influences. For example, more efficient transportation facilities make for better distribution of food and thereby increase population size, as illustrated by the role of the English-built railway system in India. Everywhere also better roads and more convenient means of transport facilitate rapid migrations of people from one region into another, and especially into urban centers, resulting in population increase where the economic opportunities are a little greater and the food supply a little less erratic. The modern technologist, rather than the physician and public health officer, has played the largest role in accelerating population growth.

It seems appropriate to quote here from a prescient paper written by Benjamin Franklin in 1751 and published in Boston in 1755, entitled *Observations Concerning the Increase of Mankind and the Peopling of Countries.* Here Franklin estimates that the population of the American Colonies doubles every 25 years and that within 100 years more Englishmen will be west of the Atlantic Ocean than east of it. He also made the sound but still not understood observation that emigration does not reduce the population of a country and that immigration does not increase it. In Franklin's words, "If you have room and subsistence enough, as you may, by dividing, make ten polypuses out of one, you may of one make ten nations, equally populous and powerful; or rather increase a nation tenfold in numbers and strength."

The only generalization that can be safely made at the present time is that population growth is the fastest in some of the areas that have the least adequate food supplies, the worst sanitary conditions, and especially the most deficient medical facilities. It is important to emphasize that modern medicine and public health have played but a limited part in the population avalanche because, as we shall see later (Chapter XI.3 and 4), health improvement is in fact one of the essential conditions of birth control.

c. Population size vs. population density

Even in areas where the *recent* phase of population growth has been accelerated by medical advances, there is no proof that the economic and social difficulties posed by overpopulation have been as greatly increased thereby as is commonly believed. The special case of Ceylon is worth considering in some detail because it is the one most often quoted as evidence that the population avalanche is a simple and direct consequence of disease control, in this case the virtual elimination of malaria through the widespread use of insecticides.

It should be emphasized at the outset that the lowering of death rates in Ceylon antedates the control of malaria by DDT spraying. The Ceylon death rate (all causes) has been falling almost continuously since 1905, the downward trend having been interrupted only twice, once in 1935 after a disastrous drought, and once again during the latter part of the Second World War. Admittedly, the rate of change became more striking as a result of the systematic DDT campaign. Mosquito control was followed by a sharp reduction in the incidence not only of malaria but also of infant diarrhea. As a result, the overall death rate dropped almost one half; and furthermore the birth rate, already high, became even higher. The population of Ceylon had risen from 2.4 million in 1871 to 8 million in 1950, corresponding to an average yearly increase rate of 1.7 per cent, but this rate jumped to 2.8 per cent after the DDT campaign.

While the figures mentioned above leave no doubt that insect control accelerated markedly the rate of population growth, the interpretation of this fact is not as simple as is usually assumed. The population figures that are quoted, such as those given in the preceding paragraph, refer to the island as a whole, and not to the restricted areas that were already heavily populated two decades ago. However, the antimosquito campaign did more than increase the population density in the areas that were already settled; it had the even more important effect of making once more habitable parts of the island that had been very productive and populated in the past but had been abandoned during the last century because of malaria. In other words, the DDT campaign reopened new lands into which the population could expand, much as drainage had done by facilitating human habitation in the malarious forests and lowlands of Europe during the Middle Ages.

In the light of what has happened in Ceylon and elsewhere, the population avalanche should be described in terms somewhat different from those commonly used. To a very large extent, it represents the acceleration of a process that has been going on ever since the emergence of man, namely the progressive filling up of all areas of the world by human populations. Throughout history, new agricultural and industrial processes have made it possible to produce food, process raw materials, and therefore establish civilization in areas previously considered out of bounds. The unpleasant fact that the whole globe will soon be completely inhabited is not so much a consequence as a cause of the population avalanche.

Drainage and then insecticides have rendered new regions available for man by ridding them of malaria and of other insect-transmitted diseases. Because water can be brought almost everywhere under sanitary conditions and because any remote region can be readily supplied from the outside with food and the raw materials required for specialized industries, deserts now provide acceptable and even pleasant conditions for work and for everyday life. The recent population increase in the American Southwest was made possible by the technological advances that made water available everywhere and thus permitted the transformation of desert lands into productive areas both for agriculture and for industry. Other technological innovations are now beginning to make the frozen North also habitable for man.

Thus it is grossly misleading to identify the increase in world population with increase in population density. Large congested cities have long been part of human civilizations; in fact, it is probable that congestion has existed ever since the beginning of recorded history. But it was a highly localized congestion, as illustrated by the life of the cliff dwellers and especially of the artisans and laborers in the medieval walled cities. The striking and really new phenomenon of our time is that the *totality* of the globe is rapidly becoming populated. Needless to say, this process cannot go on forever; indeed, it will not continue much longer since such a large percentage of the earth is already occupied. A biological disaster would therefore be inevitable if the population growth were to continue at the present rate. The çolonization of oceans and of outer space need not be considered here, since it is in the domain of science fiction rather than a practical possibility.

Despite the dire affirmations of the prophets of gloom, however, there are several reasons to be hopeful, the most important being that, progressively but surely, most human groups are becoming industrialized and urbanized. In the past, industrialization and urbanization have had a two-phased effect upon population growth —first accelerating it, then retarding it. As we shall see, the retarding influences are now beginning to operate in some of the overpopulated areas of the world where industrialization is in the process of changing the social order, just as happened in Europe a century ago. Before discussing the social and economic determinants of population control in human societies, however, it will be useful to consider some of the biological factors that affect the size of animal populations in nature.

2. The homeostatic regulation of animal populations

a. Fluctuations in animal populations

Man occupies a unique place in the animal kingdom with regard to changes in population size. His numbers have been increasing almost continuously since the time of his appearance on earth, whereas animal populations seem to reach approximately constant levels in a given environment. Some of them, true enough, fluctuate widely from season to season and from year to year. But on the whole these fluctuations are around an approximately constant value. When environmental conditions change, animal populations do also change in size, but then they stabilize once more around a new level.

The population crashes discussed in Chapter IV.2 constitute a dramatic aspect of the tremendous but transient changes that can occur in the population size of many plant and animal species under natural conditions. Thus, reindeer populations continue to build up even while their habitat is being impoverished by overgrazing; then suddenly they crash either from starvation or infection, or through the agency of the metabolic disorders discussed in

Chapter IV.2. About 40 years ago, a pair of Sika deer were put on a small island of about 150 acres in Chesapeake Bay, and the population that developed from them was kept well supplied with food. The colony grew until it reached a density of about one deer per acre. Then the animals began to die off, and this in spite of adequate food and care.

The records of the Hudson's Bay Company constitute a classical source of documentary information for those who believe that animal populations vary according to a more or less regular cycle. Over the long period of time they cover, these records provide evidence for large fluctuations in the numbers of fur animals and of the rodents on which they feed. Similar fluctuations have been observed in the course of many recent wildlife surveys all over the world. For example, the numbers of lemmings, voles, and hares in the far north are said to exhibit 4-year and 10-year cycles. Even more striking evidence of cycles has come from the harvesting of fox pelts in the Canadian arctic and subarctic zones, and from observations of breeding and large-scale dispersals of birds of prey. In many cases, however, the figures on which the claims of periodicity are based might correspond to random fluctuations rather than to true periodicities. Because of this fact, the doctrine of population cycles must still be considered *sub judice.*

Whether or not they occur according to a true periodicity, fluctuations in the size of plant and animal populations are probably determined in part at least by changes in the physical environment. Many efforts have been made to relate them to certain cosmic events, particularly to sunspots or to changes in radiation, but the evidence on this score is still very meager. In contrast, more precise information is available concerning the many varied effects climate exerts on the distribution, abundance, and even the characteristics of several animal species in the wild.

It can be taken for granted that climatic factors act as a selective force and thereby affect population structure through genetic mechanisms. The fact that animals living in cold climates are usually larger than those of related species living in warmer climates probably constitutes an example of this selective effect. Another illustration is provided by the history of rabbit populations in Australia. The European rabbit *Oryctolagus cuniculus* was introduced into Tasmania at the beginning of the nineteenth century and it rapidly spread over much of the island. Color variations with a genetic basis soon became established at different frequencies in different areas, the black rabbits reaching a frequency of 20 per cent in places of highest rainfall. In this case, the variation was genetic in nature and the mutants were probably selected by climatic factors. More generally, however, the biological responses to climatic changes occur so rapidly that they must depend on phenotypic rather than genetic mechanisms.

Much knowledge concerning the possible role of climatic factors on populations and their characters has come from the comparative study of the thickness of tree rings. Sections of very old trees collected in the Southwest of the North American continent show such great variations in the thickness of the rings that the rate of plant growth must have differed markedly from one period to another during the past 2,000 years. Variations in temperature and atmospheric precipitation were probably responsible for the differences in the growth rate of trees. As these climatic factors certainly influenced also agricultural crops, it can be assumed that indirectly at least they have played an important part in the life of man. They probably determined, for example, the location, size, and duration of the settlements of the Pueblo Indians in the American Southwest. In fact, it can be taken for granted that climatic changes have influenced the development, performance, and decay of other civilizations all over the world.

The mechanisms through which climatic factors cause population changes are at times rather simple. For example, plankton varies regularly in abundance and in composition from season to season and from year to year. In 1925, however, the warm equatorial countercurrent off Colombia and Ecuador (El Niño) shifted its course so strongly to the south (as it does approximately once every seven years) that the population of plankton, fish, and water birds normally found off the Peruvian coast fled or died and was replaced by warm-water species.

Similarly, when the current from the Atlantic predominates on the English coast it brings water rich in phosphate, which favors one species of glassworm on which the herring feeds. In contrast, current from the Channel brings water poor in phosphate, resulting in failure of the herring fisheries. The "red tide," which swept immense numbers of dead fish onto the Florida beaches in 1946 and again in 1952, as it does approximately once a decade, was caused by a microscopic flagellate always present in the waters off the Florida coast, but in numbers too small to be harmful; its population reaches toxic levels only when atmospheric circumstances bring about the local stagnation of low-salt brackish water in certain areas.

In most cases, however, the mechanisms of population fluctuations have not yet been unraveled. For example, wildlife surveys carried out in Iowa leave no doubt that the populations of ruffed grouse, snowshoe hare, and especially muskrat have fluctuated enormously and repeatedly over the past four decades. Yet there is no indication that population changes, disease states, or patterns of behavior of the animals can be traced to any of the ordinary meteorological factors such as heat, humidity, water levels, and other obvious variables customarily recorded by the Weather Bureau.

The greatest difficulty in explaining population fluctuations is that they occur so often through very indirect channels, involving complex food chains. For example, the lemmings and other microtine herbivorous mammals constitute the chief source of food for the various carnivorous predators living in the arctic. As already mentioned, the fluctuations in the numbers of these predators have provided some of the best evidence for the theory of population cycles. It has long been suspected, furthermore, that the size of herbivorous populations is dependent on qualitative changes that affect the nutritional value of the particular types of vegetation on which the animals feed.

In the case of Iowa muskrats there is evidence that their numbers are correlated with the intensity of ultraviolet radiation, but the changes are not explainable in terms of quantities of staple foods available to the animals. It is possible, therefore, that radiation acts indirectly through some effect on the qualitative characteristics of the food supply rather than on its amount. In any case, most population cycles depend upon the ultimate outcome of complex predator-prey interrelationships, in which the herbivores prey on the vegetation and the carnivores on the herbivores.

Increasingly, man introduces still further complexities in the interplay between the various determinants of population growth by interfering with natural ecological systems. Because so many leopards have been killed in Africa, there is now a plague of baboons, which destroy the crops in certain areas. When rabbits became scarce in England following introduction of the myxoma virus, the foxes and birds of prey began to raid poultry yards. Contrary to what might be believed, the destruction of predators is not an unmixed blessing even for the populations on which they prey. Extermination of the wolves and mountain lions on the Kaibab Plateau has proved unfavorable in the long run to the deer by removing one of the checks to their multiplication and thus increasing their tendency to overgraze their feeding areas.

The observations on wolves and caribous recently reported by L. Crisler in her book *Arctic Wild* provide a classical example of the importance of indirect mechanisms in the maintenance of natural populations.

> The fit and healthy caribou, even the quite young fawns, are apparently able to outdistance the hunting wolves with comparative ease. In almost all cases investigated, it was found that the caribous killed were those hampered by disease, old age, or injury of some kind. The natural role of the wolf, in fact, is as the culler of the unfit. As a result wolf "control" has had disastrous effects on the Nelchina herd of Alaskan caribou. From an estimated total of 4,000 animals in ten years the figure, without the limiting factor of wolf predation, had risen to the startling total of about 10,000. The available winter range was no longer sufficient, and the consequent trampling and over-grazing threatened the livelihood of the excessive caribou population.

The complexity of the interrelationships between food supplies, animal life, and human societies can best be illustrated perhaps by quoting from a very serious book the following facetious account of animal ecology in England.

> The cats were kept as company by maiden ladies, and the number of these in Britain was large because so many Englishmen were away at sea keeping the British Navy strong. That the circle of relationships was complete came from the food habits of Englishmen in the British Navy—eating the beef raised on the clover pollinated by the bumblebees which were protected from mice by the house cats kept by the spinsters at home. . . . Everywhere we turn, a living fabric of food relationships links nature into a web [Milne and Milne, 1960].

It is because all living things, including man, are part of such a web that disturbances in the order of nature are likely to have unexpected and dangerous consequences. Ecological studies have proven beyond doubt that the larger the number of species in a natural community, the greater the chance that this community is able to absorb shocks from the outside without damage. Man, like other living things, can become adapted to a kind of monoculture civilization, but this form of adaptation is precarious because it lacks the stability of more complex natural systems.

b. Physiological mechanisms of self-regulation in animal populations

The dependence of population size on the quantity and quality of food seems to provide support for the doctrine advanced by Malthus a century and a half ago that living things in general, and human beings in particular, multiply until their numbers are held in check by starvation, disease, and conflicts. In reality, however, field observations and experimental studies have shown that the Malthusian doctrine is certainly incomplete and often misleading, at least as far as animals are concerned. In nature many animal species maintain their population size at a fairly constant level, year after year, and even century after century. In a natural environment undisturbed by man, this level is low enough to be compatible with the maintenance of health because it permits conservation of food resources. While such homeostatic population control was first recognized in the course of wild animal surveys, recent experiments have established that it can occur also under laboratory conditions.

Several colonies of mice were started, each from one male and two females in one enclosure provided with several nest boxes and excess food and water. Fighting occurred between sexually mature males as soon as the populations began to increase in size, and this resulted in the formation of social hierarchies with dominant and subordinate animals. The distribution of the mice between the nest boxes soon became strikingly uneven, and it was common to find that one nest box contained more than half of the final population while others were empty or were occupied by solitary males. The occurrence of empty nest boxes was due to the patrolling activity of dominant males. After 18 months, all colonies contained about 150 mice and reproduction had almost completely ceased. The populations had reached a stable level.

When the females of these stable populations were removed from the restricted enclosure, the majority soon became fecund again; most of them conceived within about one week after being placed in a larger area. It is clear therefore that the populations within the restricted enclosures had been controlled by self-regulatory mechanisms, which did not impair potential fecundity.

External factors such as food shortages, disease, predation, weather, etc. are usually assumed to be the agencies that limit population size under natural conditions. Yet the results of laboratory experiments agree with observations in the field in showing that internal factors, intrinsic to the population, are at least as important as external factors. The operation of such internal factors is illustrated by the unexpected results of the following study.

Several independent groups of rats were exposed to exactly the same conditions: they were placed in identical cages, had equal access to food, and harbored the same kinds of pathogens. Yet, in each group the pattern of population growth was different; one population would shrink while the other would grow and a third remain more or less constant in size. Evidence of internal mechanisms of growth control has been obtained with other animal species in the wild, such as muskrats in Iowa, great tits in Dutch forests, and Levant voles in Palestine. Other laboratory experiments have provided further evidence that adjoining but independent populations of one given species can differ profoundly in their pattern of growth. For example, six groups each consisting of four pairs of mice were placed in six large enclosures of identical shape and size, and were allowed to multiply unrestricted. Although every effort had been made to arrange that the environmental conditions be identical for all the six groups, the descendant populations after two years were found to vary more than fivefold, ranging from 24 to 130 animals.

When a detailed census is made of experimental vertebrate populations placed in various types of environments, it becomes clear that the variations in their sizes are not random. They are conditioned by specific physiological and behavioral traits characteristic for each population. The self-regulatory processes that control the size of each individual population operate through several different mechanisms affecting, in particular, sexual activity, production of eggs and sperm, implantation of fertilized eggs, survival of embryos and the newborn, rate of sexual maturation, health of individuals, and mobility of adults. A few examples will serve to illustrate the operation of some of these self-regulatory mechanisms.

A mouse colony was placed in an enclosure and allowed to feed and multiply freely on a constant but limited daily food allotment. The population expanded rapidly at first, but its growth stopped when the point of incipient food shortage was reached. (See for example Figure 13.) At this time there was a sharp rise in the mortality of young animals and shortly thereafter a striking and generalized depression of sexual activity. Eventually reproduction ceased altogether. There was no rise in mortality, nor any fighting among adults; furthermore, the physical appearance of the latter indicated that self-regulation had occurred at a time when their nutritional state was still adequate. The depression of sexual and reproductive activity was a reversible phenomenon. The females

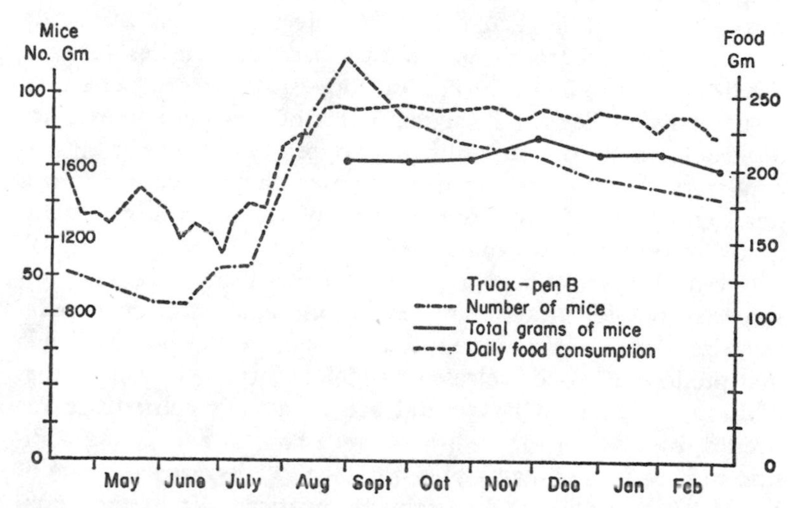

FIGURE 13. A population of mice, *Mus musculus,* was provided with a constant food supply. After a period of growth during the summer, the numbers of animals declined but the biomass and the food consumption remained stable. (Reproduced from R. L. Strecker and J. T. Emlen, 1953.)

could be mated successfully soon after being removed from the regulatory influences that had operated while the population was on a restricted food allowance.

Self-regulation of population size has been observed also among wild animals. Commonly in areas browsed to the limit, the outcome is not so much an increase in death rate as a decrease in birth rate. Abortion and even failure of conception increase in frequency as food becomes more scarce. As a result, a new state of equilibrium becomes eventually established between the size of the population and the amount of food at its disposal.

In general, the self-regulatory mechanisms do not act uniformly throughout the population; instead, they affect differentially its individual members. A population of Norway rats was started from a single pregnant female and maintained for a two-year period with ample food in a quarter-acre enclosure. The colony eventually stabilized its size at the low level of 120, even though the area was sufficiently large to accommodate 5,000 rats. Stable and small in size as it was, the population nevertheless was not uniform. It consisted of eleven subgroups, each with different numbers of males and females differing in reproductive characteristics. Some subgroups had few males and many females, and the latter were successful in rearing litters or were pregnant when the study ended. Other subgroups in contrast had more males to each female, and these females were less successful in reproduction, or even totally ineffectual from this point of view. Homosexuality was common in the subordinate groups. In other words, self-regulatory mechanisms had differentiated the population into a small group of actively reproductive animals and a larger group of low or no reproductive ability.

Recent discoveries have made it clear that many heretofore unsuspected physiological factors have a profound effect on population size. For example, some species of flour beetles possess glands that produce a gas, the release of which is increased by crowding. This gas is lethal to larvae and acts as an anti-aphrodisiac for beetles. Even when the supply of fresh flour is ample, the addition to it of flour contaminated with the gas decreases the rate of beetle multiplication and thereby automatically limits the population.

The sexual cycles of animals are affected by many environmental variables, such as light and temperature, as well as by psychological and social factors. Certain odoriferous substances are of such importance in various aspects of the social life of mammals that they can act as regulatory agents of their population size. In the mouse, olfactory stimuli excite neurohumoral responses that affect profoundly oestrus, pseudopregnancy, and pregnancy. For example, the grouping together of female mice leads to mutual disturbance of oestrus cycles, by the intervention of pseudopregnancies if the groups are small, or of anoestrus if the groups are large. (See also the effect of crowding on sexual cycles in ducks, Chapter II.2.) Even more interestingly, pregnancy block occurs if a newly mated female is caged with a strange male within a few days after mating. The pregnancy block is caused by an odorous substance released by the male, which persists in the cage after removal of the animal that produced it. Pregnancy block depends on a neurohumoral chain of events initiated by a continuous olfactory stimulus that must last two to three days for maximal effect. Thus, the odorous substance excreted by the male can be considered as an *exocrine*—in contrast to an endocrine—which is capable of acting as a chemical messenger helping to integrate the individual members of a population.

In most mammals, if not all, odor plays also an important role in territory marking, in defense, in identification, in sex attraction, in sexual and aggressive behavior. The widespread use of musk, civet, and castoreum in the perfumes of high quality makes it obvious that man is not qualitatively different from animals in his response to odors.

c. Behavioral and adaptive aspects of population control

While it is obvious that the homeostatic processes of population control in animals operate through restriction of sexual activity and interference with reproduction, it is also true that they have social determinants based on complex behavior patterns. Territoriality and hierarchy within the group play a fundamental part in limiting the size of animal populations.

Generally, animals living in a state of freedom engage in social competition rather than in the more direct and destructive competition for food and mates. The croaking of frogs, the song of birds, the howling of monkeys, the roaring of carnivores or seals, and even the "language" of the dolphins have been interpreted as methods of communication to establish status and territorial claims. The deposition of odorous substances from special glands or in excreta at characteristic spots also serves to delineate territorial boundaries. Territoriality, caste formation, peck order, and the various forms of social display constitute as many conventions that help animal societies to organize themselves and thereby regulate their numbers. Animals defend their home grounds from intruders or force several members of their group to emigrate when the population becomes too large. Some of the animals allowed to remain in the group become unable to reproduce when their access to mates or food is limited by the more dominant animals. Often, as we have seen, a few dominant males account for the largest percentage of reproductive activity.

It can be taken for granted that territorial and hierarchial patterns of behavior have an adaptive value, if not for the individual members of the group, at least for the population as a whole. By limiting the number of animals allowed to breed in a given area, these behavior patterns make it possible for the population to utilize the available resources in such a manner as to maintain a satisfactory state of health. Fitness to the environment is thus achieved not only for a short period of time but more lastingly. As already mentioned, many vertebrates living in the wild regulate their population size to a level so low that they never completely use their food resources. This behavior is consonant with the belief of conservationists that the exploitation of natural resources is optimal only when it remains somewhat below maximum utilization. In contrast, total destruction of food supplies often ensues when consumption exceeds a certain critical level. The ability of vertebrate populations to regulate and limit their size even amidst plenty corresponds therefore to a kind of ecological wisdom.

There is a paradoxical aspect to the view that the homeostatic regulatory mechanisms discussed above are a form of adaptation since their result is that many members of the group are deprived of the ability to reproduce or even are sacrificed altogether. The Darwinian individual fitness must be subordinated in such situations to a higher order of fitness in which the whole population is considered as the biological unit; its future is more important than the welfare of its individual members. It is in this regard that the growth of the human population is so completely out of line with that of most animals. The innate biological and social wisdom that keeps animal populations from multiplying to the extent that they destroy their habitat no longer seems to operate in man. Yet the time has come when he too finds himself in the absolute necessity of regulating his numbers.

The dynamics of all living populations are influenced of course by the external factors that compose the total environment, but, as we have seen, they are controlled most effectively by internal regulatory mechanisms which are the expression of complex interactions between the individual members of the group. In all cases, these mechanisms involve biochemical, physiological, and social processes. There is every reason to believe, however, that the purely social aspects of the phenomenon become of increasing relative importance as one ascends the evolutionary scale.

Despite what might be inferred from the rapid increase in the world population, it is certain that self-regulatory processes based on social patterns of behavior operate at least to some extent in human societies. It is even probable that the biochemical and physiological regulatory mechanisms that exist in animals still persist in man, even though they are usually inhibited or masked. Although the so-called rural-urban fertility differential is usually attributed to social, economic, and cultural factors, high population densities may still exert a subtle biological influence that accounts in part for the lower reproductive rate of city dwellers. In particular, population density probably affects the reproductive physiology of man through his neurosensory receptors.

Thus, the problems posed by the internal self-regulatory mechanisms that control population dynamics in animals are of direct relevance to the future of mankind. For this reason, it is urgent to determine not only how animal populations in the wild manage so often to maintain their size at an optimum level, but also why these regulatory mechanisms fail to function in other cases. Man is one of the species in which homeostatic population control is obviously failing at the present time. More extensive knowledge of animal life might help him to avoid the biological disaster exemplified by the exodus of the Norwegian lemmings, and by other types of population crashes.

3. The Determinants of Human Birth Rates

a. Social factors

The biological necessity to limit population size may not have been as acute for Paleolithic man as it is for most animals now. While he was still in the food-gathering phase of his evolution, and as long as his numbers were small, ancient man was probably free to emigrate and start new colonies when his local food resources became scarce. One may assume, however, that as soon as he became more numerous, he began evolving a number of tribal practices and taboos to keep population size under control.

Prolonged breast-feeding, prohibition of sexual intercourse during the period of nursing, the use of various contraceptive techniques, abortion, and infanticide are among the many methods used by primitive people in the past and still used by them today. Such customs, along with a high rate of infantile mortality, have usually been sufficient to keep population density in balance with the amount of food available. According to anthropologists who have lived among them, the pygmies of the Congo practice strict population control even though they never lack food in the rain forest where they live. It seems that the populations of the South Sea Islands had also remained fairly stable for long periods of time before the arrival of the Europeans. And a similar state of affairs still exists today among the Xavante Indians.

The agricultural revolution that occurred some 10,000 years ago during the Neolithic period probably changed man's attitude toward population size as it did toward most of his other tribal problems. It seems legitimate to assume that the greater availability of food progressively weakened the biological and social checks on population growth in the areas where the production of plants and domesticated animals was abundant and fairly dependable.

As we have seen, Paleolithic human sites have yielded many statuettes emphasizing sexual characteristics as well as representations of pregnancy in women and animals (Chapter I.1). Goddesses of fertility are even more common in Neolithic settlements, and all primitive people still have today elaborate fertility rites. These rites played a large role in ancient times among Chinese, Hindus, Egyptians, Greeks, etc., and they still exist in many places even though in a distorted form. The Bible teaches that men should go forth and populate the earth; Oliver Goldsmith's *The Vicar of Wakefield* begins with the statement: "I was always of the opinion that the honest man who married and brought up a large family did more service to the state than he who continued single and only talked of populations." Even today the patriarch surrounded by large numbers of descendants is still regarded as the living symbol of near holiness.

The dedication of mankind to the cause of fertility was probably increased by social attitudes derived from the advances in agricultural production during ancient times. Abundant food provided an inducement for large populations by giving power to the groups that were the most numerous. It made possible the emergence of towns and cities, thus providing a biological environment in which urban man became progressively selected for the ability to live and reproduce under conditions of crowding. Whatever the validity of these assumptions, it is certain that several of the most characteristic aspects of human life now work in opposition to the regulatory mechanisms of population size that exist in animals.

Man is an extremely social creature, and generally prefers to live in close association with large numbers of other men. His racial survival in the historical past has depended on the production of many children, in part because infant mortality was then very high, and also because the social and economic structure made large families a positive asset. It can be surmised, therefore, that evolutionary development has favored the selection of genetic traits associated with large family size. In any case, human population control can no longer operate effectively through the kind of innate physiological mechanisms that operate in animals. It depends ultimately on willful limitation of family size.

The large fluctuations in birth rates in the countries of Western civilization during modern times bear witness to the fact that family size is determined by a multiplicity of social factors involving willful action, but they are so ill defined that they have not yet been identified. Surveys made among middle-income populations living in prosperous suburbs have revealed that the average number of children desired by young couples has almost doubled during the past three decades. In fact, unexplained fluctuations in birth rates have repeatedly occurred all over the Western world during historical times.

In France, the population level and average family size reached a plateau during the late nineteenth century and remained essentially unchanged during the first four decades of the twentieth century. Then immediately at the end of the Second World War (but not after the First) the birth rate increased markedly and remained high until 1950, even though France was then in a state of political confusion, and the country's economic future looked very bleak indeed.

All predictions concerning demographic changes have so far proved erroneous, but a few general remarks concerning past events may be justified in order to illustrate how complex and obscure are the human factors that determine birth rates. In the countries of Northern and Western Europe, and in the other areas peopled by their emigrants, the trend toward smaller families began almost a century ago and has persisted until very recently. During the early postwar period, however, birth rates increased almost everywhere in the Western world, and reached levels much above those of the 1930s. In some countries, including the United States, Australia, New Zealand, and Canada, they have remained close to the postwar level, although a new downward trend is now becoming evident. In other countries, especially France and Norway, birth rates seem to have become stabilized for the present at a position intermediate between the immediate prewar and postwar levels. In most other European countries, they have drifted down to the 1940 levels. In Finland, for example, the birth rate fell from 28 per 1,000 population in 1947 to 18 in 1958.

The rapid and marked shifts in birth rates during recent decades in the countries of Western civilization can hardly be explained by changes in the availability of food, in knowledge of contraception, or in religious beliefs. For lack of these customary explanations, efforts have been made to search for more subtle determinants of human behavior.

Certain sociologists have attributed the remarkable stability of the population level in France from the late part of the nineteenth century until the Second World War to the essential stability of the social and economic structure; the bourgeois desire to transmit an unchanged state of affairs to the following generation made it undesirable to divide the family estate among many children. This conservative social attitude, it is said, vanished when the disasters of the Second World War shattered faith in the French national economy as well as in the bourgeois social values, and when inheritance taxes made it almost impossible to pass on accumulated wealth to following generations. The deeper biological urges then took the upper hand and children became a value in themselves, irrespective of any concern for the uncertain future.

Some American economists have suggested a different social motivation for the postwar baby boom in the United States and for its continuation longer than had been expected. In the words of one of them, "Americans have behaved in the past decades as if diminishing relative marginal utility sets in, after a point, for durable consumers goods; and they have chosen, at the margin, larger families" (Rostow, 1960). In other words, automobiles, washing machines, and the various forms of conspicuous wealth having lost much of their symbolic value, children have taken the place of material goods as symbols of prosperity and happiness. Furthermore, so the claim goes, the American population shows a tendency to "reimpose the strenuous life by raising the birth rate." It has been suggested also that when birth control is still a new idea, keeping up with the Joneses means having fewer children. Now that small families no longer convey an invidious distinction in prosperous countries, having a large family has come to represent a new kind of status symbol, and to take on a Veblenian value as a form of conspicuous expenditure.

The various explanations offered to account for the higher postwar birth rates in the countries of Western civilization are entertaining and suggestive, even though not entirely convincing. Whatever their validity, however, they serve to illustrate the range of human factors that can be invoked to explain choices with regard to family size. The unadorned facts, in any case, clearly establish two points of great practical importance. One is that birth rates can be profoundly affected by factors that are neither biological in the ordinary sense of the word, nor related to simple economic necessities. The other is that population growth cannot be checked merely by further education in contraceptive techniques. In the world as a whole, one of the most guilty social groups from the point of view of the population avalanche is the prosperous American suburbanite—hardly an innocent in matters of birth control.

The problem is different in the underprivileged countries. There the birth rate is much higher as a rule than in the Western world; furthermore, it has not yet shown any significant tendency to decrease among rural people, except in a very few special situations. As infant and childhood mortality is decreasing (although not as much or as fast as generally assumed), high birth rates naturally result in very high growth rates. From 1950 to 1960, the excess of births over deaths averaged 31.4 per 1,000 per year in Taiwan, 26.8 in Ceylon, 32.1 in Malaya, 26.7 in Mauritius, 27.7 in Albania, 31.8 in Mexico, 33.9 in El Salvador, and 37.3 in Costa Rica. Such figures acquire their full and somewhat frightening significance when it is realized that, with an annual increase of 30 per 1,000, a population will double itself in 23 years!

It is widely assumed that the rate of population growth in the underprivileged parts of the world would be rapidly slowed down if inexpensive and convenient contraceptive devices were made available to low-income populations. In reality, however, and granted of course the urgent need for improving the techniques and knowledge of contraception, it is unlikely that this approach alone can have a significant effect on birth rates. What is needed first is to create a genuine desire for smaller numbers of children, and a belief that this goal is both desirable and possible. In other words, there has to be a basic change in attitudes before couples are willing to reduce fertility and to exercise the care required for its control.

The magazine *Wisdom* has recently given, unwittingly, an enlightening pictorial representation of the difficulties in establishing the state of mind required for reducing family size. In 1960 the magazine dedicated its June issue to "The Wisdom of India." The presentation obviously had the official blessing of the Indian Government since the two leading articles were signed by the late Jawaharlal Nehru, who was then Prime Minister, and since copies of the magazine were widely distributed by the Indian Embassy in Washington. As is well known, India is one of the countries where the overpopulation problem is most acute, and also where government agencies have been most active, imaginative, and bold in developing birth control programs. Yet several pages of the special issue devoted to "The Wisdom of India" consist of endearing photographs illustrating the theme that children are the country's greatest wealth; their charm and the love bestowed on them constitute the most important source of happiness; their presence makes life bearable and enjoyable even in the most impoverished villages.[1]

Advocates of birth control naturally find it difficult to enlist the cooperation of destitute people who regard children as the one worthwhile aspect of their lives. Anyone inclined to criticize the Africans, Asians, and Latin Americans for their failure to make the efforts required for birth control might ponder on the unwillingness of Western people to change their own ways with regard to cigarette smoking, overeating, or lack of physical exercise,

1. In the June 7, 1853, entry of his journal, Eugène Delacroix presented a more earthy interpretation of the large numbers of children in the poor families of France:

> Dined at Madame Barbier's; all evening, the talk was only of love and its singularities. She had the drollest idea in the world when we were speaking of the number of children that one finds at Soisy. "As a matter of fact," said she, "what could they do in such a dull place? There is no view. They have got to amuse themselves somehow."

despite the wide publicity given to the relation between these practices and various forms of illness. And it bears repeating that the United States is one of the countries where the population is increasing the fastest. Among us, as in the underdeveloped countries, it is difficult to sacrifice the pleasure of the present or the immediate future for the sake of a distant future which is but dimly perceived.

Population control programs are plagued by the fact that the poorer the community, the more important is the role played by children in its daily life. Under restricted conditions of existence, children usually constitute the only hope of reward. They provide labor on the farm and insurance against the solitude and trials of old age; they are the most reliable and often the only source of deep emotional satisfaction; finally and probably most important, they symbolize hope and eternity. Men and women playing with children amidst poverty or destruction proclaim faith in mankind, whatever the present trials of life. As Plato wrote in *Symposium:* "Marvel then not at the love which all men have for their offspring; for that universal love and interest is for the sake of immortality."

b. The willful control of birth rates

The first need and greatest difficulty in population control is therefore to create values and motivation based on the ability to visualize the future. Human beings must be taught to sacrifice the satisfaction of the present for the sake of a form of life vaguely envisioned, but not factually experienced. In this light, the prospects of population control appear rather discouraging since nothing is more difficult to change than values and motivation. On the other hand, history shows that industrialization always brings about a rapid fall in birth rates, irrespective of religious allegiance. Italy and Japan are two of the countries in which the industrial upsurge after the war was immediately followed by a marked fall in birth rate.

Through mechanisms that are ill understood and complex, but highly effective, industry brings about a non-Malthusian check on birth rate, a check based not on poverty and limitation of food but on economic and social progress. Rev. Richard Jones, who was Malthus' immediate successor at the East India College, was one of the first to suggest that the most effective break in population increase would be not primary wants of food, but the secondary wants stimulated by civilized life, which he thought capable of indefinite extension. Between 1900 and 1962 the birth rate in Italy, a Catholic nation, fell from 33 to 19 per 1,000, and there is reason to believe that the change was related to marked gains in literacy and technological development. In fact, the curtailment of childbearing all over Europe might have occurred even somewhat faster if it had not been slowed down by the official opposition of governments, churches, and all "right thinking" people. Individual couples began to practice birth control against the advice of their leaders as soon as the social and economic structure put a premium on smaller families.

For historical reasons, the population problem in the underdeveloped countries today has little in common with what happened in Western countries during the first Industrial Revolution. In the West, mortality rates declined and population increased only after industrialization was well on the way; wealth was being created faster than people were born. Moreover, the empty lands of America and Africa were providing additional sources of food as soon as larger amounts were needed. In contrast, the underdeveloped countries are now in a state of accelerated population growth, but efforts to modernize and industrialize them have hardly begun. The result is a conflict between the demands of the people for higher living standards and the need to set aside capital for agricultural and industrial development.

The pessimists believe that the conflict between demographic growth and the economic requirements for industrialization cannot be resolved before the population has reached a catastrophic level. The optimists, while admitting that excessive population growth is a hindrance to industrial development, nevertheless believe that there is still enough time before an irreversible stage is reached. According to them, living standards can be raised through industrialization even while the world population continues to grow. Historical precedents, and a blind faith in the human condition, are my only justification for selecting from this controversy examples that support the optimistic attitude.

The voluntary control of family size that has accompanied industrialization in the past is not a phenomenon peculiar to people of European origin. It occurred also in Japan, and the present population trends in Lebanon give a picture of what will probably happen in other parts of the world. Until very recent times, the Lebanese family, whether Moslem or Christian, exhibited the traditional high, uncontrolled fertility found in all rural areas of low socioeconomic status. Lebanon has now begun a process of urbanization, and the traditional family pattern is rapidly being replaced by the Western pattern, with controlled lower fertility brought about by later marriages, recourse to induced abortion, and the earlier use of contraception. Although the trend is most pronounced among educated city women, it is gradually spreading through the other sections of the Lebanese community as their economic and social status approaches that of European people.

A study carried out among Indian villagers in 1954 revealed that fertility first tended to increase when people on a bare subsistence level moved to a slightly higher standard of living. Fertility reached a maximum at a "critical level" of living, then fell rather rapidly with increasing economic status.

Thus, it seems likely that the population experience of Western Europe will be repeated in the rest of the world. Europeans utilized every possible means of birth limitation, including abortion, abandonment of children, and even infanticide, once they were sufficiently motivated to limit the size of their families. Admittedly, interest in family control is still lacking today among most underprivileged peoples. Frustrating poverty and shortages of food tend to make for an unplanned existence, focused almost exclusively on the experience of today with little thought of tomorrow.

But aspirations for a new kind of life are now spreading everywhere in the world, and this will almost certainly change attitudes toward family size. There is even a possibility that the change will happen sooner than anticipated, and that other people will be bolder than Western countries have been in their acceptance of various methods for conception control. For example, several national groups have no ethical or religious scruples against voluntary abortion, witness its widespread use in Japan and probably in many other countries. Voluntary sterilization of men and women is popular in Puerto Rico.

It is necessary at this point to open a parenthesis and discuss attitudes toward the various methods used for the limitation of births, since these methods constitute at present the *sine qua non* of population control. Ignorance of contraceptive techniques or unwillingness to use them is usually held responsible for excessive population growth wherever it occurs, especially in the underdeveloped parts of the world. Religious objections to the very concept of artificial contraception, among people of Catholic and other religious faiths, are also considered an important contributing factor. There are many reasons to believe, however, that these explanations deal with but minor aspects of the overpopulation problem.

Many forms of contraception have long been known and practiced among primitive people. The Ebers papyrus (1550 B.C.), the oldest known Egyptian compendium of medical writings, describes the formula for a medicated tampon designed to prevent conception; it includes honey and the tips of the shrub acacia, a mixture now assumed to yield the spermicidal agent lactic acid as it undergoes fermentation. In Europe, the use of the condom was already well known during Malthus' time, and became so widespread in the early nineteenth century that it was banned by the Pope in 1826. When Francis Place launched the birth control movement in 1822, he recommended several of the contraceptive techniques still in use today. Finally, conception control can be achieved by several methods other than contraception, and most of these have long been known and continue to be used all over the world.

The influence of religious doctrines on birth control, while real, is probably much smaller than usually assumed. The demographic data for different countries of Western civilization, listed in Table 25, reveal no relation whatever between the dominant religion of a country and its birth rates. For example, Italy, Belgium, and France, three countries where Roman Catholicism is the domi-

TABLE 25. Birth Rates per 1,000 People *

	1945–49	*1950–54*	*1955–59*	*1960–61*
Belgium	17.3	16.7	17.0	16.9
France	20.3	19.5	18.4	18.2
Italy	21.1	18.4	18.1	18.6
Ireland	22.5	21.4	21.1	21.3
Denmark	21.6	17.9	16.8	16.6
Finland	27.0	22.8	19.9	18.4
Northern Ireland	21.9	20.9	21.4	22.5
Scotland	19.4	17.8	18.8	19.5
New Zealand	26.5	25.8	26.3	26.8
South Africa (white)	26.4	25.1	25.0	25.3
United States	23.4	24.5	24.6	23.5
Chile	35.7	33.8	35.5	35.4
Ceylon	38.2	38.5	36.6	37.0

* Adapted from United Nations Demographic Yearbook 1961.

nant faith, have birth rates significantly smaller than those of Northern Ireland, Scotland, New Zealand, and the United States, where the non-Catholics vastly outnumber the Catholics. In Holland the average birth rate among Catholics is smaller than among Protestants. True enough, birth rates in Canada are higher among Catholics than among Protestants, but religious faith is not the only determining factor in this particular case. A large percentage of Canadian Catholics live under a type of rural economy that is associated everywhere in the world with birth rates higher than those prevailing in urban industrial areas, irrespective of religious faith. Moreover, the desire of the French Canadians to achieve the "revenge of the cradle" by outbreeding the English-speaking majority may also play some role in their higher birth rates. As to the Catholic countries of Latin America, their birth rates are extremely high, but not more so than those of many Oriental countries where other religions prevail.

Surveys carried out in the United States indicate that at least 90 per cent of white couples with no fertility impairment use some form of contraception when the wife is over thirty; to a very large extent, this applies to Catholic families as well as to those of other religious faiths. Extending birth control to the remaining small minority of couples who do not exercise effective voluntary limitation of fertility would therefore have at most a very slight effect on the national birth rate. Fertility is high at present in the Western world, and especially in the United States, simply because couples deliberately choose to have more children than they did several decades ago. To a very large extent, the rapid population increase in the Western world since the end of the war had its origin in social mores and fashions, rather than in failure to control fertility through lack of convenient contraceptive techniques or through neglect of their practice.

Many different contraceptive devices and other methods of conception control have long been available and many new ones are being introduced at the present time. But whatever their convenience, safety, and effectiveness, they will not help in checking population growth unless individual couples really want to limit the size of their own families. The problem of excessive population growth is not peculiar to people of darker skins or of lower economic status. It will not be solved unless it becomes the concern of *everybody, everywhere,* white as well as dark-skinned people, not only in the tropical slums but also in our own luxurious suburbs.

4. Population density and human life

a. Overpopulation and ways of life

Of the many dangers created by overpopulation, the most frequently discussed are those arising from shortages of food and raw materials; even water may soon come to be in short supply in our communities. Famine is commonly regarded as the first likely consequence of overpopulation. It is true indeed that approximately one-third of the people now living do not have enough to eat, and even more have a qualitatively inadequate diet (Chapter III.2). As we have seen, however, food production could be increased quantitatively manyfold, and improved qualitatively, by many different technological procedures. Just like synthetic materials in industrial processes today, proteins, amino acids, and vitamins produced by an industrialized form of agriculture, or even by chemical techniques, might come to play an essential role in the nutrition of the crowded world.

Present-day agricultural methods, even when carried out with modern equipment, still remain close in principle to the ancient practices of land utilization developed almost 10,000 years ago during the Neolithic period. The real change in food production will probably depend more on the introduction of truly new technologies unrelated to Neolithic farming than on more powerful tractors to till the land, or more widespread use of fertilizers and pesticides. The extraction of concentrated proteins from crude and otherwise inedible plant materials, a more efficient utilization of solar energy in photosynthesis, and the application of chemical techniques to the production of synthetic foodstuffs may come to occupy with regard to traditional agriculture the same position that automated work occupies today with regard to the skill of the eighteenth-century craftsman. Farming and gardening will of course persist, but less from economic necessity than for the satisfaction of traditional values. To a large extent, the modernized forms of ancient agricultural practices will survive as hobbies, just as do deer-hunting, trout-fishing, and home cabinet work, which are still actively pursued only because man finds it hard to cast away the habits and skills on which his survival depended for so long.

According to recent estimates, the application of present scientific knowledge to agriculture would permit producing enough food to support the existence of 50 billion human beings on earth—almost 20 times the present level! Barring unforeseen technical difficulties and natural disasters, it is therefore unlikely that the food supply will soon constitute a limiting factor in the growth of the world population. Man will need simply to adapt himself, as he certainly can, to consider that roast beef, sweet corn, tender lettuce, and tasty fruit are not essential to good nutrition, but need be used only as adornments to diets made up of foodstuffs designed and produced for their nutritional value, rather than for their taste or traditional meaning.

However, the necessity to modify nutritional habits is only one of the many qualitative changes in human life that can be expected to result in the near future from the population avalanche. In fact, the most disturbing problems that will arise from a larger world population may not be amenable to technological solutions. As he populates more and more of the earth, man will have to eliminate all forms of wildlife that would compete with him for space and food; he will increasingly have to flood deserts and fell forests in order to create more farm land, factories, houses, and roads; in brief, he will tolerate wild animals, wild plants, and wild landscapes only to the extent that they serve his needs. Highways, factories, and dwellings will occupy much of the scenery; all natural resources, including water, will have to be carefully husbanded.

Most disturbing perhaps are the behavioral consequences likely to ensue from overpopulation. The ever-increasing complexity of the social structure will make some form of regimentation unavoidable; freedom and privacy may come to constitute antisocial luxuries and their attainment to involve real hardships. In consequence, there may emerge by selection a stock of human beings suited genetically to accept as a matter of course a regimented and sheltered way of life in a teeming and polluted world, from which all wilderness and fantasy of nature will have disappeared. The domesticated farm animal and the laboratory rodent on a controlled regimen in a controlled environment will then become true models for the study of man.

Thus, it is apparent that food, natural resources, supplies of power, and other elements involved in the operation of the body machine and of the industrial establishment are not the only factors to be considered in determining the optimum number of people that can live on earth. Just as important for maintaining the *human* qualities of life is an environment in which it is possible to satisfy the longing for quiet, privacy, independence, initiative, and some open space. These commodities will be in short supply long before there is a critical shortage of the materials and forces that keep the human machine going and industry expanding. In the words of Stuart Mill a century ago: "A population may be too crowded, though all be amply supplied with food and raiment. It is not good for man to be kept perforce at all times in the presence of his species. A world from which solitude is extirpated, is a very poor ideal."

The question then is not so much how many mouths can be fed, or how many bodies can be accommodated, as what can be done to maintain the qualities that give to human life its peculiar characteristics. In other words, the optimum size of the world population cannot be discussed without consideration of predetermined objectives. There are many possible choices of goals open to a wealthy and enlightened community. The most difficult problem is to select what to do among all the things that could be done, many of which cry out for attention.

From the simple materialistic point of view, the creation of wealth is the most obvious goal; but the provision for continued growth, or rather for continued change, is equally important because happiness depends more on variety of outlook than on wealth itself. The development of military and other forms of power is probably justified, because it is required for the protection of national freedom, in the present state of the world at least. The conservation and improvement of resources for future generations is also an essential part of social responsibility, unfortunately much overlooked at the present time. Improvements of human health, in quality and longevity; the development of social institutions, of education, and of culture; a deeper understanding of man's nature and his place in the cosmos; all these criteria must be considered in deciding the optimum population density.

Because the importance of family control appears so self-evident, it is at first sight difficult to understand why certain populations continue to grow after reaching a density that results not only in shortages of food and other resources but also in a hopeless outlook for their children. The fundamental reason is that human beings regard life as the supreme good, even though it involves privations and suffering. The love for life, and the desire to expand it through the creation of children, is not a rational attitude based on thought or learning, but a motivation more powerful because antecedent and spontaneous, indeed primeval, in nature.

The two large obstacles that stand in the way of fertility control are therefore independent of familiarity with contraceptive techniques. One is the primeval urge for the creation of new life, and the other a deep uncertainty as to the future of the human condition. This uncertainty calls to mind the question from *Alice in Wonderland:* "Cheshire Puss, would you tell me please, which way I ought to go from here?" To which question the very proper answer was, "That depends a great deal on where you want to get to." The great weakness of the campaigns for population control is that everyone knows where one should *not* go, but no one has a clear notion of what to substitute for the satisfactions that human nature derives from sharing the experience of children, and for the feeling of safety afforded by the company of numerous human beings. Birth control techniques are only means to an end, and they will be used only if they serve a purpose judged worthwhile. For this reason, campaigns for the limitation of family size will make little headway until people are given a convincing and clear vision of goals as appealing as the creation of more life.

b. The human qualities of life

Everyone realizes that quality is a better criterion than quantity in formulating population goals, but the qualities judged most desirable for human existence depend on ill-defined and highly subjective value judgments. Human being readily agree on the need for improvements in health, longevity, comfort, and conformity with social mores. However, other aspects of the qualities of life raise questions that cannot be answered conclusively, in part because they involve a kind of scientific knowledge not yet available, and more importantly because they raise philosophical and ethical considerations that have not yet been made objective.

In a community where most children survive into adulthood, population stability demands of course that family size be strictly limited. For example, with a life expectancy at birth of 60 years, the population will stop growing only if the birth rate does not exceed 16.7 per 1,000. With the death rates now prevailing in most of the Western world, even as low a figure as three children per family would cause an approximate doubling of the population every 50 years. Clearly then, population control will demand a limitation of family size much more stringent than the average practice in the United States at the present time. It is not unlikely that such strict limitation would interfere somewhat with the play of the selective forces that operated in the past, which helped to minimize the spread of certain genetic defects in the population. Unfortunately, knowledge of human genetics is not sufficiently developed to warrant predictions as to the distant consequences of real population control.

Concern with the possibility of genetic deterioration and a desire to improve the qualities of the human stock have led some geneticists to advocate a program of eugenics aimed at favoring the spread in the general population of traits they consider beneficial. But it would take a soothsayer to design the ideal human being, since the only thing known about the future is that it will certainly differ greatly from the present (Chapter XVI). Science cannot specify the goal toward which evolution should proceed or the kind of environment best for human life, because this involves value judgments that transcend objective knowledge. The wiser course for the time being is probably to favor the greatest possible diversity among the human population, in the form of genetic polymorphism, so as to make it easier for the human race to take best advantage of unpredictable conditions. In a changing world, it is more important to be adaptable than to be perfectly adapted.

Fortunately, the medical aspects of the population problem are somewhat simpler and clearer than those faced by geneticists and sociologists. As emphasized earlier and contrary to what is so often stated, the population avalanche is not the consequence of medical action (Chapter XI.1). It began before modern medicine could exert any significant influence on either death or birth rates, and it is reaching its most disturbing intensity in areas where medical services are the most deficient. Indeed, improvements in public health might well be the most effective way to slow down population growth. The people in primitive and poor societies have learned through long experience that most of their children will die before reaching adulthood, and that the surviving progeny will be sufficiently large only if many more children are born than are desired. However, it is now within the power of preventive medicine to change this passive acceptance of fatality, by showing that the majority of newborn can be made to survive through the use of relatively inexpensive techniques applicable everywhere.

Modern medicine can also help in changing attitudes toward the population problem by improving the general state of health. As everyone knows, planning for the future demands a kind of stamina rarely possessed by the sick and malnourished. For social groups as well as for individual persons, mental apathy is a common consequence of malnutrition and debilitating diseases. Improvements in nutrition and health will go very far toward helping people in low-income countries to change their ways of life and in particular to raise their sights beyond the dismal present. Larger hopes and aspirations will render people who are now lethargic and passive more receptive to birth control teaching, by providing them with the vision of a future different from the dreary past they have known, in which the mere experience of being alive was the only reward of life.

One can take it for granted that man will not rest content until he has established permanent settlements all over the earth. For this reason, increases in the world population are inevitable. A larger population, however, need not imply true overpopulation since the increases can come about through the occupation of new territories, as DDT made possible in Ceylon, and as canalization of water made possible in the American Southwest.

It is equally certain, on the other hand, that the demographic expansion cannot continue very long. If men allow themselves to continue breeding like rabbits, their fate will inevitably be to live like rabbits, a precarious and limited existence. The greatest dangers posed by overpopulation do not arise from shortages of food and resources for industrial development. Immensely complex as it is, the world population problem can be seen nevertheless as a fairly simple pattern in which the material aspects of living are interwoven with the aspirations of man's nature. Long before the world population has reached the maximum level for which the earth can provide material support, there will be other forms of scarcity, less obvious but more important in the long run. The most critical commodities are not the material ones, but those which condition the qualities human life must preserve if it is to remain above the brutish level and retain its superiority over the rest of animal life.

The undisciplined increase in population is obviously an evil, but as Malthus himself stated in the final paragraph of his first essay, "Evil exists in the world, not to create despair, but activity." At the present time, activity in this regard should be directed not only to the development of birth control techniques but also to creating a motivation, based on social goals, appealing enough to substitute for the emotional satisfactions derived from the mere creation of more life. Health improvement will facilitate population control by giving underprivileged people the confidence that children can be saved from early death. It will give them the vigor necessary to visualize and create the future, instead of passively accepting life as the dreary experience of the day.[2]

2. After I completed this chapter, the Honorable Quintin Hogg, then Minister of Science and Education of Great Britain, published an article in which he discussed the "foreseeable impacts of science on political life." The following quotation is taken from the section of his article dealing with the world population problem:

> even though it [a magic contraceptive pill] were found to exist, its use in the conditions of Indian, African or Amer-Indian village life is problematical. What is certain to my mind is that the high birth-rate in the developing countries is less the product of the absence of such methods (since cruder methods of control have existed since the dawn of history) as of the simple desire for children in a society not yet used to low infant mortality rates and accustomed to regard a quiverful of offspring as a poor man's old age pension, a personal bodyguard or the answer to the servant problem. Social insurance, monogamy, later marriage, improved living and housing conditions, and a higher degree of personal security will, in my judgment, do more in the long term to lower the birth-rate than all the contraceptive pills in the world. These conditions will, I believe, follow on the development of agriculture and industry and an intelligent social policy.

XII. Hippocrates in Modern Dress

1. The hippocratic doctrines

All ancient people of whom we have records formulated theories to account for disease and developed practices of prevention and treatment. Many of these practices certainly operated by suggestion and were most effective against disease states of psychic origin. But others had a more physical basis and dealt with the ailments of the body. Baths, massage, blood-letting, dry-cupping, cauterization, the setting of fractures, trephining, the use of many drugs derived from plants or from the earth, even vaccination against snake bites and infectious diseases, are among the many therapeutic procedures primitive man discovered empirically and learned to use skillfully. It is possible that some of these procedures originated from an innate health instinct. Just as wild animals instinctively choose the right food, eat certain herbs when they are sick, treat themselves by licking their wounds or by extracting foreign bodies, so did primitive man probably have "medical" instincts which were essential for his survival. Eventually some of these instincts may have evolved into magical practices incorporated into primitive doctrines of health and disease.

Medical theories have changed greatly in the course of time, but not according to a continuous and orderly process of evolution. Rather the changes have occurred as discontinuous steps, the direction of which was determined by the prevailing view of man's nature and of his relation to the cosmos. Emphasis has been placed at times on the whole man and on his relationship to demons or to God. In contrast, the focus under other circumstances has been on fragments of man's nature, either on his mind or on the components of the body machine. Throughout the history of medicine, the "ontological" doctrine, which regards diseases as specific entities, has alternated with the "physiological" view, according to which disease is simply an abnormal state experienced by a given individual organism at a given time. In one form or another, the ontological attitude has probably been the most consistently dominant among both laymen and physicians. It assumes that disease is a thing in itself, essentially unrelated to the patient's personality, his bodily constitution, or his mode of life. This concept reasserts itself repeatedly in everyday language when it is said that the patient has *a* disease or that the physician treats *a* disease.

To think of disease as an entity separate from oneself and caused by an agent external to the body but capable of getting into it and thereby causing damage seems to have great appeal for the human mind. This attitude is perhaps a consequence of projection, a process whereby what is felt or experienced as uncomfortable, painful, or dangerous is ascribed to a bad influence from the outside. In prescientific medicine, such explanations took the form of demonological concepts, disease being regarded as resulting from the malevolent influence of taboo violation, sorcery, revengeful ghosts, witchcraft, hostile ancestors, or animal spirits. A man became ill because he had an enemy who cast a spell or because he was being punished for breaking a taboo or for some other transgression.

Similar psychological concepts still influence today the interpretation of phenomena established by scientific medicine. Regardless of their level of education and sophistication, patients are prone to blame their illness on something they "caught" or ate, or which happened to them. They think of disease as something apart from themselves, and many account for it in terms of punishment. "Physicians also find attractive such ways of thinking, particularly if they can see the 'cause' of disease as something which they can attack and destroy" (Engel, 1962).

Irrespective of medical philosophy, ancient man acquired many practical skills for the treatment of disease. Empirical mastery of a wide range of medical techniques can be observed even today among people who are still in the Stone Age culture, for example, the pygmies of Central Africa or the Australian aborigines; their medicine men perform therapeutic feats that do not depend on suggestion alone. Indeed, so great were the practical skills of ancient people that only during very recent times has medicine come to differ significantly from what it was in the distant past. The differences in medical and surgical procedures as practiced in ancient Egypt, in post-Socratic Athens, in a medieval palace, at the Versailles court, or in a primitive African tribe today correspond not so much to differences in factual knowledge as to differences in medical philosophy of the physician. In one case he believes or pretends to believe that disease comes from demonic forces. In another case he regards health as a gift of God, or as the expression of a harmonious relationship between man and his environment.

Primitive societies developed very early a large stock of practical experience having to do not only with the treatment but also with the prevention of disease. All the great leaders of men, such as Moses, formulated strict sanitary regulations for camp life. In the words of the medical historian Field Garrison, "The ancient Hebrews were in fact the founders of prophylaxis, and their high priests were true medical police." The admonitions in Leviticus, the various kinds of taboo that rule and restrict the activities of primitive people, and so many codes of behavior based on folklore that persist in a concealed form among us are the perennial expressions of empirical practices which served a useful purpose at the time when they were first enacted in the countries of their origin.

In all ancient civilizations, medical wisdom and skills have been symbolized by the attributes of certain gods and goddesses of health. For example, the complex structure of ancient Greek medicine was personified by Asclepius, who was first regarded as the perfect physician, then as a god of health. His cult persisted longer than that of any other Greek god and, instead of being destroyed by Christianity, it progressively was integrated in Christian tradition. The two Christian saints of medicine and pharmacy, Cosmas and Damian, appeared as did Asclepius in the dreams of those who sought help in the temples for the ills of the body and the mind.

While it is customary to emphasize Asclepius' knowledge of the healing arts, his activities in the Greek legend went beyond that of a healer. From the third century on, Asclepius is shown in most of Greek iconography with his two daughters, on the right Hygeia (Hygiene) and on the left Panakeia (Cure-All). The two maidens symbolize the main attitudes modern medicine has retained from this ancient heritage, which it is trying to convert into exact sciences. Panakeia was a true healing goddess, learned in the use of drugs derived either from plants or from the earth; her cult is still alive today in the universal search for a panacea. In contrast, her sister Hygeia, one of the many personifications of Athena, was the goddess who taught the Greeks that they could remain in good health if they lived according to reason, with moderation in all things. Men still honor her memory in the use of the word hygiene. They pay lip service to the wisdom she symbolizes by attempting, even though without much success so far, to formulate ways of life conducive to health yet compatible with their appetites, inertia, and lack of discipline.

Eventually, all the attributes of Asclepius, Hygeia, and Panakeia came to be synthesized in the legends attached to the name of Hippocrates. Although it is likely that the several treatises attributed to Hippocrates were written in reality by different persons at different times, there is no doubt that a distinguished physician by this name did exist and practice medicine on the island of Cos. He was then so well known that Plato mentions him by name in two of his dialogues.

The many stylistic genres and doctrinal contradictions in the Hippocratic Corpus make it certain that its multiple fragments were composed at very different times by physicians of different schools. This multiple origin naturally constitutes a source of great complexities for the historian and detracts somewhat from the fame attached to the man Hippocrates. On the other hand, it accounts for the multifarious and lasting influence the Hippocratic Corpus has exerted on Western medicine (Chapter II.1). The Corpus, in fact, does not represent merely the doctrine of the school of Cos, with which Hippocrates is identified, but also that of the rival school of Cnidos. More broadly it defines an attitude and states a series of facts and concepts that constitute the summation of Greek medical knowledge, derived in large part from Egyptian sources. The Hippocratic Corpus constitutes even today the essential basis of modern medicine because it encompasses the various complementary aspects of medical science and practice. These can be summarized as follows:

> Disease is not caused by demons or capricious deities but rather by natural forces that obey natural laws. Hence, therapeutic procedures can be developed on a rational basis. These procedures include the use of regimens, drugs, and surgical techniques designed to correct the ill effects of natural forces.
>
> The well-being of man is under the influence of the environment, including in particular air, water, places, and the various regimens. The understanding of the effect of the environment on man is the fundamental basis of the physician's art.
>
> Health is the expression of a harmonious balance between the various components of man's nature (the four humors that control all human activities) and the environment and ways of life.

> Whatever happens in the mind influences the body and vice versa. In fact, mind and body cannot be considered independently one from the other.
>
> Health means a healthy mind in a healthy body, and can be achieved only by governing daily life in accordance with natural laws, which ensures an equilibrium between the different forces of the organism and those of the environment.
>
> Medicine is an ethical profession and implies an attitude of reverence for the human condition.

Paraphrasing Whitehead's remark on the debt of European philosophy to Plato, one might say that modern medicine is but a series of commentaries and elaborations on the Hippocratic writings. The comprehensiveness of their coverage is such that they have retained a universal appeal for 2,500 years. The scientist recognizes in them the first known systematic attempt to explain disease phenomena in terms of natural laws, on the basis of objectively made and carefully recorded observations. The clinician admires in them a shrewd description of the signs and symptoms characteristic of each disease, a knowledge of prognosis based on clinical experience, and a penetrating concern for the patient considered as a whole, integrated organism. The student of public health points to the clear statement in the Hippocratic doctrine that diseases are determined by the total environment and by the ways of life. The student of man's natural history is impressed by the statement in the Hippocratic Corpus that the general characteristics of populations in the normal state are conditioned by the topographic and climatic factors of the locality.

Thus, the Hippocratic Corpus has a significant message for almost every thinking person concerned with man in health and in disease. But in reality, each person and each age has emphasized some particular aspect of this message at the expense of the others. The practicing physician with the gold-headed cane, the sanitarian concerned with public health, the student of physiological functions and chemical structures, can all trace their scientific ancestry to the Hippocratic teachings, but otherwise their attitude and professional activities have little in common.

2. Specific etiology and host response

The belief in natural causation of disease has led to a multiplicity of scientific attitudes which differ so profoundly that they appear at first sight to be almost incompatible. Some schools of thought have emphasized the unique importance of particular organs or functions, whereas others have claimed that pathogenesis cannot be understood, or disease treated, without considering the patient as a whole. Some believe that the key to medical progress is a knowledge of molecular biology, while others are just as convinced that the most urgent task is to study the response of the body-mind complex to the total environment. Depending upon the time and place, the center of the medical stage has been occupied by anatomy, physiology, cellular pathology, molecular biology, concern with the unique experience of the individual patient, or with the responses of the social group considered as an integrated unit. All medical faiths continue to proclaim allegiance to Hippocrates even though the Hippocratic Corpus, like the Bible, is often invoked but hardly ever read.

Whatever their differences of opinion, all the physicians of Greco-Roman culture have long accepted the concept that diseases are caused by natural forces. But the history of this faith could be written as a debate that has lasted more than two thousand years, between the proponents of the doctrine of specific etiology and those who regard disease as the outcome of a constellation of factors acting simultaneously.

It would be historically naive, of course, to trace the doctrine of specific etiology to any one particular person. For the sake of convenience, however, it may be justifiable to identify its triumph during the nineteenth century with Pierre-Fidèle Bretonneau (1778–1862), because this astute clinician clearly recognized that certain clinical entities with which he had great practical experience were caused by particular agents. Bretonneau's formulation of the doctrine of specific etiology was celebrated in 1962 by an international congress held in Tours, the city where he practiced medicine, on the occasion of the centenary of his death. The doctrine of specific etiology has been so influential in the evolution of modern medical science that it seems justified to trace here some aspects of its historical development.

There is no more spectacular phenomenon in the history of medicine than the rapidity with which the germ theory of disease became accepted by the medical profession. This acceptance was due in part to the vigor of the personalities of its two leading proponents, Louis Pasteur and Robert Koch, but in part only. The triumph of the germ theory would not have been as immediate and decisive if the medical public had not been prepared for it by the observations and reflections of several great clinicians throughout the early part of the nineteenth century. From the haphazard accumulation of clinical facts, these clinicians derived generalizations that made it possible to classify diseases as if they were defined biological entities. Bretonneau extended clinical analysis and classification still further and especially managed to make order out of the incredibly confused field of febrile diseases. Through his efforts, diphtheria and scarlatina emerged as well-defined diseases out of the chaos of the "fevers."

The importance of Bretonneau's achievement can best be evaluated by recalling the statements of one of his French contemporaries, the physician-epidemiologist J. A. F. Ozanam (1773–1837), whose books were widely influential at the time. Between 1817 and 1823, Ozanam published a history of epidemic diseases in five volumes, with a German edition in 1820 and a second French edition in 1835. Its impressive title was *Histoire médicale, générale et particulière, des maladies épidémiques, contagieuses et épizootiques, qui ont régné en Europe, depuis les temps les plus reculés, et notamment depuis le XIVeme siècle jusqu'à nos jours.* In his *Histoire* Ozanam presented not only a nosography of epidemics inspired from Sydenham but also a detailed discussion of "epidemic constitutions." According to him, all epidemic diseases prevailing at any given period had "the same origin and the same diathesis." Whatever their characteristic differences, they represented, in his view, a single disease expressed through different forms and symptoms. Ozanam did not even consider the possibility that each type of disease might be caused by a characteristic etiological agent. While accepting that there were epidemic contagious diseases, he explicitly qualified as "absurd" the hypothesis of a "contagium animatum."

Bretonneau claimed in contrast that each specific disease is caused by "a miasmatic organism which is incorporated in morbid secretions" and which thus behaves as a "specific transmissible agent." This statement acquires special interest from the fact that it was written several decades before the formulation of the germ theory of disease. Furthermore, Bretonneau went as far as devising experimental models to illustrate this visionary concept, by showing that various vesicant substances had different effects on the mucous membranes of animals.

By relating clinically and conceptually the characteristics of each disease to a particular cause, Bretonneau prepared physicians to accept the teachings of microbiology, and more generally made it easier for the doctrine of specific etiology to take root in medical thought. The time was right, because this intellectual departure was consonant with two great themes of early nineteenth-century biology. On the one hand, the Linnean classification of plants and animals was then becoming generally accepted, and it seemed natural to extend it to other natural phenomena, including those of disease. In addition, cause-effect relationships had come to be taken as a matter of course in biological thinking; the principle of causality was replacing vitalism at least among experimenters. On this dual basis, it became possible to construct the theory that each particular disease has its particular cause, and that each noxious agent exerts a characteristic pathological effect.

Historical reasons account for Bretonneau's special interests in acute infectious diseases. These diseases, then known as "the fevers," were extremely prevalent in Europe; in particular there had been disastrous outbreaks of diphtheria and scarlet fever among army recruits and in the population near Tours, where Bretonneau had his practice. The overwhelming importance of acute microbial diseases during the nineteenth century thus constituted the historical accident that gave its first form to the doctrine of specific etiology.

Furthermore, these diseases provided ideal models to demonstrate the validity of the concept of specific causation. Nutritional deficiencies, which were then also widespread, constituted another type of pathological material, which soon proved well suited for studies on specific etiology, and in many cases for the development of specific methods of control. Once proven in these particular situations, the doctrine extended progressively to all fields of medicine. Biochemical lesions, molecular pathology, congenital anomalies, and genetic disorders are direct linear descendants of the doctrine of specificity. Even certain mental diseases are being shown to result from identifiable chemical disturbances or from specific life situations.

The doctrine of specific etiology has thus constituted the most powerful single force in the development of medicine during the past century. But there is now increasing awareness that it fails to provide a complete account of most disease problems as they naturally occur, and especially of those which are most important in our communities today. In general, pathological states are the consequence of several determinant factors acting simultaneously; moreover, the manifestations of any given agent differ profoundly from one person to another. Causality and specificity are therefore much less apparent in clinical situations than they are in the laboratory. A few acute infectious diseases and nutritional deficiencies of course present clinical pictures so characteristic that they can be identified without difficulty; but few are the pathological states that can be classified as true biological entities. What the patient experiences and what the physician observes constitutes generally a confusing variety of symptoms and lesions rather than a well-defined entity. In most cases, a syndrome such as anemia, cardiac insufficiency, gastric disturbance, and depression is more in evidence than the unique pathological manifestations of a specific etiological agent.

Furthermore, each noxious agent can express itself by a great variety of different pathological states. The phthisis studied by Laënnec had little in common with the primary infection that is now the most common form of tuberculosis in our communities; yet both are caused by tubercle bacilli of the same virulence. Syphilitic aortitis, paresis, and tabes, syphilitic roseola, a gumma or a chancre, can all be caused by identical forms of *Treponema pallidum*. A traumatic accident that would be fatal for an aged person might have but trivial consequences for a young adult. Thus, the characters of a disease are determined more by the response of the organism as a whole than by the characteristics of the causative agent.

To complicate matters still further, different agents can elicit similar reactions. Bacterial endocarditis in man and mastitis in cows used to be caused almost exclusively by streptococci. Now that streptococcal infections can be successfully treated with penicillin, other microbial species commonly become established in the lesions and thus give a new microbial etiology to these ancient diseases. Similarly, congestion and hypersecretion of the nasal mucous membranes can be caused by viral and bacterial infections; by inhalation of smoke, dust, allergens, or cold air; by migraine of vascular origin; by the administration of parasympatomimetic drugs; by sorrow and tears. Or again, urticaria can be caused by sunlight and cold, by contact with wool, by many types of foods and drugs.

Since the body has only a limited range of reactions, its response to assaults of very different origin and nature is rather stereotyped. For example, an intestinal pathology which mimics that of typhoid fever and shows the typical lesions of the Peyer patches can be produced by injecting into the mesenteric nodes almost any irritating substance, even a rose thorn! More generally, a stimulation of the neurovegetative system can produce important lesions not only in the viscera directly affected but also in others with indirect and distant anatomical connections. The General Adaptation Syndrome seems to constitute an irrefutable denial of the doctrine of specificity.

As is well known, furthermore, the response of the human organism to many noxious agencies is profoundly conditioned by adrenal and other hormones. The secretion of these in turn is affected by psychological factors and by the symbolic interpretation the mind attaches to environmental agents and stimuli. This interpretation is so profoundly influenced by the experiences of the past and the anticipations of the future that the physicochemical characteristics of noxious agents rarely determine the characters of the pathological processes they set in motion.

These facts might lead to the conclusion that the doctrine of specificity has been discredited, but a more useful interpretation seems to be that its meaning must be enlarged to include not only the operations of external agents but also factors that govern the responses of the organism. In its original form, the doctrine of specificity was focused on external noxious agents—microbes, poisons, nutritional deficiencies, ionizing radiations. In its more sophisticated form, it concerns itself not only with the direct effects of these noxious factors on the target organs but also with the organism's responses.

The ancient belief in the healing power of nature, Hippocrates' Naturae vis Medicatrix, is an expression of the faith that the integrated organism can respond in an adaptive manner to environmental insults. But this large philosophical concept of biology will not become part of scientific medicine until precise knowledge is available of the physiological, immunological, and psychological mechanisms through which the healing power operates. Similarly, while the concepts covered by the words "diathesis" and "epidemic constitutions" denote real characteristics of the organism and real ecological situations, they will not evolve beyond the level of understanding reached by Hippocrates and Sydenham until their determinant factors have been identified. Thus the doctrine of specificity still remains the indispensable key to the understanding of all problems of pathogenesis; it must be applied now to the factors of the internal environment as well as to the various cosmic forces that affect mankind.

The experimenter, the epidemiologist, and the clinician approach the phenomena of disease with different attitudes. The experimenter selects from these phenomena a few aspects closely related to his scientific interests; in fact, he must learn to eliminate all others as completely as possible, even while remaining aware of the complexity of phenomena as they naturally occur. Such an attitude involves choices which are intellectually difficult and often painful, but which are the inescapable price of scientific progress. The epidemiologist in contrast must observe and study the biological phenomena in all their natural complexity; he tries to identify the various components of ecological situations and to recognize their interrelationships in the hope of being able to select among the various control methods that are available to him those which are most practical and effective in a given environment at a given time.

Whereas the experimenter selects a simplified model for detailed analysis and the epidemiologist deals with large human groups, the clinician must appreciate the subtleties that determine the response of each individual person. Physicians have long known, for example, that healing tends to be slow and relapses to be frequent in the anxious and irritable person. Experience with diabetic subjects has taught that bereavement and disappointment in love or business can increase insulin requirements as much as does an infection or a physiological disturbance. A person may blush, perspire, tighten his fist, or even collapse following a trivial incident that goes unnoticed by his companion. The underlying mechanisms of these differences in response are poorly understood, but this does not decrease their importance for the physician. They illustrate that each person exists as it were in a private world, and furthermore responds to environmental forces according to a highly individual pattern. These complexities pose to human medicine problems that can usually be ignored by the student of general biology or veterinary medicine.

Finally, the clinician has the responsibility to decide the relative importance of the many different factors that impinge simultaneously on his particular patient, and what aspects of the external and internal environment he can modify usefully. As most clinical decisions must be made on the basis of insufficient information, the practice of medicine retains necessarily some of the characteristics of an art.

The complexity of the problems encountered by the clinician may help to forecast the future trends of the doctrine of specificity. Simple cause-effect relationships involving only one variable are rarely sufficient to account for the natural phenomena of disease. The total environment and the *milieu intérieur* constitute a multifactorial system, each component of which must be studied with regard not only to its own characteristics but also to its effects on the other components of the system. There is reason to hope that a better understanding of multifactorial systems would eventually give concrete meaning and practical usefulness to concepts such as Naturae vis Medicatrix, defense mechanisms, constitution, telluric factors, etc., which medicine has honored since the time of Hippocrates, but which have remained too vague to be practically useful. The concept of multifactorial causation is in reality but an extension of the doctrine of specificity that brings scientific understanding a little nearer to the complexities of the real world.

Theories are like living organisms that can survive only by evolving in order to adapt themselves to new demands. If the doctrine of specificity were restricted to its initial narrow formulation, it would experience the fate of other theories and wither away or at best become mummified. It would constitute another illustration of Thomas Huxley's statement that new truths commonly begin as heresies and end as superstitions. Fortunately, the doctrine is acquiring new life and becoming even more fruitful of understanding, because its scope is being widened. At first focused on a few noxious factors of the external world, it is now taking cognizance of the multiplicity of internal mechanisms through which the body and the mind attempt to respond adaptively to environmental stimuli and stresses. Seen from this broader point of view, the doctrine of specificity will stimulate the development of methods for the study of the response of the organism as a whole, in other words for a truer apprehension of disease. "Science," Pasteur wrote, "advances through tentative answers to a series of more and more subtle questions, which reach deeper and deeper into the essence of natural phenomena."

3. Organismic and Environmental Biology

The very process of living is a complex interplay between the organism and the environment, at times resulting in injury and disease. To study this interplay and its effects, medical scientists have naturally adopted the general method that has been used by experimental biologists since the seventeenth century, namely to devise models as simplified as possible, preferably involving only one variable.

The beginning of the modern era in scientific biology is commonly traced to René Descartes. This is not because his name is associated with discoveries or scientific generalizations of biological importance. In fact, some of his biological ideas were rather primitive or even outright erroneous, as when he challenged Harvey's theory of the blood circulation and proposed instead that the heart worked as a heat engine. Descartes opened a new era in medical science simply by asserting with logical force and literary skill that all the structures and operations of the human body are reducible to mechanical models, while the soul is a direct gift of God and is therefore out of the range of scientific understanding. These assertions encouraged scientists to focus their efforts on the body machine and to study it by the methods used for studying the inanimate world. Removing the mind from scientific concern greatly simplified the study of life and particularly of man.

One of the essential principles of Descartes' famous method was to divide each of the difficulties presented by the system under consideration into as many parts as possible, and then analyze these parts separately, in the faith that knowledge of the more complex aspects would eventually emerge from the reductionist analysis. The obvious objection to this method is that one-variable systems can never represent the complexity of living processes. Human life particularly, in health or disease, is the resultant of countless independent forces impinging simultaneously on the total organism and setting in motion a multitude of interrelated responses. In this light, most medical problems appear so complex as to be beyond the range of experimental science.

However, the Cartesian analytical approach is so powerful that innumerable discoveries immediately emerged from its application to medicine. The immensely complex problems posed by living man were converted into much simpler questions focused on the mechanical aspects of structures and functions isolated from the body machine. To a very large extent the history of modern medical science consists in an attempt to pursue the reductionist analysis until it reaches into smaller and smaller fragments, or simpler and simpler functions. The study of life has thus become almost identified with the study of the molecules of which the body is made.

While the operations of the mind have not yet proved amenable to description in terms of molecular events, it seems that Descartes was too timid when he placed this field outside the realm encompassed by his method. On the one hand, nerve action can be described in much the same terms as muscular action. Even more striking is the fact that the most glamorous achievements in the study of behavior bear the stamp of the Cartesian scientific philosophy. Pavlov and the behaviorists after him have reduced complex responses into simple conditioned reflexes. And Freud taught that the mind of each individual person could be explained if one tried with enough skill and patience to identify the particular events that had shaped it early in its development.

Unfortunately, the very success of the reductionist approach has led to the neglect of some of the most important and probably the most characteristic aspects of human life. Starting from a question singled out for study because of its medical importance, the modern medical investigator is likely to progress seriatim to the organ or function involved, then to the single cell, then to the cellular fragments, then to the molecular groupings or reactions, then to individual molecules and atoms; and he would happily proceed, if he knew enough, to the elementary particles where matter and energy become indistinguishable. There is no doubt, of course, that fascinating and important problems continuously emerge at each step in the analytical disintegration of a complex biological phenomenon. But experience shows that all too frequently the original phenomenon itself is lost on the way. Many are the investigators who descend from man to molecules, but few are those who ever try the more difficult task of using molecular knowledge to deal with the problems of real life.

From a more practical, even if narrow, point of view, there is the obvious limitation that life is too short to make it possible to reduce all problems to their elementary constituents. As stated by Latham in his *Aphorisms:*

> It is all very fine to insist that the eye cannot be understood without a knowledge of optics, nor the circulation without hydraulics, nor the bones and the muscles without mechanics: that metaphysics may have their use in leading through the intricate functions of the nervous system, and the mysterious connections of mind and matter. It is a truth, and it is also a truth that the whole circle of science is required to compre-

hend a single particle of matter: but the most solemn truth of all is that the life of man is threescore years and ten.

Many scientists who dedicate themselves to medical research tend to shy away from the problems peculiar to man's nature, and even from those posed by other complex *living* organisms. They fear that the uniqueness of each individual person makes the kind of generalizations on which science thrives all but impossible, and for this reason they deal by preference with questions pertaining to lifeless fragments of the body machine. Yet while it is true that the responses of living man are extremely complex, they exhibit nevertheless patterns that can be described in the form of scientific laws. Such patterns, however, can be recognized only if scientists devote to the study of man's responses to his environment the intellectual effort and technical skill they presently devote to the analysis of fragments isolated in a lifeless form.

Awareness of the limitations of the reductionist approach in medicine expresses itself in the ever-increasing emphasis on the "whole man." But in practice this interest has not yet been converted into an active field of science. One of the reasons for this failure is the ambiguity of the word "whole." It is used on the one hand to denote those aspects of the organism that make it function as an integrated structure; on the other hand, it refers also to the summation of all the constituents and properties of the organism, including their individual relations to the total environment. Shrewd judgment based on experience will be needed to decide the extent to which the "whole" can be investigated by scientific methods. But unfortunately the research activity in this field is so limited that there is little chance for developing the experience needed.

It is certain in any case that, in comparison with the enormous effort devoted to the components of the body machine, living as a process has hardly been studied by scientific methods. The reason commonly given for this failure is that the proper techniques are not yet available and that such study must await the completion of more "fundamental" steps. But the truer reason is that this field of research does not fit in the reductionist philosophy that has prevailed since the seventeenth century. Many phenomena that have long been known empirically have been neglected by biological and medical investigators, not for lack of techniques but because their study was not fashionable. Such is the case with adaptive processes, conditioned reflexes, the subconscious manifestations of the mind, the effect of sensory deprivation, imprinting, and other phenomena of behavior.

Pavlov, Freud, Frisch, and Lorenz opened the scientific analysis of the responses made by man and animals to various situations, not so much by introducing new techniques as by accepting the fact that many aspects of life cannot be studied except when the organism functions as a living entity. Most of the techniques they used could have been developed long ago. The areas of knowledge to which they devoted themselves could have blossomed into full-fledged sciences long before the physicochemical sciences. Similarly, techniques could now be developed to study the living process in the full complexity of its manifestations, without waiting for further advances in the knowledge of the unit structures and reactions through which the body machine operates. One single example will be mentioned here to illustrate how prevailing scientific fashions lead to the neglect of important facts which have been carefully observed and recorded.

The first extensively documented account of the existence, composition, and natural history of chylomicrons is more than 40 years old. The original paper described the dramatic change in numbers of chylomicrons in relation to food intake. As is well known, an immense amount of detailed knowledge has now been accumulated concerning these bodies with regard to their submicroscopic architecture, as revealed by electron microscopy, their chemical composition with particular emphasis on the different fatty acids, and their fate following fat consumption. The original paper on chylomicrons described also other facts that have received much less attention. It reported in detail and with striking time curves that the fate of chylomicrons was profoundly influenced by emotional states. The prospect of having to give a lecture on anatomy was sufficient to alter profoundly the rate at which chylomicrons disappeared from the blood (Figure 14).

Here then was a phenomenon, objectively measurable, that could have been used to relate emotional states to certain aspects of fat metabolism. In addition to its intrinsic physiological and psychological interest, the phenomenon may also have an important bearing on the pathogenesis of certain diseases. Yet it has been essentially ignored. If the scientific effort and ingenuity devoted during the past 40 years to the chylomicron as a morphological

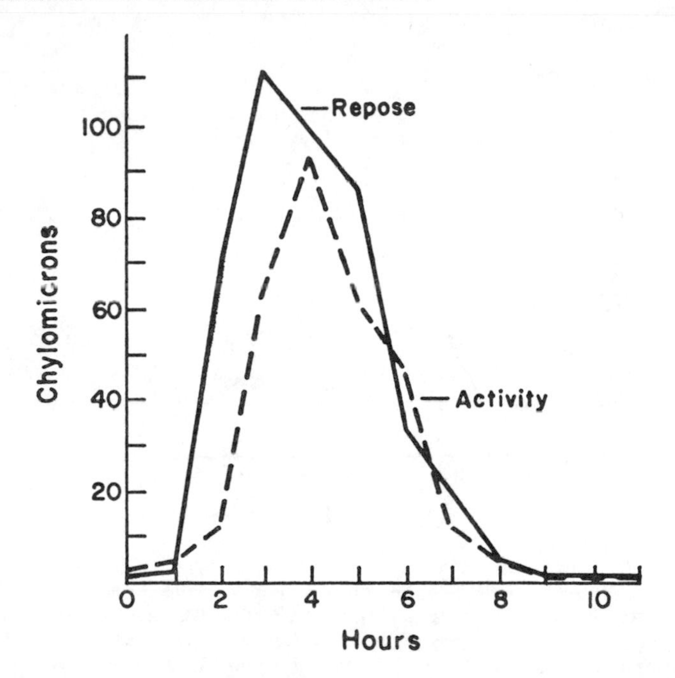

FIGURE 14. (a) Chylomicron curves, showing the digestive cycles of the same person in activity (A) and in repose (R). The area within curve A is 30 sq. cm. and within curve R 36.36. In this case the digestion was more perfect in repose than in activity. While the peak was reached one hour earlier with R than A, the entire period of the digestive cycle was the same in both—8 to 9 hours.

and chemical entity had been directed to the fate of chylomicrons as affected by life situations, it is probable that much progress would have been made in converting psychosomatic medicine from a vague word into an objective science.

Living man presents of course very special obstacles to those who attempt to investigate him scientifically. By the exercise of free will, he constantly introduces unpredictable complications into the study of his behavior and the effect that environmental factors exert on him. For these reasons, knowledge of human life cannot reach the level of precision and predictability achieved with regard to the inanimate world, or even to other living organisms in which the exercise of freedom can be kept under control. Granted these difficulties, man's responses to his environment pose problems of such urgency that scientists cannot long remain indifferent to them. The social pressures building up all over the world bid fair

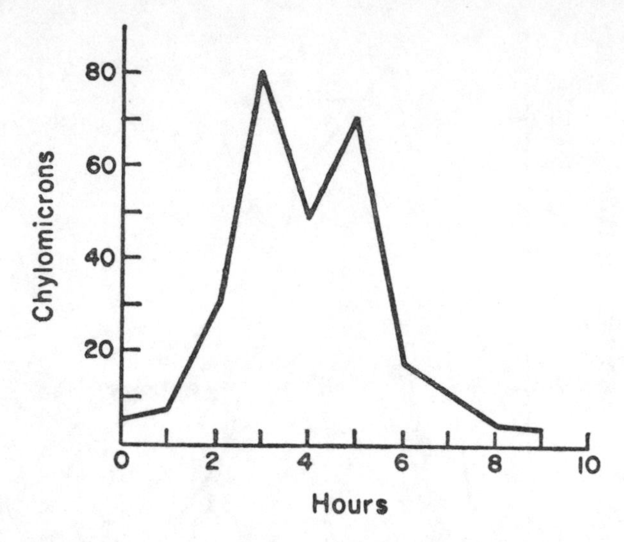

FIG. 14. (b) Chylomicron curve, showing cessation of digestion during severe mental effort. The meal consisted of 20 gm. of butter with toasted bread, coffee, and apple sauce. The subject was a teacher in the forties (K). Between the third and fourth hours a lecture was given. It is believed from the observations of Beaumont and many later workers that the corresponding depression on the curve at the time of the lecture indicates cessation of digestion (see also c).

to force biological and medical sciences into new directions, focused on the manner in which individual persons and human populations respond to their total environment.

Until the nineteenth century, most educated men believed that it was within the power of analytical science to reach a complete understanding of life, and to provide health and happiness for all. Few are those who have retained this euphoric attitude and believe that the accumulation of detailed knowledge can provide in any foreseeable future a reductionist explanation of man's special attributes. Expression of skepticism on this score might be regarded as a form of the antiscience movement, but the worst form of anti-intellectualism may turn out to be the unwillingness to acknowledge the present limitations of science, with its vested ideas and its neglect of certain problems of human life. Indeed, this self-satisfied attitude is likely to retard the development of scientific methods applicable to many of the problems that have direct relevance to the future of mankind, especially in the areas of human biology, psychology, and sociology.

There is no need to belabor the obvious truth that, while mod-

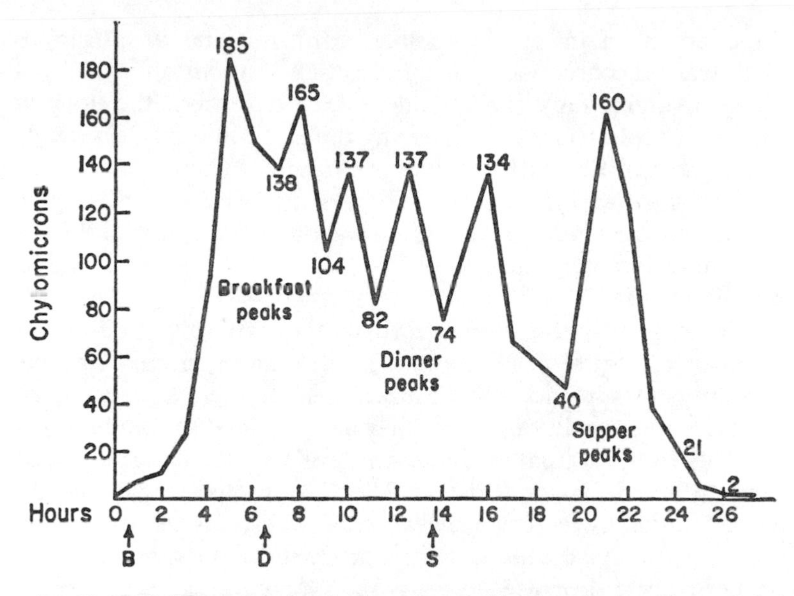

FIG. 14. (c) Chylomicron curve, showing effect on digestion of repeated mental and emotional states (3-meal cycle, full activity). Each of the 3 meals consisted of 40 gm. of butter with toasted bread and apple sauce. The letters B, D, S along the base indicate the time of each meal. During the middle part of the day there were severe interruptions; these account for the sawtooth appearance of the curve. In the morning before the interruptions and at night when they were all over, the curves are regular, typical curves of fat digestion, absorption, and assimilation. (Graphs and legends reproduced from S. H. Gage and P. A. Fish, 1924.)

ern science has been highly productive of isolated fragments of knowledge, it has been far less successful in dealing with the complexity of human problems. The high degree of specialization required for professional effectiveness accounts in part for this difficulty, since no one person can give thorough attention to the multiple facets always found in any human situation, or can control its multiple determinants. But above and beyond these techical complexities, the life sciences present other difficulties of a more philosophical nature, which do not fit readily in the usual conceptual attitude of scientists.

In the most common and probably the most important phenomena of life, the constituent parts are so interdependent that they lose their character, their meaning, and indeed their very existence, when dissected from the functioning whole. In order to deal with problems of organized complexity, it is therefore essential to investigate situations in which several interrelated systems function in an integrated manner. Multifactorial investigations will demand conceptual and experimental methods different from those involving only one variable, which have been the stock in trade of experimental science during the past 300 years. It is widely acknowledged that such methods are needed to bring sociological problems within the scientific fold; but it is less frequently recognized that the need is just as great for other biological problems. The most important aspects of life fall outside the net of reductionist analysis.

Two examples, discussed in earlier pages, will serve to illustrate types of biological problems that require an organismic and environmental approach. As we have seen, normal human beings placed under conditions where they are sheltered as completely as possible from external stimuli soon develop abnormalities in perception and behavior (Chapter I.3). The profound pathological effects caused by sensory deprivation, as well as by certain psychotomimetic drugs, demonstrate beyond doubt that the maintenance of personality structure depends upon, and indeed is an expression of, the responses that the organism as a whole makes to the bombardment of stimuli that is a constant feature of normal life. A similar conclusion emerges from an entirely different kind of facts. Animals raised under germ-free conditions, or deprived of their indigenous microbiota by prolonged treatment with antimicrobial drugs, develop gross histological and physiological abnormalities (Chapter V.3); the structural and functional integrity of many essential organs and structures depends, in other words, upon the constant stimulus exerted on them by the presence of microorganisms.

These two examples, so different in nature yet so concordant in their implications, illustrate the fundamental fact that organisms cannot be understood unless they are studied in their integrated responses to the environment. Oversimplified systems, involving only one variable, may be adequate for the analysis of the structures and chemical reactions found in the isolated components of the body machine, but they are not sufficient to study the complex manifestations of the processes and experiences of living. Yet these manifestations are the very subject matter of life, in health and in disease.

The physician ministers of course to the human body, but he is concerned also, and even more perhaps, with life as experience—physical, social, and mental. In final analysis, the success of his action is measured by the happiness and performance of his patient in a certain environment. One of the responsibilities of medical science is therefore to study the effects on the body and the mind of the new threats created by technological civilization; the constant exposure to environmental stimuli and pollutants; the estrangement of the human organism from the natural cycles under which evolution occurred; the solitude and emotional trauma of life in congested cities; the monotony and the boredom of regimented existence; the compulsory leisure ensuing from automation. These are the influences that are now at the origin of many medical problems characteristic of Western civilization. Most of the disorders of the body and the mind are expressions of inadequate response to environmental influences.

The scientific knowledge of man's environment and of his response to it is much less developed than the precise knowledge of body structure gained by one century of scientific research or than the empirical art of the healer gained from age-long experience. Like other biologists, any physician worth his salt knows that all the features and all the manifestations of an organism are influenced as much by the environment as they are by the genes. Genes determine not characters or traits, but reactions and responses. Health and disease are manifested in the phenotype of the organism. And in practice the phenotype is modifiable, or controllable, much less by the alteration of the genotype than by manipulation of the environment and by adaptive efforts.

A few examples based on subjects discussed in earlier chapters will be briefly reviewed once more in the following pages to illustrate what kind of knowledge is required to foster the growth of organismic and environmental medicine.

Laboratory methods exist or can be developed for studying the effects of most substances or stimuli on tissue cultures, biochemical systems, or isolated cells. In other words, certain biological aspects of the influence that technological innovations exert on the constituents of the human body can be brought within the field of cellular and molecular biology. This analytical approach leaves unexplored, however, the effects that have the most direct bearing on the actual life of man, as well as of animals. It does not provide the kind of knowledge that would be relevant to the various forms of response that the intact living organism makes to the actual conditions of exposure over long periods of time.

One of the reasons for this failure is that most effects of environmental forces are extremely indirect and delayed; they are transformed and magnified through a chain of reactions in which almost every organ is involved (Chapter VIII.4). Time is an essential component of the system, because exposure to a stimulus or to a substance that appears innocuous today may result after many months or years in crippling secondary effects, whether these be allergic reactions, malignant neoplasms, or psychotic states. Man fortunately has a wide range of adaptive potentialities and thus can achieve some form of adjustment to many different stressful situations (Chapter X.1). There are limits, of course, to the range of his adaptabilities, but they are unknown.

The problems posed by indirect and delayed effects, as well as the questions raised by adaptability and its limits, involve the responses of the organism as a whole. The effects on man of the conditions created by urban and industrial life in modern societies thus transcend the phenomena that can be recognized by the study of simplified experimental systems using only cells or isolated chemical reactions. They demand the use of more complex biological models observed over long periods of time.

The increasing concern with iatrogenic diseases, especially those caused by drugs, has the quality of a caricature in accentuating the new kinds of threats to health arising from technological innovations. It is a painful but richly documented paradox that each and every drug of proven worth in the treatment of disease can itself become a cause of disease, even when used with understanding, skill, and moderation (Chapter VIII.4).

Some of the toxic effects exerted by drugs are direct and can be studied by the orthodox toxicological techniques, using short-term tests in simple laboratory systems. In most cases, however, the toxic effects are extremely indirect and delayed. They result from disturbances in the physiological and ecological equilibrium of the organism. Their mechanism does not reside in chemical or physiological reactions involving direct cause-effect relationships, but rather in complex interrelated responses made by the whole integrated organism, including its indigenous microbiota. Clearly these forms of toxicity can be understood only through an organismic and ecologic approach.

The problems of behavior are of course among those which

require complex organisms for investigation. Many different kinds of substances are known to modify behavior not only in human beings but also in animals; the effects of the nutritional state and particularly of amino acid metabolism on mental processes are especially well documented. Furthermore, while behavioral response to any given situation is under genetic control it is also conditioned by various forms of deprivation during early life. Early associations and experiences, training, and crowding are among the many factors that modify behavior in animals and thus provide experimental techniques for behavioral studies (Chapter I.2).

It goes without saying that such investigations acquire their full scientific significance and practical usefulness only if carried out under a wide range of conditions, for long periods of time, even extending into several generations.

As discussed earlier, the size of the world population is conditioned by many biological and social factors that are poorly understood. It is known, on the other hand, that many animal species automatically adjust their populations to levels low enough not to overtax their resources. However these homeostatic mechanisms of population control do not always operate effectively. The understanding of the factors that control population dynamics is still very meager (Chapter XI.2).

Another aspect of the population problem is that any association of living things creates new properties that transcend the properties of each one of them. The individual members of a population interact with each other, and the consequences of this interplay affect their anatomical, physiological, and behavioral characteristics (Chapter IV.2 and XI). For example, the degree of crowding affects reproduction, growth, learning, and resistance to stress. The magnitude and direction of the effects cannot be predicted from even the most detailed knowledge of the individual components of the biological system. Crowds and even small groups respond differently from isolated organisms to almost any kind of stimulus.

Progress in the different aspects of population problems can be made only by studying experimental groups maintained under a wide range of conditions. By necessity such studies must be focused not on the component parts of the organisms, not even on the individual organisms themselves, but rather on their interplay as affected by the environment.

Harvey Cushing is reported to have taught that "a physician is obligated to consider more than a diseased organ, more even than the whole man—he must view the man in his world." So sweeping a statement expresses a lofty medical and scientific ideal, which is difficult if not impossible to fulfill in practice. But it is consistent with the view that medical sciences must concern themselves with complexities that are not encountered in most laboratory disciplines.

It has been observed, for example, that certain people within a given social group are much more vulnerable than others to almost any etiological variety of disease—whether it be infections, neoplasms, gynecological troubles, or mental disorders. Among these vulnerable people, diseases are more frequent and severe during periods when the environment is regarded by them as threatening, depriving, or overdemanding. Such facts obviously transcend the orthodox concept of etiological specificity (Chapter XII.2). They cast doubt furthermore on the view that there exists a consistent relation between certain kinds of bodily disease and personality type. As stated by Hippocrates 2,500 years ago and by Harvey Cushing more recently, the physician "must view the man in his world."

One of the most puzzling but also most important aspects of the healing art is that the physician can exert a healing effect by his very presence, even without using any objectively effective method of treatment. The maintenance of health also requires certain kinds of human experience and contact above and beyond the biochemical and physiological conditions required for cellular metabolism. Despair, grief, and other trials certainly have a part in the genesis of disease, and many physicians have stressed the role played by faith and hope in recovery. To a large extent, however, this kind of dependence of the organism on psychological forces has been dealt with exclusively in religious or spiritual terms.

Attempts have been made during recent years to accumulate evidence that helplessness, hopelessness, unresolved griefs, and "giving up" are attitudes likely to generate or aggravate many illnesses, even neoplasms. (Reviewed by Engel, 1962a and b; Schmale, 1958; Wolff, 1960; Wolf, 1961, 1963.) But despite improved insights into the workings of the mind, the problems of psychology and their bearing on health have remained on the whole outside the main channels of biological thinking and experimentation. Indeed, there is a widespread feeling that these problems cannot be studied by the usual scientific approach because each human being is unique with regard to his life experience, his attitude, and his values.

In reality, however, most members of any given culture share a number of attitudes, values, and modes of thought, which make their behavior largely predictable; furthermore, many fundamental traits are common to all of mankind. It should be possible, therefore, to base the study of man's responses on a large body of working assumptions and thus to assess the effect of certain environmental conditions on his health and performance.

The scientific exploration of these fields is rendered even more promising by the fact that most aspects of human life have their counterparts in some animal species. The response of experimental animals to a variety of insults and stimuli, and the incidence and severity of disease among them, can indeed be profoundly modified by loneliness, separation of the young from their mothers, and other conflict situations that present close analogies to the stresses of human life. Even the uniqueness of the individual is not peculiar to man, as every pet owner knows (reviewed in Engel, 1962). Since examples of changed responses brought about by isolation or crowding have been presented earlier (Chapters I.2 and IV.2), it will suffice merely to recall here the profound and varied physiological and behavioral disturbances that occur in monkeys raised by surrogate mothers.

Thus, observations in human beings and experiments with animals establish beyond doubt that health and disease are influenced by life situations that transcend the direct impact of physicochemical forces. The mechanisms involved are different from those revealed by the concepts and operations of physicochemical biology, and consequently different experimental methods are required for their investigation.

XIII. Man Meets His Environment

1. The mirage of health

Benjamin Franklin once expressed in a letter to the chemist Joseph Priestley his belief that a time would come when "all diseases may by sure means be prevented or cured, not excepting that of old age, and our lives lengthened at pleasure even beyond the antediluvian standard." Depending upon one's prejudices, this statement may appear as a prophetic view of modern medical achievements, or as a naive expression of wishful thinking. More probably, however, Franklin simply reflected the optimism that permeated life during the Age of Reason. Condorcet, for example, went even further than Franklin in his optimistic prediction of the future. In the *Sketch for a Natural History of the Progresses of the Human Mind* he wrote in 1794, "I will show . . . that Nature has set no limit to the perfecting of the human faculties, that the perfectability of man, henceforth independent of any power that might wish to arrest it, has no other limit than the duration of the globe on which Nature has placed us."

During the past two centuries, social reformers and natural scientists have gone far toward converting into reality the concept of the Golden Age, so vividly imagined by Franklin and Condorcet. Indeed, it is truly a prodigious feat that the era of plenty, which was but a vague utopian dream at the end of the eighteenth century, has come to pass within less than 200 years of that time in several countries of Western civilization. If the social philosophers of the Age of Reason had spelled out in detail what they considered to be the essential requirements of health and happiness, their imaginings would probably prove to have become part and parcel of everyday life in the United States. Yet it is plain that health and happiness have not necessarily followed in step with social, economic, and medical advances. Political freedom, abundance of worldly goods, and the miracles of modern medicine obviously are not sufficient to deal with the problems of the body and the mind that continue to plague man even under conditions of peace and prosperity.

One of the paradoxical aspects of the health picture is that despite the improvements in sanitation and nutrition, despite effective protection against heat, cold, humidity, and physical fatigue, an increasingly large percentage of the population depends on medical help for its daily existence. Most of the great plagues of the past have been brought under control, and more people than ever enjoy safety and comfort; yet the need for medical care and for hospital facilities is increasing everywhere. This paradox is not peculiar to the United States; it is encountered in each and every prosperous country of the Western world. The history of the National Health Service in England has provided a striking demonstration of the fact that general availability of medical care does not necessarily decrease the incidence of illness.

In 1942, *The Beveridge Report on Social Insurance and Allied Services,* which prepared the ground for the National Health Service in Great Britain, estimated that the annual cost of a comprehensive health service in that country would be £170,000,000. It was predicted in the Report that the development of the National Health Program and the consequent reduction in its need would cancel out, leaving the cost unchanged in 1965. The fact is, however, that the annual cost of the program had risen to £400,-000,000 as early as 1952; furthermore, it has continued to soar ever since. Only the imposition of a ceiling by the Treasury prevents it from reaching figures much higher than the present ones, and there is no indication that the upward trend will level off in the foreseeable future.

On January 13, 1949, within six months after the inception of the National Health Service, Mr. Ian Mikardo, who was then a member of Parliament, assured his readers in the *Daily Mail* that "the Health and National Insurance Services have settled down and are making us a healthy and unworried people." Although Mr. Mikardo's assertions now appear somewhat ridiculous in the light of the present health situation in Great Britain and elsewhere, the fact is that they expressed the utopian views of universal health through the eradication of disease that were then and are still widely held all over the world. Even greater hopes than those expressed by Mr. Mikardo are being fostered by the present-day flowering of scientific knowledge. There is a very general tendency to believe that the time has come to raise the level of our sights and that we must plan for the absolute eradication of certain diseases instead of being satisfied with control (Chapter XIV). Many physicians and public health officers claim furthermore that health should be regarded not merely as absence of disease but as a positive attribute, enabling mankind to take advantage of all its physiological resources as well as of its emotional potentialities.

The concept of perfect and positive health is a utopian creation of the human mind. It cannot become reality because man will never be so perfectly adapted to his environment that his life will not involve struggles, failures, and sufferings. Nevertheless, the utopia of positive health constitutes a creative force because, like other ideals, it sets goals and helps medical science to chart its course toward them. The hope that disease can be completely eradicated becomes a dangerous mirage only when its unattainable character is forgotten. It can then be compared to a will-o'-the-wisp luring its followers into the swamps of unreality. In particular, it encourages the illusion that man can control his responses to stimuli and can make adjustments to new ways of life without having to pay for these adaptive changes (Chapter X.2). The less pleasant reality is that in an ever-changing world each period and each type of civilization will continue to have its burden of diseases created by the unavoidable failures of adaptation to the new environment.

Nothing illustrates better the collective loss of objectivity created by the mirage of health than the constant reiteration that recent advances have brought about a spectacular increase in the expectancy of life. It is true, of course, that the expectancy of life *at birth* has been increasing steadily during the past few decades, but as emphasized earlier, this increase is due almost exclusively to the virtual elimination of deaths during infancy, childhood, and early adulthood (Chapter IX). In contrast, the life expectancy of adults is not very different now from what it was a few generations ago, nor is it greater in areas where medical services are highly developed than in less prosperous countries. True longevity has increased very little, if at all.

The present clamor for better medical care and more hospital facilities makes it clear that today as in the past we live in a disease-ridden society. Needless to say, the increase in medical needs does not necessarily imply a parallel increase in the prevalence of disease. To a certain extent, it comes from the fact that we are more exacting than our ancestors in matters of health, and especially are we less willing to accept the infirmities, pains, and blemishes, the catarrhs, coughs, and nauseas that used to be regarded as inevitable accompaniments of life. But granted that health criteria are constantly rising and that we are less and less willing to accept passively physical and physiological imperfections, it is also true that the modern ways of life are creating problems of disease that either did not exist a few decades ago or are now more common than in the past.

A few examples contrasting the medical problems of recent history with those characteristic of the present times will suffice to show how the pattern of diseases rapidly changes with the conditions of life.

Whereas microbiological pollution of water used to be responsible for much disease among our ancestors, chemical air pollution is now taking the limelight. Chemical fumes from factories and exhausts from motor cars are causing a variety of pathological disorders that bid fair to continue increasing in frequency and gravity and may become serious health handicaps in the near future (Chapter VIII). There is reason to fear also that various types of radiation will soon add their long-range and unpredictable effects to this pathology of the future.

During recent decades we have gone far toward controlling microbial spoilage of food, but some of the new synthetic products that have become part and parcel of modern life are responsible for an endless variety of allergic and other toxic effects.

Nutritional deficiencies have now become rare in wealthy countries; but a new kind of malnutrition is arising from the fact that the nutritional regimens formulated for physically active human beings are no longer suited to automobile-borne and air-conditioned life in the twentieth century.

In the past many human beings suffered from physical exhaustion; now labor-saving devices and especially automated operations threaten to generate a type of psychiatric disturbances that will greatly complicate the medicine of tomorrow; boredom is replacing fatigue.

Who could have dreamed a generation ago that hypervitaminoses would become a common form of nutritional disease in the Western world; that the cigarette industry, air pollutants, and ionizing radiations would be held responsible for the increase in certain types of cancer; that the introduction of detergents and various synthetics would increase the incidence of allergies; that advances in chemotherapy and other therapeutic procedures would create a new staphylococcus pathology; that patients with various forms of iatrogenic disease would occupy such a large number of beds in the modern hospital; that some maladies of our times could be referred to by an eminent British epidemiologist as "pathology of inactivity" and "occupational hazards of sedentary and light work"?

There is reason to believe that the changes in ways of life and in technology are responsible in part for the degenerative diseases characteristic of our civilization (Chapter IX.3). But while this state of affairs demands serious attention, it need not create panic. The increased prevalence of certain diseases is not a new phenomenon. Disease presents itself in different forms today simply because the world is changing and demands new adaptive responses from human beings.

2. Health as ability to function

The complex nature of man's response to his environment accounts for many of the difficulties experienced in developing methods for the prevention of disease. It is also responsible for the fact that the precise meanings of the words health and disease differ from one social group to another or even from person to person. Furthermore, the meanings change with time as well as with the environment and ways of life.

The definitions of health and disease that are commonly given by encyclopedias and academies remind one of Diafoirus' statement in Molière's play *Le Médecin malgré lui* that "opium induces sleep by virtue of the fact that it possesses a sleep producing property." In words that bring to mind this empty definition by which Molière poured ridicule on the ignorant and pretentious physicians of his time, modern dictionaries define disease as any departure from the state of health and health as a state of normalcy free from disease or pain. In 1946 the medical experts of the World Health Organization tried to sharpen and to enlarge the meaning of the word health by affirming that it implies not merely the absence of disease but rather a *positive* attribute, like a kind of primeval euphoria that would enable men to take advantage of all their potentialities for vigor and happiness. The introductory paragraph of the Constitution of the World Health Organization states: "Health is a state of complete physical, mental, and social well-being and is not merely the absence of disease or infirmity."

Health will be considered in the following pages from a more prosaic point of view. Instead of assuming that an ideal state of positive health can be achieved by eradicating all diseases from a utopian world, I shall take the view that the man of flesh, bone, and illusions will always experience unexpected difficulties as he tries to adapt to the real world, which is often hostile to him. In this light, positive health is not even a concept of the ideal to be striven for hopefully. Rather it is only a mirage, because man in the real world must face the physical, biological, and social forces of his environment, which are forever changing, usually in an unpredictable manner, and frequently with dangerous consequences for him as a person and for the human species in general. In the picturesque words of an English public health officer, "Man and his species are in perpetual struggle—with microbes, with incompatible mothers-in-law, with drunken car-drivers, and with cosmic rays from Outer Space. . . . The 'positiveness' of health does not lie in the state, but in the struggle—the effort to reach a goal which in its perfection is unattainable" (Gordon, 1958).

The evaluation of health and disease varies from person to person because it is conditioned by highly individual requirements and subjective reactions. In consequence, the words health and disease cannot be defined universally or statically. A Wall Street executive, a lumberjack in the Canadian Rockies, a newspaper boy at a crowded street corner, a steeple chase jockey, a cloistered monk rising during the night to pray and chant, and the pilot of a supersonic combat plane have very different physical and mental needs. The various imperfections and limitations of the flesh and the mind do not have equal importance for them. As pointed out by Ruth Benedict in *Patterns of Culture,* social aggressiveness is socially unacceptable and regarded as a form of disease by the Pueblo Indians, whereas this attribute is desirable in the countries of Western civilization because it facilitates the achievement of power and the acquisition of wealth.

Health is an even more elusive concept in the case of women.

A farmer's wife with several children and a New York fashion model of the same age have very different physical requirements and therefore have different concepts of health. Furthermore, it is apparent from the history of fashions and from contemporary tastes that ideals of the feminine figure and complexion have undergone a wide gamut of changes in the course of time, and still differ at present from one country to the other. The fleshiness of the Paleolithic Venuses or of Ruben's goddesses reflects an attitude toward womanhood oddly different from the tastes that generated the slenderness of the English pre-Raphaelite models or of the American flapper in the 1920s. Entertainingly enough, sexual selection of hyperthyroidism occurred in much of Southern Europe during the sixteenth and seventeenth centuries, because this disease was then considered to enhance the attractiveness of young women. Lest this should appear nonsensical to us, we should remember that the dictates of publicity certainly affect our own concepts of health and physical attractiveness.

It must be acknowledged that history records many situations in which human beings seem to have achieved a state of physical development that can be regarded as healthy according to any criterion. In the account of his travels, for example, Christopher Columbus expressed admiration for the beautiful physical state of the natives he had found in the West Indies. Captain Cook, Bougainville, and the other navigators who discovered the Pacific islands also marvelled at the vigor of the Polynesians at that time; and similar reports came from the European explorers who first saw the American Indians in the Great Plains and in the Rio Grande valley. Even in modern times, people like the Xavantes Indians or the Mebans of East Africa generally remain vigorous as long as they live in isolated communities and retain their ancestral ways of life (Chapter IX.3).

In practically all cases, however, primitive people have undergone physical decadence within one generation after having had extensive contacts with the white man. It would seem therefore that the health of primitive people, like that of animals in the wild state, depends upon their ability to reach and maintain some sort of equilibrium with their environment (Chapter X). In contrast, all of them are likely to fall prey to disease when their ancestral conditions of existence suddenly break down. For them, and certainly for us also, health can be regarded as an expression of fitness to the environment, as a state of adaptedness.

The societies of Polynesians, American Indians, Eskimos and other people who appeared so vigorous when first seen by European explorers provide examples of what Arnold Toynbee has called "arrested civilizations." They represent societies that lived for long periods of time under fairly stable physical and social conditions and had little if any contact with the outside world. New diseases appeared among them as soon as their *status quo* was disturbed, because changed conditions made new adaptive demands for which they were not prepared.

The examples mentioned above, among countless others which could have been selected, illustrate that it is not possible to define health in the abstract. Its criteria differ with the environmental conditions and with the norms and history of the social group. The criteria of health are conditioned even more by the aspirations and the values that govern individual lives. For this reason, the words health and disease are meaningful only when defined in terms of a given person functioning in a given physical and social environment. The nearest approach to health is a physical and mental state fairly free of discomfort and pain, which permits the person concerned to function as effectively and as long as possible in the environment where chance or choice has placed him. "Work is more important than life," Katherine Mansfield confided to the last pages of her journal. Searching for a definition of health as she was dying of tuberculosis, she could only conclude: "By health I mean the power to live a full, adult, living, breathing life in close contact with what I love—the earth and the wonders thereof. . . . I want to be all that I am capable of becoming."

3. Hygeia: a city of health

During the late eighteenth and early nineteenth centuries, there was a strong revival of the Hippocratic doctrine which taught that man had a good chance of escaping disease if he lived reasonably. This doctrine encouraged for a while the illusion that civilized man could recapture physical well-being and primeval vigor simply by returning to the ways of nature. Despite its romantic appeal, however, the literature based on the naive belief in the virtues and perfect health of the noble savage had little influence in modifying behavior. It did not transform civilized Europeans into children of nature. But the new attitude was important nevertheless, because it created an intellectual climate favorable for the Sanitary Revolution. In this atmosphere, the philosophers of the Enlightenment and the practical sanitarians who followed them found it possible to raise the environmental concept of health from the individual to the social level. The social reforms that provided a partial solution to the health problems created by the Industrial Revolution emerged from this new medical philosophy.

Around the middle of the nineteenth century, laymen as well as physicians became increasingly concerned with the fact that disease and physical frailty were most common among the poor classes. In *Conditions of the Working Man in England* Engels wrote of the "pale, lank, narrow chested, hollow-eyed ghosts," riddled with scrofula and rickets, who haunted the streets of Manchester and other manufacturing towns. If ever men lived under conditions completely removed from the state of nature dreamed of by Rousseau and his followers, it was the English proletariat of the 1830s. Public-minded citizens came to believe that, since disease always accompanied want, dirt, and pollution, health could be restored only by bringing back to the multitudes pure air, pure water, pure food, and pleasant surroundings.

The sanitary ideal developed at first without any support from laboratory science. As mentioned earlier, it emerged from the conviction that the high rates of disease and death were largely preventable because they were due to filth, dirt, crowding, and other social factors that could be corrected. Simple as this concept was, however, it would not have become a creative force in medicine and public health if it had not been publicized and instrumented by intensely dedicated social reformers.

In Germany the most picturesque and influential of the pioneers in sanitation was the chemist Max von Pettenkofer, who regarded hygiene more as an all-embracing philosophy of life than as a laboratory science. In Munich he persuaded the city fathers to bring clear water in abundance from the mountains, to clean streets and houses, to cart away refuse and garbage, to dilute the sewage downstream in the Isar, and even to plant trees and flowers, which he regarded as essential to the mental well-being of the population because they satisfy esthetic longings. Following these measures, the typhoid mortality in Munich fell from 72 per million in 1880 to 14 in 1898. The city soon became one of the healthiest in Europe, thanks to the efforts of this imaginative and enterprising hygienist who was not a physician and did not believe in the germ theory of disease. Max von Pettenkofer's lectures *The Health of a City* conveyed to the general public the view that collective cleanliness was the surest approach to health.

In England the frontal assault on filth with the goal to create a healthier world became a national commitment involving all social groups. The campaign was carried out by members of the aristocracy like Lord Ashley, statisticians like William Farr, clergymen like Charles Kingsley and Frederick D. Maurice, physicians like Southwood Smith and John Simon, bureaucrats like Edwin Chadwick, and by thousands of anonymous zealots who belonged to the Health and Towns Association, a movement that constitutes the prototype of the present-day voluntary health associations throughout the world.

The aim of the Health and Towns Association was to

> substitute health for disease, cleanliness for filth, order for disorder . . . prevention for palliation . . . enlightened self-interest for ignorant selfishness and bring home to the poorest . . . in purity and abundance, the simple blessings which ignorance and negligence have long combined to limit or to spoil: *Air, Water, Light.*

The association undertook a program of education in all phases of welfare and recommended steps for improving the attractiveness of habitations and surrounding areas. It even encouraged the development of gardens and the maintenance of lanes about large cities for the enjoyment of the public. Faith in the healing power of pure air, with much contempt for the germ theory of disease, was also the basis of Florence Nightingale's reforms of hospital sanitation during the Crimean War. "There are no specific diseases," she wrote. "There are specific disease conditions." According to her beliefs, these conditions could be removed by improving the environment. John Simon was the standard-bearer of this philosophy of public health in the field of governmental administration. The official outcome of the Sanitary Movement in England was the Public Health Act of 1875.

In the fall of the same year, the physician and sanitarian Benjamin Ward Richardson outlined before the Health Section of the Social Science Association a utopian picture of *Hygeia: A City of Health.* His purpose was "to show a working community in which death . . . is kept as nearly as possible in its proper or natural place in the scheme of life." Richardson's Hygeia is noteworthy among the long list of utopias for being devoted almost exclusively to health problems and for stressing that preventive measures are of greater social importance than curative treatments. His emphasis on the need for social control of the environment was the more remarkable because it was in such obvious conflict with the political philosophy of extreme individualism advocated by Herbert Spencer and so widely accepted during the Victorian era.

Richardson's description of Hygeia was received with much enthusiasm and reported at length in British, American, and Continental newspapers. Hygeian city planning, Hygeian residences, and Hygeian household products achieved wide popularity on both sides of the Atlantic Ocean and became the symbols of a new kind of social idealism. The following quotation from a recent analysis of Richardson's original text gives some idea of the scope of his vision and the extent to which it has influenced modern life and is still influencing social planning.

> Richardson limited his city to 100,000 persons. He provided, moreover, a limit of 25 persons per acre and houses which should not exceed four stories in height. For maximum ventilation and sunlight, every house had its own ample garden in back. Even public buildings were invariably surrounded by lawn and garden space.
>
> The paved streets, of course, were spaciously wide and were kept spotlessly clean. They were washed daily, and debris was carried away beneath the surface. To eliminate noise and accidents, surface streetcars were eliminated in favor of an underground railway system.
>
> Individual houses in Hygeia were built uniformly on brick arches in order to ensure good drainage and ventilation. For the same reason, wall bricks were perforated in order to allow constant circulation of air at all times and in winter to help in heating. Interior wall bricks, which came in various colors, were glazed for ease of washing and so that unsanitary wall papers with their poisonous pastes could be avoided.
>
> There were no dank cellars or evil basement habitations about which the Dickens of the City of Health could moralize. All rooms were above ground. Reversing normal practice, Richardson placed the kitchens on the upper floors. There they were lighter and cleaner, and odors disappeared

more quickly than when they were on the bottom floor. Bedrooms were large, airy, well lighted, and kept free of all unnecessary furniture. Similarly, unsanitary, dirt-catching carpets were never laid on living room floors. Bedrooms as well as bathrooms were provided with hot and cold water taps, while a continuous water supply and carefully designed plumbing eliminated the danger from the sewer gases which Victorians feared so much. It went almost without mention in 1875 that all houses were serviced by underground water, gas, and sewerage pipes.

The means of ensuring clean soil, clean water, and clean air for the citizens of Hygeia were far in advance of Victorian realities. The filtered water supply of the city was tested twice every day, while a large generator passed ozone into the drinking water whenever an excess of organic impurities existed. Scavengers came around every day to remove garbage and refuse in closed vans. Sewage was disposed of on sewage farms far out of town. Similarly, noisy factories, laundries, slaughter houses, and other offensive trades had to be located outside of the city. The conduct of such establishments, of course, as well as the raising of animals for meat, was carefully supervised.

In various respects Hygeia would have been a Methodist's delight. By ways which Richardson left unexplained, the use of tobacco had totally disappeared. Similarly, the citizens miraculously seemed to have lost their thirst for alcoholic beverages. This presumably had eliminated alcoholism, while it had also caused a great decline in the number of the insane.

If Richardson accomplished these results in Hygeia largely by wishful thinking or by unspecified administrative fiat, he was much less vague about the actual presence of sickness and how to prevent or deal with it. Drawing from the experience of the sanitary revolution, he could picture the "principal sanitary officer" of Hygeia at the head of an entire hierarchy of medical officers, registrars, sanitary inspectors, chemists, and scavenging personnel. This body of men enforced all of the meticulous sanitary rules of the city, which began with providing for the basic cleansing needs and went on to cover such things as the use of public buildings, the conduct of funerals and burials, and working conditions in various trades. Textile workers, for one example, were not

> allowed to take any of their work to their homes, so that diseases in their families would not spread to the purchasers.
>
> To inhibit the high infant mortality which was the plague of the nineteenth century, Richardson suggested a large number of public homes where trained nurses would care for infants and small children. For the rest of the populace of Hygeia there was a system of model hospitals, unique in some respects, but reminiscent of the plans of More, Mercier, and Cabet [Cassedy, 1962].

Von Pettenkofer's *The Health of a City* and Richardson's *Hygeia: A City of Health* are predicated on the nineteenth-century creed that "Pure Air, Pure Water, and Pure Food" constitute the necessary ingredients of a healthy life. On the other hand, these two books also mark the end of the first phase of the sanitary era because, as they were being published, laboratory science had begun to take the limelight on the health scene. The new laboratory knowledge was rapidly rendering obsolete the Victorian philosophy of health and was taking its place in medical schools and departments of sanitation. The general cleansing of the environment was almost forgotten as disinfection, vaccination, vitamins, drugs, and diagnostic laboratories became the new themes of the health slogans. Charles V. Chapin, the dynamic Health Commissioner of Providence, Rhode Island, went as far as to state that it mattered little hygienically whether the city's streets were cleaned or not, provided microbes were kept under control and people were protected against infection by the proper vaccines.

4. Environmental health

Experience has now proved that it is not sufficient to clean up the environment, control pathogens, and provide abundant nutrition to bring the City of Health into being. Everywhere in the countries of Western civilization, the chronic diseases of degenerative, metabolic, or neoplastic nature have come to take the place of infectious and nutritional diseases. There is as yet no solution in sight for these new plagues of civilized life, and the disturbing fact is that the climate of opinion in the lay and scientific public discourages any vigorous social effort to develop preventive measures against them.

Under the alluring title "Progress in Cures," a popular magazine quoted an "American scientist" in 1962 as stating: "The five most-wanted cures are . . . a drug to cure cancer; a drug to maintain emotional equilibrium; a penny-a-day birth-control pill; a 'virus killer' to kill viruses as antibiotics kill bacteria; a 'youth pill' to delay man's aging process and prevent such degenerative diseases as arthritis" (*Ladies Home Journal,* October 1962). This statement might be dismissed as merely a trivial example of shallow and irresponsible journalism, but in fact it symbolizes an attitude that is unfortunately very common not only among laymen but also among scientists, and dangerous for the future.

As we have seen, the disease problems created by the Industrial Revolution stimulated in the Western world a crusading spirit that engendered great social reforms and a healthier environment. In contrast, as the magazine article mentioned above illustrates, the solution to the medical problems of our times is sought not in environmental or social reforms, but in pills. A few weeks after publication of the Surgeon General's Report on cigarette smoking, a survey among high school students in Westchester County, New York, revealed that awareness of the danger had not affected their cigarette consumption. The reason was plain from the answer made by a White Plains high school senior who was a heavy smoker: "When I reach the age to develop lung cancer, *they* will have discovered a drug to cure it" (italics mine).

This answer is significant on two accounts. First, it expresses an attitude that tends to substitute for personal responsibility a hope that somebody else—symbolized by the anonymous *they*—will find a painless and effortless solution for all problems. Secondly, it illustrates how the approach to health through reform of the ways of life has been replaced by a formula that requires no greater effort than a telephone call to the drug store. Yet it is certain that now as in the past the only real solution to any disease problem is prevention rather than cure, and that prevention demands both concerted social effort and personal discipline.

Like the lay public, the scientific community has paid little heed to the fact that the diseases characteristic of our times are to a large extent the consequence of changes in the ways of life and in the environment. The search for the environmental determinants of disease is not a fashionable topic and carries little scientific prestige. As a result, so little effort has been devoted to the subject that the relevant information available is at best only suggestive and usually not sufficient to provide a basis for social action. The lack of interest in the role of the environment in the causation of disease came to light, for example, at the hearings held before the Subcommittee of the Committee on Appropriations, House of Representatives, Eighty-Sixth Congress, Second Session on "Environmental Health Problems." In answer to a plea for enlarged Federal support to the program, Congressman John E. Fogarty, chairman of the Subcommittee, pointed out to Surgeon General Leroy Burney that "environmental health doesn't seem to ring a bell with people. . . . To the average person, if you start talking environmental health, they are just not interested."

It is of course extremely difficult to determine the precise origin of any environmental insult and to attribute the guilt to any particular person or groups of persons. For example, every member of the community has a share of guilt in air and water pollution. Collectively we discharge into the environment the immense amount of chemical and organic refuse of our industrial and urban society, as if we hoped that somehow or other the collective garbage would get lost and disappear without trace. But in fact it lingers with us, polluting the air we breathe, the water we drink, and the food we eat. A few decades ago, pollution meant a dead animal, putrefying garbage, or ill-smelling water, which could readily be detected and therefore stimulated public protest. In contrast, we cannot smell the ionizing radiations, the nitrogen oxides in the air, the chemicals in the water, or the penicillin in the milk. Yet none of us can escape exposure to the pollutants that are the products of our collective life. It is estimated that by 1970, 75 per cent of the people of the United States will be huddled on 10 per cent of the land area, where all the environmental pollutants will also be concentrated.

Fifty years ago, when the horse was the chief source of energy for transportation, there were stables on every block and countless flies. The automobile has done away with this breeding ground for flies and has generated many forms of social advance. The price, however, is chemical pollution of the air with the unknown consequences that it may entail, in addition to the more than 40,000 deaths and the millions of persons maimed every year on the highways. Bumper crops and the countless conveniences of daily life go hand in hand with the yearly introduction of some 500 new chemicals, many of which survive the destructive forces of nature and therefore accumulate in our environment and our bodies. These paradoxes of modern life point to a kind of social problem that had not been anticipated either by the sanitarians or by the laboratory scientists who so earnestly believed they knew how to build the City of Health. The flaws in their plans arise from inconsistencies and conflicts in our value systems. Crudely stated, the question is to decide whether health or economic growth should have priority in determining the type of environment in which we live.

It is important to acknowledge that conflicts exist between commercial profits and prevention of disease, because awareness of this fact leads in many cases to technical solutions that are compatible with both kinds of values. In 1906, for example, three-quarters of the dairy herds of the New York City milkshed contained tuberculous cows; living tubercle bacilli could be recovered from more than 10 per cent of the milk samples examined. The public health demand for pasteurized milk was at first opposed with the arguments that heating would destroy the value of milk and that the capital and maintenance costs needed for pasteurization plants would price milk off the market. When the dairy industry was compelled to meet the challenge, however, it developed efficient and inexpensive techniques even while it was fighting the very principle of pasteurization. Similarly, there is reason to believe that if the public were really concerned, it could compel the various industries to eliminate many types of environmental pollutants, and to investigate more thoroughly than is done at present the potential health dangers of technological innovations.[1]

Making studies of the health aspects of technological innovations a part of development research would naturally increase costs. and might even retard economic growth in some cases, but the overall results would be of benefit to the community in the long run. A kind of crusade similar to that conducted by the nineteenth-century sanitarians is needed to create the social atmosphere in which the study of long-range health consequences is considered an

1. There is as yet no evidence that the general public is much concerned with the dangers to health posed by modern technologies. On the other hand, several isolated groups of scientists have taken it as their social responsibility to discuss in public the health and social implications of science. The Greater St. Louis Citizens' Committee for Nuclear Information has been a pioneer in this movement. Its journal *Nuclear Information* has recently been renamed *Scientist and Citizen* to reflect the broadening of its field of interest, especially with regard to environmental pollution. The activities of the various local groups concerned with informing the general public on the implications of science are now coordinated through a national Scientists Institute for Public Information (SIPI).

essential part of technological research. In fact, a precedent and a model for this attitude is provided by the large-scale efforts that have been made during the past few decades to determine the biological effects of ionizing radiations long before they were used in industrial practice.

Granted that the threats to health posed by the present-day technologies are often exaggerated, there are many facts that justify, nevertheless, anxiety for the future. One of the alarming aspects of environmental pollution is that despite all the new powers of science, or rather because of them, man is rapidly losing control over his environment. He introduces new forces at such a rapid rate, and on such a wide scale, that the effects are upon him before he has a chance to evaluate their consequences. A few examples will suffice to illustrate how technological developments can have dangerous effects that are recognized only after the biological damage has been done.

The photochemical conversion of hydrocarbons and nitrous oxides into the toxic products responsible for the Los Angeles type of smog was recognized only after the California economy had become dependent on an excessive concentration of industries and automobiles. The absorption of radioisotopes in the human body became known only several years after the beginning of large-scale nuclear testing. The resistance of synthetic detergents to bacterial decomposition became evident only after their universal use as household items had led to their accumulation in water supplies. Many drugs have been used indiscriminately before their potential toxic effects could be recognized. In this regard, the thalidomide tragedy served to highlight a situation of far more general significance; for example, more than 10 years of widespread use of the tetracyclines elapsed before it was noticed that these drugs accumulate in certain tissues and can interfere with bone growth.

In brief, the techniques available for developing new means of action are far more powerful than those available for recognizing the long-range biological effects of new technological innovations. Even when thorough studies are carried out in advance, and elaborate precautions are taken, some new technologies are bound to result in accidents, and in manifestations of chronic toxicity that cannot be foreseen. This is the price of change and of industrial growth. To require a certificate of absolute safety as a condition for the acceptance of a new product, or technological procedure, would be to paralyze progress, indeed would be incompatible with social growth. (See Chapters XV.6 and XVI.2 for further discussions of this topic.)

5. The limitations of environmental control

Richardson and von Pettenkofer believed that a City of Health would automatically emerge if people were given a chance for a moral life in a sanitary environment providing pure air, pure water, pure food, and esthetic surroundings. Two decades later, Elie Metchnikoff expressed a similar faith when he coined the word *orthobiosis.* According to him, men could avoid degenerative diseases and become centenarians by eating the proper kind of food, controlling the microbiota of their intestinal tract, and behaving reasonably. Being thus able to fulfill their biologic potential during a long life, they would no longer fear death, but instead welcome it as one does rest at the end of a full day.

Richardson, von Pettenkofer, and Metchnikoff were representatives of a medical tradition long in vogue among social planners. In ancient China, this tradition was codified in the Yellow Emperor's *Classic of Internal Medicine,* which taught that "the ancients followed Tao and the laws of the seasons under the guidance of their sages who were credited with the realization of the value of education in the prevention of disease." In classical Greece the goddess Hygeia personified physical and mental health achieved through a life of reason—*mens sana in corpore sano.* Plato asserted in his *Republic* that the need for many hospitals and doctors was the mark of a bad city. Physicians, according to him, were of use only for the treatment of wounds and during epidemics. "To stand in need of the medical art through sloth and intemperate diet . . . obliging the skillful sons of Asclepius to invent new names of diseases, such as dropsies and catarrhs—do you not think this abominable?" he asked. In Imperial Rome, Tiberius declared in a similar mood that anyone who consulted a doctor after the age of 30 was a fool for not having yet learned to regulate his life properly without outside help. J. J. Rousseau, the nineteenth-century sanitarians, and the nature worshippers of our era have kept alive in the form of a few reasonable practices and many futile if romantic fads the belief that health can be achieved through a sensible way of life in harmony with nature.

The medical utopias of our times are very different in spirit from those of the past. They assume that the achievement and maintenance of health depend not upon living in accordance with the laws of nature, but upon medical care based on scientific biotechnology. It is this philosophy that governs the World Health Organization in its efforts to establish health on a global scale. In the words of its Director General, Dr. M. G. Candau,

> If the great advances gained in science and technology are put at the service of all people of the world, our children and their children will live in an age from which most of the diseases our grandparents and parents took for granted will be banished. . . . It may no longer be utopian to envisage a new chapter in the history of medicine.

And another official of the World Health Organization stated on the occasion of its tenth anniversary, "Man seems to have found out how to make his dreams of a paradise on earth come true."

In all countries that have adopted the Western ways of life, especially of course in the United States, the emphasis is on controlling disease rather than on living more wisely. In fact, the general assumption throughout the world is that health cannot be achieved, or maintained, without vaccines, drugs, pesticides, genetic guidance, and the general availability of medical and surgical services.

There are valid reasons, of course, for placing greater reliance in scientific medicine than in the healing powers of nature. However wisely we manage our lives, we have to cope with physiological imperfection; the environment is forever changing and makes on us adaptive demands to which we cannot always respond successfully even when social conditions are favorable. The more complete the human freedom and the more rapid the technological and social changes, the greater is the likelihood that new and unexpected causes of disease will appear. Richardson's Hygeia and Metchnikoff's orthobiosis are ideals that imply a static view of the world. They are inapplicable to dynamic societies, in which living conditions are in a constant flux and eradication of disease is therefore impossible.

H. G. Wells was probably the first to state explicitly that the utopias of the past were static institutions incompatible with evolutionary concepts. The evolutionary point of view that inspired his book *A Modern Utopia* is applicable not only to social institutions and practices, but also to matters of health. As Wells pointed out:

> The Utopia of a modern dreamer must needs differ in one fundamental aspect from the Nowheres and Utopias men planned before Darwin quickened the thought of the world. Those were all perfect and static States, a balance of happiness won forever against the forces of unrest and disorder that inhere in things. . . . Change and development were dammed back by invincible dams forever. But the Modern Utopia must . . . shape not as a permanent state but as a hopeful stage, leading to a long ascent of stages.

Although evolutionary concepts and faith in science are usually identified with a belief in continuous progress, H. G. Wells eventually lost the robust optimism of his early adult years. His *Mind at the End of Its Tether* expressed the gloomy view of his later years that "everything was driving anyhow to anywhere at a steadily increasing velocity. . . . The pattern of things to come faded away. . . . The end of everything we call life is close at hand and cannot be evaded." H. G. Wells' *Mind at the End of Its Tether* thus constitutes a precursor of E. M. Forster's *The Machine Stops,* Aldous Huxley's *Brave New World,* George Orwell's *1984,* and the many other anti-utopias that are such a characteristic expression of social skepticism in the most prosperous countries of the world today.

There is as yet no outspoken anti-utopian movement directed against scientific medicine, but there are many sceptics among laymen and scientists concerning the possibility of eradicating disease and of creating a state of "positive health" through a scientific management of life. The public health services themselves, despite their misleading name, are concerned less with health than with the control of the specific diseases considered important in the communities they serve. They do little to define, recognize, or measure the healthy state, let alone the hypothetical condition designated "positive health." The fact that physicians and medical organizations are more interested in disease than in health might be regarded as an expression of professional bias, but this attitude corresponds in reality to a widespread human trait. Most persons take it for granted that disease constitutes an inevitable accompaniment of life; and many derive great pleasure indeed from recounting in detail their physical ailments and mental disturbances.

Whatever may have been the physical condition of the human race during primitive existence amidst untamed nature, disease is today and has been for a long time a living experience for the man of flesh and bone in civilized societies. This universal aspect of human life accounts for the worship bestowed upon Asclepius the healer in the Greco-Roman world; for the dramatic use of illness by playwrights, poets, and novelists of all times; for the appeal of pictures showing the doctor by the side of his patient; for the popularity of charlatans and of medical programs on television.

In contrast, health is a disembodied concept. It stimulates no emotional response and inspires only dull official speeches and allegoric paintings, which do not touch the heart because they deal only with an inhuman and fleshless abstraction. True enough, health in the abstract did acquire some human warmth for a short time during the fifth century B.C. in Greece. It then became personified in the handsome maiden Athena-Hygeia, who taught the Greeks that they could escape disease by living wisely. But Hygeia never captured the hearts of men. She became a real heroine loved by the multitudes only after being transfigured in Rome from an abstract health symbol into the more human Minerva Medica, who healed the sick in her sacred groves.

The healer, in fact, remains to this day the public symbol of the contribution medicine makes to human welfare. But while it is obvious that the therapeutic triumphs of medicine and surgery are snatching many lives from the grave every day, it is even more certain that the greatest contribution of medical science to the welfare of mankind has been prevention rather than cure of disease. Life expectancy has increased and health has been improved, not so much because of therapeutic advances as because we manage to escape some of the diseases that were most common and most destructive in the past.

The forces contributing to our relative freedom from disease are many and diverse. Even when he acts as a healer, any physician worth his salt practices a form of preventive medicine. A large part of his responsibility is to try to prevent the development of complications of already established disorders. Ideally, he hopes also to prevent certain diseases from occurring in the first place. However, the medical problems of the people as a whole are usually beyond the range of what the medical practitioner can do, and they must be attacked by community effort. The enteric infections transmitted through water and food were typical of the population problems of disease that were effectively dealt with by community approach during the late nineteenth century. The various forms of chemical pollution and other environmental threats posed by modern technology are among the new unsolved problems of preventive medicine that are characteristic of our times.

Although the body of scientific knowledge that constitutes the basis of modern preventive medicine is of very recent origin, a definite improvement in the general health of the public became apparent in much of Europe and also in North America early during the second half of the nineteenth century. The improvement began therefore several decades before the medical discoveries of the scientific era could be converted into practical policies. There is no doubt that a large share of this achievement can be credited to the very active role played by boards of health and other municipal bodies during that period. They did much for public health by organizing antifilth campaigns and by helping to improve the availability and quality of food.

It was not until the end of the nineteenth century that the decrease in death rates from infectious diseases and malnutrition became obvious to everyone. For this reason, scientific medicine and the germ theory of disease in particular have been given the credit for the marked improvement in the general state of health. The present generation goes still further and now believes that the control of infectious and nutritional diseases dates from the widespread use of antibacterial drugs and from the availability of vitamins and processed foods. So short and parochial are our memories!

The truth is, however, that improvements in the general nutritional state began when prosperity and greater facilities for the transportation of food made it possible for many people to afford at least one square meal a day; likewise, the mortality of many infections began to recede in a dramatic fashion in Western Europe and North America long before the introduction of specific methods of therapy, indeed before the demonstration of the germ theory of disease. No medical discovery made during recent decades can compare in practical importance with the introduction of social and economic decency in the life of the average man. The greatest advances in the health of the people were probably the indirect results of better housing and working conditions, the general availability of soap, of linen for underclothing, of glass for windows, and the humanitarian concern for higher living standards.

The magic formula of the nineteenth-century sanitarians, "Pure Air, Pure Water, Pure Food!" proved so effective in controlling many of the worst problems which had plagued our ancestors that it might encourage the belief that raising living standards still higher will inevitably result in better health. This extrapolation is unfortunately unjustified. A more accurate statement would probably be that man is becoming more and more dependent on environmental control for the maintenance of his health, because the ways of civilized life are causing a progressive atrophy of his natural mechanisms of defense against environmental noxae. In the distant past, man depended entirely on biological adaptability for his survival. Increasingly in the course of time he has supplemented his innate adaptive powers with social practices designed to control the environment. The human condition can certainly be still somewhat improved by further environmental control, but it is likely that the rate of progress in this direction will be from now on slower than it has been in the past. Furthermore, there are biological and social limits beyond which it may not be wise to tamper with the environment.

Sanitation improved rapidly during the past hundred years because the improvements dealt with factors external to man that could be manipulated through community action. The bacteriological purity of water and food, the quality of clothing, and the comfort of habitation are factors of great importance for health that can be controlled, so to speak, outside of man. They do not interfere appreciably with his freedom of action and do not require from him any personal effort. In contrast, there are many other aspects of the environment that are uncontrollable unless man is willing to make a personal effort and change his ways of life. He may have to curtail his smoking, drinking, or overeating; walk instead of riding; refrain from becoming dependent on drugs; stop polluting his air and streams, even though this means slowing down industrial expansion. Environmental control through collective and anonymous social measures has proved relatively painless, but physicians and hygienists have not yet succeeded in convincing human beings that they should change their ways of life for the sake of health if the change involves a continued personal effort.

Many natural forces, furthermore, cannot be controlled by raising living standards or changing the ways of life. Such forces are likely to acquire increasing importance in disease causation, because human life is becoming more dependent on modern technology. As we have seen, certain factors that are built into the very fabric of man, such as his biological rhythms and his innate emotional drives, can become a source of physiological disturbances because they are increasingly in conflict with modern life (Chapters II.2 and IV.2). Man may illuminate his dwellings at night, heat them during the winter, and cool them during the summer, but he cannot change the operation of the inner biological clocks, which impose on him diurnal and seasonal metabolic rhythms linked to those of the cosmos. He can be taught to control the outward manifestations of his temper, but not the hormonal processes set in motion by the flight and fight response he inherited from his animal and Paleolithic ancestors. Thus, disease can be caused by many kinds of natural forces that are either unrecognized or uncontrollable. And, more important, the very growth of technological civilization is creating new and unpredictable dangers. Medicine itself generates threats to health, for example through the use of drugs of unknown toxicity or by making possible the social accumulation of genetic defects.

The medical problems posed by the new environmental stimuli and insults have acquired a critical urgency because technological and social changes now occur with explosive rapidity and affect almost simultaneously all parts of the world and all economic classes. In the past, the rate of change was generally so slow that mankind, if not individual persons, could make unconsciously the adjustments necessary for survival. The genetic endowment of the population became progressively altered; phenotypic modifications helped many individuals to function effectively in their particular niche; and especially, man learned to achieve better fitness to his milieu through technological and social innovations. If the rate of change continues to accelerate, however, there will not be time enough for the orderly and successful operation of these unconscious adaptive processes. Biologically and socially, the experience of the father is becoming almost useless to the son.

The diseases characteristic of highly industrialized and urbanized societies are, to a large extent, the manifestations of the effects of new environmental forces to which man has not had a chance to become adapted. The emergence of new diseases that replace the plagues of the past as these are being controlled makes a hopeless ideal of the ancient Greek and Chinese concepts depicting health as universal and eternal harmony, as a way of life according to the seasons and in equilibrium with nature. These concepts would be valid only if man could live in a stable environment to which he would become really adapted; but such a static world cannot possibly exist.

To believe that disease can be kept at bay through environmental and medical control is to assume that change can be banished from the world. Such a situation will never exist, first because man cannot control all the cosmic forces, and also because one of his most characteristic and powerful urges is to move onward into the unknown. "Man has never sought tranquility alone," Sir Winston Churchill wrote. "His nature drives him forward to fortunes which, for better or for worse, are different from those which it is in his power to pause and enjoy."

Mankind is committed to "progress." The etymological meaning of this word is "moving forward," not necessarily improving health and happiness. Free man endlessly moves into new conditions during his restless search for adventure, and continuously changes his enviroment through technological innovations. The state of adaptedness to an ancient agrarian society is very different from that required during the nineteenth-century Industrial Revolution, and this in turn differs from the attributes that will be needed to function effectively in the Automation Age. Progress always implies the risk of encountering new dangers, and disease ensues whenever man fails, as he usually does, in making rapidly enough a perfect adaptive response to the new environments in which he elects to live and to function.

XIV. Eradication versus Control

1. Eradication of microbial diseases

a. Eradication defined

The destructive epidemics of the past become medical rarities wherever existing knowledge of control is vigorously put into practice. However, epidemics can be avoided only through an unrelenting and costly effort. Eternal vigilance is truly the price of control of microbial disease. The ever-present threat of infection requires that several different types of control measures be constantly applied. Elaborate epidemiological and medical surveillance, complex vaccination programs, skillful use of antimicrobial therapy, and other such procedures can be fully effective only under stable and fairly prosperous conditions. Moreover, their success is never complete. A few cases of poliomyelitis, smallpox, yellow fever, diphtheria, typhoid fever, plague, malaria, etc. occur every year even in the best administered communities. The incidence of new cases of tuberculosis remains discouragingly high even in countries where deaths from the disease have become rare.

All control policies are thus exacting of effort and money; furthermore, there is little hope that they can be made much easier or cheaper. Indeed, the techniques of control become more demanding, and their success less certain, as more and more people travel more rapidly from one continent to another, and as environmental conditions are more profoundly altered by social and technological changes. Because the control of microbial diseases as presently practiced is a never-ending process, some public health administrators have become convinced that a radically new approach is needed. According to them, the goal should be eradication instead of mere control.

Etymologically the word eradication means "pulling out by the roots," and this is exactly what epidemiologists have in mind when they speak of eradicating an infectious disease. They want to destroy once and for all the roots of the disease, namely its causative agent or its vector, or preferably both. Despite its apparent sharpness, however, this definition fails to define in operational terms what is actually implied in eradicating a disease. The most obvious ambiguity appears in the crucial differences between two semi-official formulations of the particular policies involved. According to some epidemiologists, eradication means "the purposeful reduction of specific disease prevalence to the point of continued absence of transmission within a specified area" (Andrews and Langmuir, 1963). According to other epidemiologists, the word eradication means "the complete extinction of the pathogen that causes the infectious disease in question; so long as a single member *of the species* survives, the eradication has not been accomplished" (italics mine) (Cockburn, 1961). In other words, eradicating malaria, tuberculosis, or poliomyelitis signifies eliminating the last malaria parasite, the last tubercle bacillus, or the last polio virus from the surface of the earth.

The crucial difference between these two definitions is that in the former case the goal of the program is merely to achieve the "absence of transmission *within a specified area,*" whereas, in the other, eradication is a meaningful goal only if applied to the whole world. As the proponents of the latter policy rightly point out, "Regional eradication implies a basically unstable situation, because at any time the infection may be reintroduced by carriers or vectors from the outside" (Cockburn, 1961, 1963, 1964).

b. Eradication of pathogens and their vectors

Eradication in its most exacting sense, namely as applied to the whole world, is not merely an armchair game of epidemiologists; it has become the official policy of several national and international organizations. The Thirteenth Pan American Sanitary Conference gave its approval in 1950 to programs for the eradication of yaws, smallpox, and malaria. In 1954 the Fourteenth Pan American Sanitary Conference agreed to strengthen the malaria eradication program by providing it with its own budget. In 1955 the Eighth World Health Assembly voted that "the World Health Organization should take the initiative in a program having as its ultimate objective the worldwide eradication of malaria." Eradication programs are also being carried out or contemplated for yaws, smallpox, tuberculosis, and poliomyelitis. The advocates of these programs claim that once worldwide eradication has been achieved the phase of "vigilance activities" will come to an absolute end in all countries.

At first sight, the decision to eradicate certain microbial diseases appears to constitute but one more step forward in the development of the control policies initiated by the great sanitarians of the nineteenth century, which have been greatly expanded since the beginning of the microbiological era. In reality, however, eradication involves a new biological philosophy. It implies that it is possible and desirable to get rid of certain disease problems of infection by eliminating completely the etiological agents, once and for all. Unfortunately, the significance of this belief is made uncertain by another ambiguity in the meaning of the word eradication. To eradicate the infectious agents most commonly responsible for the *diseases* malaria, tuberculosis, diphtheria, smallpox, yellow fever, or paralytic poliomyelitis is not the same thing as to eliminate entirely from the surface of the earth *all the kinds of microbial agents* that are *potentially* capable of causing these conditions.

Several different strains of infective agents, as is well known, can give rise to similar signs and symptoms. For example, it is impossible to differentiate on clinical criteria alone the forms of tuberculosis caused by the human, the bovine, the avian, and some at least of the so-called atypical forms of mycobacteria. In other words, the definition of tuberculosis involves etiological, pathological, and clinical criteria. The diagnostic confusion can be even greater for many other diseases. At the Seventh European Symposium on Poliomyelitis in 1961, several outbreaks of paralytic disease were described which had been thought to be caused by one of the polio viruses until isolation of the causative agent proved them to be caused by other enteroviruses. A strain of Coxsackie virus A type 7, isolated from such an outbreak in Russia, caused paralysis and poliolike lesions in the nervous system of monkeys. It had also caused paralytic illnesses among human beings in the vicinity of New York.

Because of the uncertainties posed by clinical and pathological diagnosis, programs of eradication must therefore be formulated not in terms of diseases but rather of etiological agents. However, other kinds of difficulties arise from the fact that all pathogens without exception exist in multiple variant forms. Just as several different strains of plasmodia, of mycobacteria, or of polio viruses are known to cause disease in man, so is this true for most human and animal pathogens. Furthermore, several strains of each microbial and viral species certainly exist in addition to those known at present; experience has shown that when a protozoan, bacterial, or viral species exists in more than one serological type, still other types usually turn up on further investigation. The question is thus to decide whether *all* species of a genus and *all* serological strains of a microbial species must be eradicated, or only those which happen to be considered important at the time the eradication program is organized.

It is true, of course, that a few of the most important pathogens are known to occur in only one immunological type, for example yellow fever and smallpox. In these cases, furthermore, a thorough program of vaccination with the appropriate attenuated strains of the respective viruses elicits a high level of immunity. Yet vaccination does not necessarily eliminate the viruses from the environment. In most cases, indeed, lasting immunity depends probably upon persistence of the virus in the body.

It is also certain that *all* attenuated strains of pathogens are potentially capable of reverting to the virulent state. Whether it be a protozoon, a bacterium, or a virus, each kind of microorganism can give rise to mutant forms not known to be associated with disease at the present time, yet capable of mutating to a virulent form. In addition to the attenuated strains used for vaccination against smallpox, yellow fever, poliomyelitis, tuberculosis, brucellosis, etc., there exist in nature many other mutant strains of these microbial species. Some of them persist unnoticed in a variety of hosts, but are capable of acquiring virulence for man and animals under certain unpredictable conditions.

The detection and elimination of persisters may constitute the worst stumbling block for eradication programs. As emphasized in Chapter VII.3, it is probable that all pathogens can persist in the tissues for prolonged periods of time, often in not readily detectable forms. Even in the case of infections such as smallpox, for which the persister state has not yet been recognized, it would be foolhardy to assume that it does not exist. Repeatedly in the past, a minor modification of technique has been sufficient to bring out potential pathogens that had remained completely masked. Examples of unexpected recovery of a microbial agent from a source that had been assumed to be free of it are now so numerous that persistence in a masked form can be considered the rule rather than the exception. The following familiar cases will suffice to illustrate that it is theoretically unjustified and practically dangerous to take negative cultural results at their face value.

The formol-treated vaccines used in the early days of the polio vaccination program appeared to be completely free of active virus as judged by the assay techniques then in vogue, including intracerebral injection into monkeys; yet some of the preparations proved infective when injected into children, with the tragic results that are well remembered. Human tubercle bacilli injected into rabbits fail to establish a progressive infection and seem to be destroyed in the animals' tissues; yet when fragments of the organs that appear sterile and histologically normal are incubated in vitro, these tissue cultures are killed by the multiplication of tubercle bacilli that had remained inactive in vivo and were undetectable by the best known methods of cultivation on artificial media. The persistence and occasional reactivation of rickettsia, various types of virus, or malaria parasites in persons removed from any possible contact with these pathogens for many decades should suffice to prove that recognition of the silent carriers of infection constitutes a very large unsolved problem in human pathology.

In brief, there does not exist at present any dependable technique for ruling out the presence of persisters, or for eliminating them where they are known to exist. Such techniques would have to be developed before it becomes possible to convert the concept of eradication into a really effective operational program, especially if one accepts the doctrine that "so long as a single member *of the species* survives, then eradication has not been accomplished."

The case of yellow fever illustrates difficulties of a simpler conceptual type, yet probably just as large in practice. This disease has been recognized as a definite clinical entity for at least three centuries; its transmission through mosquito vectors has long been known; its causative virus seems to exist in only one serological type; infection elicits vigorous and well-defined immunological responses. These properties of the virus have greatly facilitated epidemiological studies and have permitted the development of a very effective vaccine. As a result of this combination of favorable circumstances, it proved relatively easy to eliminate yellow fever from cities. But when this was achieved, the unexpected occurrence of so-called "jungle yellow fever" revealed that the virus has sanctuaries other than the ones that had been recognized at first.

Jungle yellow fever was first observed sporadically among men working away from settlements, and shortly thereafter the disease was found to be widespread in monkeys, of which the howler is one of the more important and also one of the more populous, with a range extending far north in Central America. The jungle mosquitoes that transmit the disease among primates are not susceptible to present eradication procedures. In other words, the discovery of jungle yellow fever revealed that the epidemiology of the disease is far more complex than the one on which the programs of "eradication" had been based. Similarly, it is probable that even if the epidemiological manifestations of smallpox that we know today could be eradicated, other unsuspected sanctuaries of the disease would soon be discovered, especially as we go on altering the environment all over the world.

A few years ago it was widely believed that chemotherapeutic agents would open a royal road to the eradication of pathogens. However, several limitations of this approach soon became apparent. One was the recognition that all drugs exhibit some form of toxicity, and their general and continued administration requires strict medical surveillance. Suffice it to mention here that it was only after several years of widespread use of chloroquine that this remarkably effective drug was found to produce ocular lesions in a large percentage of treated persons.

Another limiting factor is the rapid occurrence of mutations endowing the parasites with resistance to any kind of drug that is widely used. In several parts of the world, mutant forms of *Plasmodium falciparum* have appeared that are resistant to all the available antimalarial drugs except quinine. Likewise, the increase in resistance of tubercle bacilli to isoniazid is beginning to limit the usefulness of this drug, especially in the parts of the world where it is most desperately needed. A last example among so many others that could be quoted is that resistance of the polio and vaccinia viruses to respectively 2-(x-hydroxybenzyl) benzimidazole and 5-iodo-2^{1} deoxyuridine began to be recognized very shortly after the discovery of the antiviral properties of these drugs.

The availability of several different kinds of inexpensive synthetic insecticides seems at first sight to bring vector elimination within the range of practical possibility in a number of cases. However, experience has shown that while the problems posed by eradication of insect vectors are somewhat different from those experienced in the eradication of microbial pathogens, they are not simpler.

The first difficulty encountered was that populations of insects, like populations of microbial parasites and of all other living things, rapidly give rise to mutant forms possessing genetic resistance to the poisons with which they come into contact. More recently unexpected problems have arisen from the peculiar ability of insecticides to stimulate irritability in mosquitoes, a property commonly referred to as excito-repellency. The effect of excito-repellency is that the insect concerned avoids lethal contact with the insecticide; but this does not reduce its attacks on man, and so transmission of the pathogen continues. An example is the greater irritability of *Anopheles gambiae* as compared with *A. funestus,* which explains why the former insect is, in general, less amenable to insecticidal control than is the latter. Excito-repellency appears also to be responsible for the fact that DDT has been a comparative failure against *A. gambiae* in the savannah regions of West Africa. Finally, new problems are arising from the change of behavior of mosquitoes. In certain parts of Africa, notably Swaziland and Southern Rhodesia, the application of residual insecticides to houses has led the main vector of malaria, *A. gambiae* Giles, to so change its habits that instead of resting indoors and biting man, it now rests outdoors and feeds on cattle and other animals.

In the light of these difficulties, and of others not to be mentioned here, eradication of vectors with residual insecticides is probably not feasible even though, of course, a very satisfactory degree of control (not eradication) can often be achieved through this approach. We are engaged in an endless cycle of synthesizing *at an ever-increasing cost* more and more insecticides to which vectors become progressively resistant. Furthermore, all insecticides are toxic to man, and the fact that they accumulate in plant and animal tissue poses, as we have seen, serious problems of delayed toxicity (Chapter VIII.3).

It must be acknowledged that some agents of disease have been eradicated from certain limited areas in a few specialized situations. For instance, the Mediterranean fruit fly, which is a menace to citrus fruit crops, has been eradicated from Florida several times during the last 30 years. Similarly, hoof and mouth disease has been eradicated from the United States and Canada on a number of occasions following its introduction from abroad. In both these cases, the pathogens had been imported from the outside, and it was economically advantageous and ethically possible to use ruthless methods to eradicate them. Eradication of *Aedes aegypti* from the Americas was likewise greatly facilitated by the fact that this mosquito is domesticated and lives around dwellings, where it can be controlled by residual insecticide sprays or by larvicides.

The eradication of *Anopheles gambiae* from northern Brazil in 1944–45 is commonly mentioned to support the view that an insect vector can be completely eliminated by a concentrated effort pursued for several years. Remarkable as this achievement was, however, it dealt with a very atypical situation and therefore lacks generally convincing value. On the one hand, *A. gambiae* had been introduced into Brazil only a short time before the eradication program was begun, and therefore had not had time to establish deep biological roots in this area. Furthermore, all the breeding places of the species were readily accessible and could be treated with Paris green once a week. The same was true in Upper Egypt. The contrast with *A. labranchiae* in Sardinia is very striking. In this case, the mosquitoes are indigenous to the island and probably have been there for millions of years; they breed in uninhabited and inaccessible mountain areas, blithely independent of man, his house, his roads, and all his other artifacts. For these reasons, mosquito eradication programs failed in Sardinia.

Thus, it is possible to eliminate a particular microbe or vector from a limited area if the circumstances are favorable and if the project is pursued with enough effort, time, money, and especially ruthlessness. However, elimination from a localized area constitutes an achievement that has no bearing on the possibility of eradicating a species from the surface of the globe, except when this species exists only in that area, a very unusual situation indeed.

In the picturesque words of Dr. Walsh McDermott, "There are no rattlesnakes in the streets of Houston or Dallas, but their elimination from big cities does not mean eradication from the state of Texas, let alone from the rest of the world." It is relatively easy to create in limited areas ecological conditions incompatible with the survival of rattlesnakes, rats, mosquitoes, or lice; but these species could be "eradicated" only if the whole surface of the globe were changed.

2. Difficulties inherent in eradication programs

Malaria will be singled out to illustrate some of the theoretical and practical difficulties that stand in the way of eradication programs, because there is long and worldwide experience of the problems posed by its control. Malaria is not a single disease, but a huge congeries of diseases, occupying a large number of ecological provinces each with its particular conditions. The multiplicity of problems comprised under the single word malaria makes it therefore hazardous to conclude that a given procedure that has been effective in a particular area will necessarily prove effective in another where the climate, the topography, the plant and animal populations are different, to say nothing of the habits of the human populations.

The recent discovery that monkey malaria can be transmitted to man by mosquitoes, and thence from man to man, adds still further unknowns to the problem. *Anopheles leucosphyrus,* an important vector of human malaria in Borneo, was recently shown to be infected with *Plasmodium inui.* This observation demonstrates that transmission of monkey malaria to man is possible in nature. Furthermore, it raises the disturbing possibility that the primate strain could become a widespread human pathogen if the present human strains of *Plasmodium* were actually eradicated.

In addition to these difficulties, there are others of a more general nature originating from the fact that the ecological determinants and characteristics of an infectious disease in its descending phase are very different from those most important during its period of great prevalence. For example, even though tuberculosis is very far from being eradicated in the countries of Western civilization, its epidemiology has already changed profoundly. Familial or localized outbreaks are becoming common and in certain places human beings contract the disease from cattle. As physicians rarely diagnose tuberculosis in its early phase, the detection of these small foci of infection is often delayed, and the disease thus continues to retain a foothold even in the communities with the best public health services. There is no doubt that similar difficulties will be encountered in all other microbial diseases once control has proceeded so far that they are no longer part of the physicians' common experience.

In all cases also the problems posed by the biological and epidemiological peculiarities of each type of infection are still further complicated by financial, administrative, and political uncertainties. Even if genuine eradication of a pathogen or vector on a worldwide scale were theoretically and practically possible, the enormous effort required for reaching the goal would probably make the attempt economically and humanly unwise.

As a given pathogen or its vectors become scarce, the procedures needed to detect or destroy them become technically more exacting and also more costly. "I hazard a guess that it may sometimes cost a million times more to eliminate the last case of a disease than the first" (G. James, 1964). The resources and technical skills being everywhere in short supply, even in the most prosperous countries, excessive concentration on a particular program inevitably means the neglect of others. It has not been proved that the discovery and destruction of the last plasmodium or last smallpox particle is a more desirable objective than minimizing the damage done by the many other microbial diseases that still plague mankind today. There were only 431 cases of poliomyelitis in the United States in 1963. "As vaccination moves on, this figure should go down. But how many billions should be taken away from other health programs to make this figure read zero while certain other figures are soaring" (G. James, 1964). The difficult problem is to evaluate what things are most worth doing at a given time among all the things that could be done and cry out to be done. The popular appeal and fervid ring of the word eradication is no substitute for a searching analysis of the manner in which limited supplies of resources and technical skills can best be applied for the greatest social good.

Social considerations, in fact, make it probably useless to discuss the theoretical flaws and technical difficulties of eradication programs, because more earthy factors will certainly bring them soon to a gentle and silent death. Certain unpleasant but universal human traits will put impassable stumbling blocks on the road to eradication. For example, it is easy to write laws for compulsory vaccination against smallpox, but in most parts of the world people would much rather buy the vaccination certificate than take the vaccine; and they always find physicians willing to satisfy their request for a small fee.

It can be taken for granted, furthermore, that the public and its elected representatives will get tired of devoting enormous funds to the eradication of a particular disease once it has ceased to cause significant damage and is no longer an obvious threat, if the effort means, as it certainly will, the neglect of other and more acute medical needs. There were 697 deaths from pulmonary tuberculosis in New York City in 1962 and only 297 deaths from emphysema, but the latter disease is rapidly increasing in importance. Its death rate has shot from 8 per million nationally in 1950 to 67 per million in 1962. What will happen when emphysema (or other forms of chronic respiratory disease) passes tuberculosis as a cause of death? A straw in the wind may be seen in the fact that *The American Review of Tuberculosis,* the official journal of the American Thoracic Society, recently changed its name to *The American Review of Respiratory Diseases.*

Eradication programs can be successful only if they enlist the whole-hearted cooperation of the most deprived and least educated members of the community; for it is among them that infectious diseases are most likely to persist the longest. Genuine eradication of malaria from the Congo, or of tuberculosis from Calcutta, constitutes an undertaking that is not likely to receive active support until the bitter end from the local population, as long as so many other of their medical problems remain unsolved. In fact, medical scientists themselves show great reluctance to work on problems that have become somewhat less pressing; witness the fact that fewer and fewer American physicians or investigators are willing to work on tuberculosis or leprosy today.

Public health administrators, like social planners, have to compromise with the limitations of human nature. For this reason and many others, eradication programs will eventually become a curiosity item on library shelves, just as have all social utopias.

3. Ecological approach to disease control

Needless to say, the geological strata of the earth's crust contain fossilized remains of thousands of species that were once abundant yet no longer exist today. The extinction of the passenger pigeon illustrates the disappearance from the American continent of a species that constituted an immense population little more than a century ago. However, the complete disappearance of a species has always been the result of profound environmental changes that rendered conditions unsuited for its survival. The passenger pigeon would not have disappeared merely from being hunted to feed the pigs if the forests that were its abodes had not been cut down at the same time to create farmland. Likewise, malaria disappeared from the European lowlands during the Middle Ages when the Cistercian monks established drainage as a standard agricultural practice around their monasteries; five centuries later malaria disappeared also from the Ohio valley when the settlers cut down the trees and transformed the forests into cultivated farms.[1]

Judging from past experience, it is unlikely that the eradication of vector-borne diseases can be achieved by spraying of insecticides alone. Complete success will require a simultaneous alteration of the environment to make it less favorable for vector or parasite survival. Such ecologic approach, however, demands prolonged ecological research in order to create a sophisticated awareness of the complex secondary effects exerted by any environmental change. One of the unfortunate consequences of the simple-minded and unfounded illusion that insecticides and drugs can eradicate vectors or microbes has been a decadence in biological thinking. The experience gained with malaria control has shown that no lasting success can be achieved without resorting to the time-tested procedures of the pre-DDT phase, such as antilarval campaigns and extensive treatment of people. But this approach will not be practicable without retraining malariologists in a kind of biology more sophisticated than spraying of insecticides. Unfortunately, in the words of an experienced scholar in this field, DDT went further toward the eradication of malariologists than of mosquitoes.

The interrelatedness of things is so pervasive, and the variables are so numerous, that it would be practically impossible, for example, to eradicate rats or mice from a continent. The complexities would be even far greater in the case of microbial species. The unresolved problem is to identify those aspects of the environment that would have to be manipulated to make conditions incompatible with their survival. If eradication could be achieved, furthermore, the indirect consequences of the program would certainly reach far beyond those expected from the operations required to eradicate the species concerned. One need only recall the disturb-

1. While it is true that some living species have become completely extinct, in certain cases the possibility remains that a few specimens persist in areas outside the main avenues of travel. Some parts of the inhabited world remain unexplored and unseen even by the most hardy World Health Organization officers. Until recently the Metasequoia tree was known only as a fossil and thought to be extinct. Then in 1946 a young Chinese scientist, American trained, discovered a living specimen of the tree in an old temple of West China. Two expeditions of the Harvard Arboretum confirmed the discovery in 1947 and 1948, and found twelve more living trees. The Metasequoia, formerly thought to be extinct, has now been successfully transplanted in this country.

The proponents of the eradication program will find it difficult to make sure that a few malaria parasites, tubercle bacilli, and polio viruses have not taken refuge in Tibetan monasteries or Buddhist temples, or among people whose religious convictions make it a crime to kill any form of life.

ances in bird and fish populations that have followed upon the mild and ineffectual attempts to control a few insects in our own communities; or the variety of intestinal disorders brought about by the administration of antimicrobial drugs to man and animals (Chapter V.1).

Eradication programs are advocated on the ground that "once eradication is achieved, the infection is gone forever and the costly burden of recurring control measures may be dropped. . . . After eradication, no further effort is required" (Cockburn, 1961, 1963, 1964). These statements are grossly misleading. The biological truth is that elimination of one particular agent makes the ground available for another or even usually for several other kinds of pathogens. Control measures cannot be "dropped"; rather they have to be constantly redirected to meet the problems peculiar to each particular ecological situation. It is dangerous to claim that we can look forward to "freedom from infectious diseases at a time not too distant in the future," and that "within some measurable time, such as 100 years, all the major infections will have disappeared" (Cockburn, 1961, 1963, 1964). Man does not live in a biological vacuum; elimination of one type of microorganism simply creates better opportunities for other potential pathogens.

Infectious diseases, major or minor, are the expression of the interplay between living things, and they take different forms as environmental conditions change. Conditions will continue to change because no society remains static for long, and even if it did, natural forces bring about genetic and phenotypic alterations which cannot be predicted at the present time and which bid fair to be still unpredictable 100 years hence.

Eradication of microbial disease is a will-o'-the-wisp; pursuing it leads into a morass of hazy biological concepts and half truths. On the other hand, the effectiveness of the ecological approach to the control of disease has been demonstrated by centuries of useful empirical practices and is supported by modern biological science. Proper skillful handling of the ecological situation makes it possible to create environments in which diphtheria, tuberculosis, leprosy, malaria, poliomyelitis, smallpox, etc. are converted from major diseases into minor public health problems. For leprosy, this change for the better occurred during the sixteenth century in Europe through a variety of empirical social improvements; for tuberculosis, it is happening at present by the combined operation of public health, chemotherapeutic, and economic factors; for diphtheria, smallpox, poliomyelitis, it can be achieved by applying immunological techniques that do not require the unrealistic paraphernalia of eradication programs.

Some of the statements made by the proponents of eradication bring to mind a reader's comment published in 1962 by one of the weekly magazines: "Nikita S. Khrushchev drips honey and talks of peaceful coexistence. Does one peacefully coexist with a polio virus? The only sensible attitude is to destroy both of them." As it happens, experience has shown that mankind can achieve coexistence with the polio virus, or with any other microbe, without resorting to the fantastic programs required for genuine eradication. Among people who live without sanitary facilities, polio viruses are ubiquitous and contaminate everyone; as a result, immunization to them occurs spontaneously during the first months of life, and paralytic disease is rare. In our communities, sanitary practices minimize early contact with the polio viruses and thereby prevent the spontaneous development of immunity; but control of paralytic disease can be achieved by artificial methods of vaccination, which are even more effective than the haphazard early exposure to infection as it occurred in the past. There is no evidence that vaccination will ever eliminate all the strains included under the poliomyelitis viruses; and it is biologically wise to expect that mutant forms will now and then continue to emerge. But vaccination will go very far toward controlling at low cost the paralytic disease caused by the strains we know today. And the experience thus gained will help the following generations to cope more rapidly with the new forms of paralytic disease, or of any other kinds of microbial disease, which are bound to emerge as man continues to change his environment and his ways of life.

4. Social limitations to the application of knowledge

Eradication is an unattainable goal, not only in the case of microbial diseases but also of other environmental threats. In fact, progress in disease control does not always go hand in hand with scientific progress, because social factors often interfere with the application of knowledge. Naturally, these social limiting factors are most in evidence among populations of low economic status.

As mentioned earlier, protein malnutrition dominates the pathological picture in a large part of the underprivileged world today.

Further biochemical and metabolic studies would be useful, of course, in advancing the scientific understanding of this problem, but it is questionable whether such studies would be of any help to the health of the people concerned. Their primary needs are well understood. They consist in more food of the correct amino acid composition, presented in a physiologically and socially acceptable form. The situation is similar with regard to infectious diseases. Elaborate epidemiologic surveys and sophisticated studies of host-parasite relationships could sharpen knowledge of these conditions. But in practice higher standards of living and better sanitation would go very far toward disease control in the parts of the world where malaria, tuberculosis, parasitic infestations, etc. kill young people and sap the energy of adults.

In other words the scientific aspects of the medical problems that are most important in poor countries have been delineated and the remedial steps to be taken are obvious. But this knowledge remains mere intellectual exercise because social and economic factors interfere with its application. These factors range all the way from lack of facilities and trained personnel for the delivery of medical services to the shortage of suitable farmland and agricultural equipment, and to traditions and taboos that prevent the needed changes in the ancestral ways of life.

Social and economic limitations to the application of scientific knowledge also exist in prosperous and technically advanced populations, but they take a form and affect aspects of the health picture different from what is observed in poor countries. Populations that are well informed on medical matters and free to change their ways of life nevertheless are often prisoners of social strictures; they are thus unable to apply scientific knowledge to their health problems, even when they have the necessary resources. One of the greatest difficulties in disease control is to generate social action with regard to environmental factors the effects of which are remote in time. The lack of public response to the anticigarette-smoking campaign is a notorious case in point, but not the most important. The difficulties experienced in controlling environmental pollution illustrate in a painful manner that enlightened communities are often paralyzed when faced with health problems which are scientifically well understood, but not yet pressing enough to outweigh the interests of the immediate present. Social paralysis can set in even when the need for action is recognized by all.

Public action is fairly readily mobilized against soot, grit, and irritating industrial gases, as well as against the presence of detergents in water supplies, because these pollutants cause an obvious and immediately perceived nuisance. But the situation is very different with regard to other aspects of environmental pollution. The public is now as familiar with air, water, and food pollutants as it is with the microbial agents of disease. It has been warned that some pollutants that have no immediate nuisance value are nevertheless of importance because they have cumulative and remote effects. But such warning evokes little response except for a short time when it is dramatized through a vibrant symbol, such as *Silent Spring,* or by a special situation, such as the fear of carcinogenic substances in cranberry sauce at Thanksgiving time. In fact, the scientific community is almost as indifferent as the general public with regard to long-range health problems. In general, scientists hesitate to dedicate themselves to medical questions that lack popular appeal or are not likely to yield rapid results.

The largest stumbling block in the way of environmental control is therefore not ignorance but rather the difficulty of arousing public and scientific interest in problems the consequences of which appear somewhat remote. This lack of interest stands in sharp contrast to the eagerness with which both public and independent agencies supply funds for programs devoted to the discovery of drugs. The word drug evokes a rapid effect, instantly noted; for this reason, it has more appeal among the scientific as well as the lay public than long-range control programs, which appear hazy through the smog of time.

Even when interest can be aroused, and where practical control methods are apparent, the social structure often creates practical difficulties for the application of scientific knowledge. Here again environmental pollution can serve as an example to illustrate this type of special paralysis.

The origin of the most important air pollutants is well known (Chapter VIII.2). No further investigation is needed to establish that certain courses of action compatible with technical knowledge would either prevent the release of most air pollutants or at least discharge them into areas where they would not threaten human health. Improvement in design of combustion equipment and gasoline engines, increase in height of smokestacks, rerouting of automobile traffic, and displacement of certain industries are among the many measures that would go far toward minimizing the medical problems caused by air pollution. Unfortunately, these measures would disturb the economic structure of the community, and create inconveniences for many of its members. Suffice it to mention here one current example taken from the daily press.

A large corporation (International Minerals and Chemicals Corporation) recently announced that it would close its phosphate plant in Mulberry, Florida, because of "demands from the State Board of Health brought about by pressures from *non-residents* concerning the level of dust emitted from the plant" (italics mine). According to the management of the corporation, the changes required by the Board of Health were "not economically possible." The interest of this news item for the present discussion is that 270 employees lost their source of income as a result of closure of the plant, in an area where chances of relocation are small. The fact reported by management that "there have been no *recent* complaints from the *residents*" (italics mine) must be interpreted in the light of the fear of unemployment.

The health aspects of air pollution cannot be dissociated therefore from human values, which are difficult to evaluate. To what extent is society willing to sacrifice economic prosperity and certain conveniences of life in order to prevent lung cancer, cardiac failure, emphysema, chronic bronchitis, and other ailments that may or may not develop and, if they do, will become manifest only in the future? How can one balance the social performance today against human suffering at some undetermined time?

Conflicts between public health and other social values occur in practically all aspects of modern life. For example, the biological dangers of ionizing radiations must be balanced against the potential usefulness of nuclear energy in industrial processes; *complete* protection against these dangers would probably handicap the efficiency and economy of nuclear processes. One can predict that, after some hesitation, mankind will behave in this respect as it has in similar situations in the past and will accept a certain amount of risk for the sake of industrial growth.

Food production presents a similar dilemma. Many of the substances used for controlling agricultural pests and in the processing of foodstuffs do indeed have some direct or indirect toxicity for man, but like most other pollutants arising from modern life, they also have built-in beneficent values. In most cases the borderline between advantages to the consumer and dangers to his health can be so blurred that the health officer finds it almost impossible to make a fair decision based on scientific evidence. A recent partial list of food additives includes more than 500 natural and synthetic chemicals. Many of them improve the nutritive value, the keeping quality, or the appearance of the product, but some are potentially toxic. In general, the information concerning their long-range toxicity is at best uncertain, a fact especially regrettable in the case of materials added to the foodstuffs in daily use.

The fear of cancer being so widespread, it is not surprising that efforts have been made to prohibit the use in food of any substance suspected of being carcinogenic. The Delaney clause in the Food Additive Amendment of 1958 (Public Law 85-929) was designed to meet this end. It provides that "no additive shall be deemed safe if it is found, after tests which are appropriate for the evaluation of the safety of food additives, to induce cancer in man or animal."

The amendment provides for no tolerance, even with regard to substances known to be essential nutrients, if these prove carcinogenic in huge dosage in one animal species or another. And yet it is certain that several common food ingredients and even products of human metabolism can induce cancer when ingested by experimental animals. Human beings consume table salt as a matter of course even though it usually contains traces of radium, a known carcinogen. Lamb has been a part of human diet from time immemorial, and yet it frequently contains traces of selenium, another carcinogen when ingested in much larger amounts. According to a recent report, even hens' eggs contain a carcinogenic substance. Chickens and mice fed a diet made up of wheat bran and eggs grew faster and became sexually mature earlier, but developed cancer of the ovary.

Clearly, it is unreasonable to insist on zero concentrations of a given substance in food merely because administration of massive doses of it has increased the incidence of tumors in experimental animals. Chromium and selenium are of special interest here because both elements have been reported to act as carcinogens under certain conditions and yet both are essential for growth. Ironically enough, selenium has even been studied at various times as a cure for cancer! Thus, it is far from simple to evaluate whether the consequences of complete prohibition of a food additive are beneficial to society on balance.

There is no doubt, for example, that addition of the hormone diethylstylbestrol to farm feeds increases the growth rate of cattle.

Federal regulations now in force restrict the use of diethylstylbestrol in such a way that this hormone should not be detectable in beef with the available analytical techniques. This does not rule out, however, that trace amounts of the hormone can be present in the meat. If its use were to be prohibited altogether, and the present beef consumption maintained, the cattle industry would have to raise some one-third million additional head. This might cause a rise in the number of industrial accidents to workers handling the additional cattle, and it would almost certainly increase beef prices. Where, then, is the happy medium?

To single out food additives for strict control seems, furthermore, a policy of very limited usefulness as long as there is no control over discharging countless substances with well-established carcinogenic potentialities into the air we breathe and the water we drink. Even caffeine is mutagenic in *Escherichia coli* and in Drosophila. Yet it is consumed in large amounts not only in coffee and tea but also in soft drinks. If caffeine raises the mutation rate in man as much as it does in bacteria and insects, then the consumption of three cups of coffee daily over a 30-year period would have as much mutagenic action as that produced by exposure to 40 *r*! It would be an entertaining, though macabre, kind of arithmetic to compare the numbers of lives—if any—that *might* be saved from cancer by the Delaney clause with the numbers of cancer deaths that certainly result from excessive smoking, from pollution of the air and water, or from eating certain *natural* foodstuffs known to be carcinogenic.

Almost any nutrient and any substance added to food can be dangerous under certain conditions. This is true even of essential amino acids, which can exert toxic effects if employed unwisely, for example when there is too much of one in relation to the others. Whether natural or synthetic, a given food additive can be useful, dangerous, or indifferent, depending upon the conditions under which it is used. The judgment of its acceptability demands in each particular case a comparative evaluation of gains and dangers based on very sophisticated scientific knowledge; it involves questions clearly outside the competence of legislative bodies. Thus, the health problems created by the addition of chemicals to food cannot be disposed of by ruling, on legal grounds alone, that certain substances are safe and others carcinogenic or otherwise unsafe.

As mentioned earlier, it will prove difficult if not impossible to carry out on all the substances used in agricultural and food technology the long and onerous tests required to determine their potential toxic effects, especially those of delayed nature. In consequence, many lines of progress in food production would be closed if a verdict of absolute safety had to be rendered before a substance could be used. More likely, society will be willing in the long run to take chances for the sake of greater efficiency and of lower costs in food production. Complete eradication of carcinogenic agents is as unrealistically utopian as eradication of microbial pathogens.

5. Scientific possibilities and social choices

More and more rapidly, advances in fundamental knowledge are being converted into methods applicable to the prevention and treatment of disease. But the practical developments of theoretical science always require highly specialized technical skills. Even where ample funds are available, the shortage of scientific personnel constitutes a limiting factor to practical development. The production of viral vaccines is a case in point.

The recent advances in tissue culture techniques make it theoretically possible to cultivate in vitro almost every kind of virus in any desired quantity, and therefore to produce vaccines against most viral diseases. However, while the theoretical problems of vaccine production are solved, countless practical details have to be worked out for each particular case. The development of mass production techniques, the safety controls, and the clinical tests are all essential steps in vaccine production that require the work of many different specialists who will certainly remain in short supply. Thus, the most difficult problem is no longer the scientific one of producing effective vaccines, but rather the social one of deciding which vaccines should be produced.

There are limits to the number of vaccines that can be administered to a given child or adult under practical conditions. Furthermore, the use of any vaccine, however carefully controlled, entails some danger (Chapter XVI). Finally, it is impossible in practice to produce all the vaccines that can be produced. The consequent necessity to make choices involves decisions as to the kinds of disease against which protection is most desirable for the sake of individual persons and society. Should emphasis be placed on diseases of the nervous system which are fatal or crippling but affect only a small number of individuals? Or should priority be given to ailments of the upper respiratory tract, rather mild and self-limited, but of great economic importance because they affect a large percentage of the population and are responsible for much absenteeism from school and work?

The use of chemotherapy as a public health measure to control tuberculosis constitutes another example or a situation where social choice rather than medical knowledge is the deciding factor. It was recently agreed at a meeting of experts sponsored by the United States Public Health Service and the National Tuberculosis Association (Arden House Conference, 1959) that the incidence of new tuberculous infection could be greatly lowered in this country by treating spreaders of bacilli with isoniazid and other drugs. Naturally, attainment of this goal would demand a very systematic and continued search, treatment, and follow-up of persons with active tuberculosis. The scientific techniques required for such a program are available, but their application would be costly, both in funds and personnel. Moreover, even granted that such a program can be highly successful, it would not greatly decrease the burden of tuberculosis in the immediate future, but would yield dividends only during the following generation. Thus the community has to decide whether to undertake a very costly program of tuberculosis control from which it may not derive any immediate advantages; this decision is made still more difficult by the fact that there are many other problems, equally important, crying out for support. As already mentioned, for example, emphysema and other forms of chronic respiratory diseases may soon be responsible for more deaths than tuberculosis. All programs of eradication present similar dilemmas (Chapter XIV.2).

As medical science progresses, it puts into the hands of physicians an increasing number of expensive drugs, elaborate surgical procedures, and complicated diagnostic techniques. All these advances improve patient care but they continuously increase its cost, as well as the need for specialized personnel. In consequence, the time is rapidly approaching, if it has not been reached already, when economics will have to play a large part in medical decisions. In fact, as we shall see, modern societies appear to limit their medical expenditures more or less unconsciously. In all countries of Western civilization the ratio of these expenditures to the gross national income is maintained at an approximately constant level, and this level is much the same everywhere, irrespective of economic prosperity (Chapter XV.1). There is good reason to believe, however, that the medical limitations imposed by economic necessities will soon become more stringent than they were in the past.

Until recent times all technical knowledge could be applied because it was limited in range and required only a small specialized staff and fairly simple equipment. During recent years, prosperous countries have been in the happy situation of not having to worry about the economic infrastructure of medical care. The physician in our communities does all he can for an individual patient, with the confidence derived from the knowledge that whatever problems lie beyond his skills or his means will be taken care of by a colleague or some specialized organization. Rapidly, however, the situation is becoming different, for two reasons. On the one hand, we know more than we can apply, especially because of an inevitable shortage of specialized skills. On the other hand, some of our modes of action have such powerful biological effects that a particular medical procedure that is helpful for a given person may be antagonistic to the interests of the community as a whole.

The medicated survival of hemophiliacs, diabetics, or children with phenylketonuria, and their ability to beget their kind, points to the likelihood that the genetic load will continuously increase. In contrast, but for similar reasons, there will be a decrease in the frequency of those genes which made for resistance to certain specific infections once prevalent in our communities. As these infections continue to smolder in other parts of the world, and also in other species, they might become the cause of biological disasters if populations now vulnerable should ever experience unfavorable social circumstances. The increasing cost of prophylaxis constitutes one of the unfavorable legacies of modern medicine to our descendants. In other words, the very progress of medicine is creating new dangers for the human race, and the control of these dangers can be achieved only at a constantly increasing cost.

Given the overwhelming dominance of economic considerations over human problems, it is unwise to foster the view that health can be acquired without struggle simply by eradicating diseases one after the other. Health is a precarious state and can be maintained only through constant effort. One of the most important characteristics of man is to be goal-seeking, an attitude very different from mere threat-avoiding. In any case, emphasis on avoidance of danger, effort, and pain brings its own disease problems, just as real and perhaps more dangerous than those resulting from an adventurous and creative way of life. A medical philosophy which assumes a priori that the ideal is to make human life absolutely safe and effortless may paradoxically create a state of affairs in which mankind will progressively lose the ability to meet the real experience of life and to overcome the stresses and risks that this experience inevitably entails.

XV. The Control of Disease[1]

1. The purchase of health

The Yellow Emperor's *Classic of Internal Medicine,* published in China during the fourth century B.C., taught that the really good doctors are not those who treat sick persons but rather those who instruct people on how not to get ill. Employing a physician while you are well sounds like a typical piece of Oriental oddity, but in fact the social reformers and medical philosophers of Western civilization have long advocated the merits of this ancient Chinese doctrine. Indeed, what was once regarded as a utopian view of medicine is now becoming reality all over the world. In a great diversity of ways and in ever-increasing amounts, public and private funds are devoted to preventive measures designed to protect the health of the public. The ancient Chinese paid personal physicians to keep them healthy. Through taxes and insurances we pay physicians, nurses, scientists, and the employees of multifarious government agencies to provide us with sanitation and vaccinations, advise us on what to eat or how to behave, and watch over us for any sign of disease.

The process of democratization and socialization is much less advanced, of course, in therapeutic than in preventive medicine. One of the reasons for the difference is that therapeutic medicine is highly dependent on personal physician-patient relationships, which are not readily converted into mass practices and for this reason require much time. Furthermore, the prevention of disease can often be achieved at low cost, whereas therapeutic medicine increasingly involves the use of expensive techniques, equipment, and supplies.

The constant increase in the cost of individualized medical care does not result from greater financial rewards to physicians and nurses, or from inefficiency in the management of hospitals. Its

1. Many of the views expressed in this chapter are the outgrowth of extensive discussions with Dr. Walsh McDermott, of Cornell University Medical College. The articles listed under Dr. McDermott's name in the bibliography present but a small part of the ideas he has contributed to the formulation of my own attitude.

cause lies in the fact that medicine now provides new kinds of highly complex and costly services that are beyond the financial resources of most persons, yet must be made available to all. Furthermore, therapeutic medicine is probably now entering a phase of diminishing returns. Many of its most spectacular and costly achievements are of help to only a few persons. Therapeutic procedures such as open heart surgery, the maintenance of patients with artificial kidneys, and the control of phenylketonuria constitute great technological feats, but they benefit only a minute percentage of the population, while they loom very large on the national bill of health.

Economic considerations are in principle irrelevant to the right of every human being to health. And in fact disease control is rapidly becoming everywhere a responsibility of government, whatever the kind of political philosophy in power. Contrary to what is generally believed, there is nothing fundamentally new in the universal trend to make the care of the sick a collective rather than an individual responsibility. In the past, only a few people could afford the services of private physicians. Religious and other benevolent organizations were then responsible for most of medical care. What is new is the universal acceptance of the philosophy that the prevention of disease is the responsibility of government. This change in attitude did not become obvious until the second half of the past century.

In August 1848, Rudolph Virchow published an article that can be regarded as one of the manifestoes of social medicine. He wrote:

> History has shown more than once that the fates of the greatest empires have been decided by the health of their peoples or of their armies, and there is no longer any doubt that the history of epidemic diseases must form an inseparable part of the cultural history of mankind. Epidemics correspond to large signs of warning which tell the true statesmen that a disturbance has occurred in the development of his people which even a policy of unconcern can no longer overlook.

Virchow made this statement after returning from Upper Silesia, where he had studied an outbreak of epidemic typhus among the labor classes.

At the same time, other epidemics were affecting all parts of

Europe undergoing industrialization, and also Ireland as a result of the potato famine. Thus, Virchow had good reasons to believe that epidemic diseases have their primary origin in economic misery. As the rich contracted infection from their servants and from other poor people through inevitable social contacts, it was clear even to the most selfish that infectious diseases could be combatted only by educating the public and improving the general standards of living.

In the United States the public involvement in medical matters was illustrated by the motto selected for his department by Hermann Biggs when he was Commissioner of Health of New York State: "Public Health is Purchasable. Within natural limitations, any community can determine its own death rate." All countries have tried, each in its own way, to live up to the challenge of this statement. The general willingness to support the World Health Organization and its various programs indicates furthermore that freedom from disease is now universally regarded as one of the fundamental birthrights of man. However, there is growing evidence that economic limitations will inevitably interfere with disease control, even in countries enjoying great material prosperity. An article recently published in one of the national magazines under the title "Can We Afford to Be Healthy?—The Impossible Trick" illustrates that economic limitations exist even in the United States.

> We are spending as individuals more than $16 billion a year—or almost 6 per cent of all the money we use for personal consumption—on the pursuit of health. Another $5 billion is spent by the government. Because medical care has all the earmarks of a "seller's market," the ante is still rising, and a full package of an adequate standard of medical care may reach 9 per cent in today's market. . . . We are finding ways to prevent and cure disease faster than we can find dollars to pay for them. Since what we want seems to be beyond our present means, we have become a nation of medical windowshoppers—the goods are on display, our motivation to buy has been stimulated, but our pocketbooks won't meet the price tag [Strauss, 1960].

Needless to say, general and individual economic considerations affect the percentage of the income devoted to medical care. In the United States, for example, medical expenditure rose from 3.3 per cent of the gross national product in 1929 to 4.5 per cent in 1933, and then fell again in the late 1930s. However, while the gross national product during the latter period was reduced by almost half, expenditure on medical care dropped by less than a third, despite the profound and prolonged economic depression. It seems that the proportion of its income a given community spends on medical care is much less variable than the proportion devoted to other items of economic life.

Even more remarkable is the fact that all the developed countries tend to spend about the same percentage of their national income on medical care, no matter how this is financed, as shown by the following official figures: 4.06% for Belgium (1954); 4.41% for Canada (1953); 3.67% for Denmark (1952–53); 4.43% for France (1952); 3.76% for the Netherlands (1953); 4.56% for New Zealand (1953); 4.48% for Norway (1955); 4.48% for the United States (1953); and 4.05% for England and Wales (1953–54).

It is obvious that, even under optimum conditions, the present programs of prevention and treatment of disease are grossly inadequate. The national health survey now being carried out by the United States Public Health Service has revealed that

> 8 out of 10 noninstitutionalized persons aged 65 and over had one or more chronic conditions. Some of these, of course, are minor (allergies, bronchitis, and the like) but many others involve serious illnesses such as heart disease, arthritis, and diabetes. The activity of more than half the aged with chronic conditions is limited to some extent. When these figures are projected against the anticipated population of 22 million older persons in 1975, it becomes apparent that the challenge to the health professions will then become staggering [Porterfield, 1964].

Every *single day* in the United States more than 1,000 new persons are added to the rank of the 17 million who are older than 65 years of age. Approximately 11,000 persons 100 years old or older are now drawing Old Age and Survivors Insurance benefits! It is gratifying to know that new medical procedures which permit prolongation of life are constantly being discovered, but frightening to realize the extent to which each achievement adds to the medical load. In the light of these facts, there is reason to fear that even the wealthiest economies may not be able to carry forever the enormous burden created by the scientific advances that permit "medicated survival."

Although it is widely assumed that the growth of medical science and the development of health services will progressively reduce the number of persons requiring medical attention, there is no evidence for this assumption. The experience gained from almost two decades of medicine under the Health and National Insurance Services in Great Britain provides at present the most factual source of information on what can be expected from the general availability of medical care in a modern society. As mentioned earlier (Chapter XIII.1), the *Beveridge Report on Social Insurance and Allied Services* predicted that the cost of the comprehensive health service in Great Britain and the consequent reduction in its need would cancel out. It was tacitly assumed that if the present generation were willing to accept a heavy financial burden, posterity would have smaller medical needs because it would enjoy better health. In the words of the *Report,* "The primary interest of the Ministry of Social Security is . . . in finding a health service which will diminish disease by prevention and cure."

The demand for medical services in Great Britain has not confirmed these optimistic expectations. The cost of the National Service has increased constantly from year to year, and in fact similar increases in medical expenditure have taken place in all the countries of Western civilization, irrespective of political system. At the present time, for example, one out of each eight persons in the United States enters a hospital during the course of a year, as compared with one out of 18 thirty years ago, and the percentage is constantly going up.

The increased demand for medical care is due partly, of course, to the fact that society is setting for itself ever higher standards of health as it becomes more sophisticated and prosperous, and as medicine advances. But there is also reason to believe that medical progress is being outstripped by a real increase in the incidence of disease, or at least by a change in the pattern of diseases. The situation is further complicated by the fact that a larger part of the population now consists of persons in the older age groups.

The continually increasing demand for medical care can thus be expected to create an intolerable drain on the financial and technical resources, and to bring about a progressive lowering of medical standards. In this respect, it is worth emphasizing that a very large percentage of the total expenditure for prevention and treatment of disease goes to professional salaries and wages, over 70 per cent according to some estimates. The practice of modern therapeutic medicine depends increasingly upon the services of highly skilled persons, who have had a long, expensive, and specialized training and who must constantly undergo further training in order to keep in step with scientific advances. Limitation of adequately trained personnel, rather than shortage of funds, is likely to constitute the Achille's heel of disease control in the future.

Outside of humanitarian considerations, many other complex factors play a role in deciding how much of its financial and personnel resources a community should devote to health programs. In many underprivileged countries or social classes, improvement in health is almost a *sine qua non* before economic development can take place. Human beings weakened by malnutrition, malaria, parasitic infestations, and other forms of infectious disorders can hardly be expected to make the sustained effort required for the development of their resources. In such situations it is economically sound and indeed humanly imperative to give priority to health control. The problem, however, is altogether different in prosperous countries.

As an argument in favor of the National Health Service program, it is stated on page 158 of the *Beveridge Report* that "disease and accident must be paid for in any case in lessened power of production and idleness." Everyone knows, however, that many minor pathological conditions have no significant effect on earning power, creativity, effectiveness, or life expectancy, yet require elaborate medical services if attended to. Economic barriers to the medical care of such minor conditions certainly existed in England and Wales before the National Health Service came into being, as shown by the fact that physician utilization markedly increased after its introduction. However, it is difficult to determine whether the increase in physician services reflected a true medical need or merely an unmet demand, and whether it improved significantly the health of the adult population. What is certain is that, despite the greater availability of medical care under the National Health Service, absenteeism from work caused by sickness in Great Britain has not decreased; it has remained at an almost constant level of about 4 to 6 per cent between 1950 and 1960.

Thus, it is not entirely justified to state that health services are "wealth producing." In many cases, "it is not the health services which produce wealth, but wealth which makes possible expenditure upon the health services, like all those other expenditures of which neither the purpose nor the outcome is economic benefit but which are the specific mark of a human society" (Powell, 1961).

In final analysis, the standards of medical care are determined by a number of unrelated and entirely independent factors. There are minimum biological needs below which a human being or a social group cannot function effectively. Other needs are determined by the state of medical knowledge, since higher criteria of health emerge from the availability of new techniques of prevention and treatment. Finally, the very advances of medicine paradoxically create new medical demands. On the one hand, prolonging the life of the aged and the sick increases the number of persons who require constant and costly care. More important, perhaps, medical advances create an atmosphere of rising expectations, in which men are no longer willing to tolerate the discomforts to which the flesh is heir, once they become aware that these evils might be mollified.

The possibility of unlimited progress toward an unattainable medical objective becomes an even more painful paradox as we approach the ideal of a classless society. Complex medical and surgical procedures, psychiatric treatment, and elaborate medical check-ups at frequent intervals are very expensive and require the services of large numbers of highly trained specialists. Whatever the income of a nation and however effective its medical establishment, there will remain services that cannot possibly be made available to all. Yet the concept of a two-class society is even more repugnant to us medically than it is from the economic point of view. Human idealism demands that the bank teller and the laborer be given the same medical privileges as the president of the firm for which they work. But there are limits to what can be done. Unless unexpected scientific developments truly revolutionize medical practice by making it far less costly and especially far simpler technically than it is now, the ethical principles that have governed medicine heretofore will have to be reformulated in the light of inescapable social limitations.

2. Human needs and human medicine

Two independent forces have completely revolutionized the scientific basis of medicine during the past century, and have shifted the interest away from concern with the individual patient to a more impersonal approach. One aspect of this revolution is the emphasis on pathological and biochemical lesions rather than on the total integrated responses of the sick person. Medical and surgical practices depend more and more on information derived from laboratory analyses. The other aspect of the medical revolution is the increasing development of public health programs focused on the population as a whole. Whether he deals with water-borne infections, working conditions, or environmental pollution, the public health officer hardly ever concerns himself with the art of medicine or with individual human beings; he functions as a biotechnologist rather than as a compassionate healer.

Granted the trend toward an impersonal approach to disease control, there persists nevertheless an aspect of medicine that is essentially independent of modern scientific developments. While the biochemical and collective aspects of medical science and practice are essentially the same for animals and men, it is obvious that human medicine remains profoundly different from veterinary medicine. The differences in professional attitude between the physician and the veterinarian have their basis not in the natural sciences, but in peculiarities of man's nature, and especially in ethical considerations that give a unique value to each human being.

The development of scientific knowledge naturally requires that generalizations be made about *Homo sapiens,* but these generalizations are rarely applicable to a given person. One reason is the highly individual manner in which each person converts experiences and events into symbols and reacts to these symbols as though they were actual stimuli. Another reason is that even when he professes to being an uncompromising rationalist, man is always influenced by irrational thoughts; the human condition has been inextricably involved in magic ever since Paleolithic times. Man never completely succeeds in shedding his biological or social past. Preoccupations that transcend plain material existence are as deeply ingrained in man's nature as is the way he uses his hands or digests his food. Indeed, psychic forces often determine the character and intensity of man's responses to the environment. The fact that imaginings, fears, and hopes cannot be readily described, let alone quantitated, does not make these forces less significant in the causation and control of disease than are the structures and reactions that can be measured by physical and chemical methods.

Each human being differs from all others in so many different traits, both genetic and phenotypic, that individual responses to the environment must be considered from two unrelated points of view. Within fairly narrow limits, all members of a given group are similarly affected by nutritional deficiencies, heat or cold, radiations, contact with parasites, exposure to toxic substances, and certain sociocultural stimuli. On the other hand, each person within the group reacts as if he or she were living in a private world, in which environmental forces take a meaning determined by his or her own peculiarities.

According to biochemical and immunological doctrine, the reason that one man's meat can be another man's poison is to be found either in inherited idiosyncracies or in specific allergies determined by prior experience with materials having a certain chemical configuration. But the response is in fact determined also by prejudices, cultural taboos, and emotional experiences totally unrelated to the chemical nature of the object to which the prejudice, taboo, or experience is directed. Likewise, a particular combination of light, temperature, and wind lowers blood pressure in one man and raises it in another, calls one to work or to battle, and puts another in a mood of relaxation or of despondency. Thus, the objective aspects of the environment acquire a significance and have effects that are determined to a very large extent by the attitude of the person on whom they impinge. This attitude, furthermore, changes with time, as the experiences of the person widen and accumulate.

The problems posed by the biological uniqueness of each human being are further accentuated by purely ethical considerations. Throughout the living world, the adjustments of the natural system are so designed as to insure the survival and success of the species. The individual plant or animal has little meaning in this large scheme of nature and indeed is usually doomed to early death for the sake of its species. In contrast, mankind places a special and unique value on the individual person. As far as we know, this concern does not exist in the same form in the rest of the animal kingdom, but its existence in man poses special problems to human medicine. In many cases, medical ethics demand that the interests of the community be sacrificed to those of an individual person, however low the status of that person. At their best, medical institutions try to reconcile individual and social values. This attempt imposes on human medicine preoccupations and directions that reach above and beyond those derived from biochemical knowledge and from the collective policies of preventive medicine.

Human medicine reflects the interplay of all these independent forces. Their influence will be illustrated in the following pages by a few facts concerning the evolution of medical services in Western civilization. As we shall see, the structure and role of the hospital were determined originally by peculiarly human attitudes, then by social forces that continuously change as society evolves.

3. FROM INSTITUTIONS OF MERCY TO HOSPITALS WITHOUT WALLS

Anyone who feels pessimistic about the human condition might find some solace in the long history of continuous efforts to make the hospital a living expression of faith in the rights and dignity of man. Originally, the Latin word *hospitalis* referred not to an institution concerned with disease but to a place where a guest was received. The *hospitalia* of the early Middle Ages were essentially guest houses for pilgrims. Then, more and more, hospitals became places where poor people were housed when they were sick or injured. People came to the hospital not so much to be treated as to be protected and decently sheltered.

The Ospedale del Ceppo (Hospital of the Stump), in Pistoia, Italy, symbolizes how hospitals thus progressively became institutions of mercy. This lovely structure was erected in 1277, presumably through the efforts of an order of lay brothers dedicated to the care of the indigent sick. Some 200 years after its foundation, toward the end of the fifteenth century, the Ospedale del Ceppo commissioned the sculptor Giovanni della Robbia, son of the more famous Andrea della Robbia, to execute a number of ceramic panels representing its various works of mercy. The six panels placed between 1514 and 1526 represent "Burying the dead," "Clothing the naked," "Aiding widows and orphans," "Lodging pilgrims," "Ministering to the sick," and "Visiting prisoners." A seventh panel, "Giving drink to the thirsty," was placed around 1585.

Thus, the institutions of mercy in medieval times were motivated less by the hope of preventing or curing disease through medical intervention than by the concept of brotherly love in God. This attitude was consonant with the view of disease that prevailed during the Middle Ages. Health was then primarily regarded as holiness and spiritual health. There was no dichotomy between mind and body; both physical and mental illnesses were treated much the same way, by ministering to the whole person, especially via the soul. Actual treatment was extremely limited. In practice, going to the hospital was, more frequently than not, merely going to a place of refuge to rest or await death, especially for those who did not have the facilities to rest or die more comfortably at home.

In association with the main hospitals, institutions progressively developed for segregating and housing special classes of unfortunates—the lepers, the blind, the aged, the orphans. Then, as the awareness of infection became more prevalent during the eighteenth century, these institutions were supplemented by the fever hospitals or pest houses, where patients were isolated in the hope of preventing them from spreading their diseases to the community. All these trends became intensified during the nineteenth century, when the Industrial Revolution created immense problems of disease in the destitute proletariat of Europe and the United States. To meet the ever-increasing problems of crowd diseases, the hospitals evolved into large municipal institutions, which progressively took over the role of the medieval houses of mercy. They served as refuges for the poor, while persons who were better situated economically still continued to be cared for at home.

Little by little the reputation of hospitals improved during the late nineteenth and early twentieth century. The greater acceptance of hospitalization by the public was due in part to control of hospital infection. It resulted also from the fact that as scientific medicine advanced, only well-equipped institutions could provide the facilities required for the new therapeutic procedures. Moreover, the intellectual mood was now different from the one that had pervaded the medieval houses of mercy. Scientific materialism had taken control of medicine. Disease was regarded as a derangement of the body machine; even mental illness was increasingly approached via the body.

Despite all these changes, however, only a small percentage of diseases, chiefly the very serious ones, ever reached the hospital, until the first decades of this century. Home care remained the general rule; the personal compassion and attention of the family physician still provided the most common form of healing. The structure of medicine really changed at the end of the nineteenth century when the doctrine of specific etiology began to yield new methods of control for many kinds of derangements affecting the body, and to some extent the mind. Individual incomes were generally increasing at that time in the countries of Western civilization, and insurance coverage was more widespread. This made it possible to direct to the hospital a larger variety of pathological conditions. Instead of being primarily an institution of mercy and a place where poor people went to die, the hospital progressively became a medical center for the whole community, geared not only for the treatment of disease but also, ideally, for providing general guidance in matters of health.

The new therapeutic technologies introduced during recent years have had such a tremendous impact on the delivery of medical services that they are bound to change the philosophy of hospital management as much as they have changed medical practice in general. For example, antimicrobial drugs now make it possible "to place in the hands of a barefoot non-literate villager more real power to effect the outcome of a child critically ill with, let us say, meningitis or pneumonia, or tuberculosis, than could have been exerted by the most highly trained urban physician of twenty-five years ago" (McDermott et al., 1960). Such technologic developments have large implications also for hospital practice.

The design and administrative structure of the famous European and American hospitals reveal even today the influence of medical practice as it was carried out when they were built. Until two decades ago, there were few really effective therapies, and the main help to the patient was prolonged nursing care. Recovery from disease was always slow, and continued surveillance by the physician was needed at every step in order to meet emergencies. For this reason, the hospital was built, organized, and administered on the assumption that the patient might be incapacitated for a long time and dependent for almost everything on the medical staff.

In contrast, many of the steps in diagnosis and even in therapy can be carried out now by auxiliary personnel, functioning under the guidance of physicians, but nevertheless acting in a certain measure on their own initiative. Whether the patient comes within the domain of surgery, internal medicine, or any other specialty, the time he can spend usefully in the hospital is ever becoming shorter. Indeed, there is increasing evidence that, except in special cases, prolonged exposure to the abnormal environment that necessarily exists even on the best hospital wards is objectionable for the subsequent welfare of the patient. This is equally true for the newborn baby, the growing child, or the adult recovering from an accident or an illness.

Another fact of importance in planning for the delivery of medical services is that a very large percentage of illness is of a kind that does not take the patient to the hospital. The young physician going into practice soon discovers that most of his patients in the real world are troubled by conditions he had little chance to recognize, let alone to treat, during his ward training. Since the kinds of patients who find their way into hospitals provide an incomplete and distorted picture of the total health problems in the community, hospital medicine should extend its ministration to the home, the school, the office, the shop, and the factory. The medical center can encompass the whole range of community problems and thus broaden the scope of its ministration only if it is enlarged conceptually and becomes, so to speak, a hospital without walls.

Another aspect of medicine that cannot be fully apprehended in the absence of intimate contact with the problems of the world outside the hospital is the relationship between the medical profession and the other professions that also play a vital role in the health fields. Until recently, medicine was almost exclusively concerned with the care of the sick, a highly personal relation between the healer and the individual patient. This role is, of course, as important today as it ever was but there are other social problems that are fundamentally medical, yet transcend traditional medical activities.

As our societies become technologically more complex and more highly organized, they generate problems that affect the well-being of the social body as a whole, by giving rise to new types of pathological disorders and to new types of collective responsibilities. In this sense, there is rapidly emerging in the modern world a set of problems that could properly be called *social* medicine, not to be confused with the very different concept designated as *socialized* medicine. The health field is no longer the monopoly of the medical profession; it requires the services of all sorts of other skills. This collaboration will become increasingly urgent as the community demands that steps be taken, not only to treat its diseases, but also to protect its health.

The danger in this inescapable trend is that the medical profession may be progressively edged out of many social aspects of medicine. While persons trained in the physical and social sciences, from engineers to general biologists and lawyers, play an essential role in the total medical picture of our society, it is usually difficult for them to comprehend all the complexities and subtleties of health and disease problems. Limited points of view are likely to generate oversimplified formulae of action, unless technical knowledge is supplemented by broad medical philosophy and guidance. Seen in this light, the care of the social body as a whole presents new and exciting challenges to the medical profession; it constitutes an enlargement of its calling.

There is no doubt, in fact, that technological civilization needs medical guidance. Medically trained persons must transcend their preoccupation with disease and concern themselves with the effects of technological and social innovations on normal healthy men. They must learn to collaborate with engineers, sociologists, and politicians to ward off insofar as it is possible the dangers to health inherent in all technological and social changes, especially when these occur rapidly, as is the case now. Only through closer contacts between the medical professions on the one hand and all specialists concerned with the social order can we hope to govern our rapidly changing technological civilization in such a way as to provide better health and more desirable ways of life. Thanks to its knowledge of the imperative but often subtle needs of the body and soul, the medical profession can help mankind to avoid sacrificing the fundamental requirements of man's nature at the altar of dehumanized progress.

Hospital medicine must therefore add to its ancient responsibilities in the healing art a new role very essential in the modern world—namely that of adapting technological civilization to the eternal needs of the human body, and of the human heart. The physician can no longer be merely a healer, in Luther's words "God's body patcher." His ultimate goal is not merely to restore the patient to physical fitness, for true fitness really implies successful adjustment to existing social patterns. The hospital can fufill its responsibilities only to the extent that it evolves beyond the status of a "repair shop." It must concern itself with man in society and therefore must overcome its traditional isolation.

Many attempts are being made to incorporate in the modern medical center the ancient attributes of the houses of mercy and of the traditional hospitals, and to supplement them with the services, teaching, and research activities that link medical practice to the various biomedical and social sciences. An ideal medical center would give to physicians and students a direct contact with the whole spectrum of these sciences, from the most abstract knowledge to its ultimate practical applications. Such a center would provide a suitable environment for the birth of an idea, the establishment of its validity, the shaping of it in the form of a usable concept, the testing of its practical utility and limitations, the teaching of its theory and practice, the concern for the moral and ethical problems that inevitably arise whenever technological innovations are applied to human beings, and finally the discriminating application of the new knowledge for the benefit of a particular human being as well as of the community as a whole. Ideally, the medical center should recapture the intimate relationship with the patient, symbolized by the traditional picture of the family physician; it should cultivate the rational approach to disease that has grown out of scientific medicine; it should be a forum for social studies, where the problems of the body, the aspirations of the soul, and the needs of society are integrated into a new science of human and social engineering.

The requirements outlined above correspond of course to a utopian concept, according to which the medical center would become all things to all men in the world of medicine. At first sight, such a wide-ranging and utopian view of medical organization seems to stand in sharp contrast to the worldly advice given by William Osler to young physicians in an address delivered before a group of Army surgeons on February 28, 1894. Said Osler, "Your business is not to *see* what lies dimly at a distance, but to *do* what lies clearly at hand." In reality, however, the modern idealized picture of the medical center is not incompatible with Osler's admonition. One of the great difficulties of academic medicine is to recognize the nature of the medical problems that *really* lie at hand, instead of being exclusively concerned with those selected from the social pattern of diseases by the artificial environment of the hospital, however up-to-date its equipment may be.

Let it be acknowledged that the modern medical center is not essential to the growth of medicine. Enterprising physicians and medical students have always made use of the clinical resources available for teaching, learning, and research, even under the most primitive conditions. To illustrate the scientific role of ancient medical establishments, one need only mention the development of wound surgery in military institutions, the knowledge of infectious diseases gained in pest houses, the emergence of morbid pathology in the nineteenth-century general hospitals.

Until recently, in fact, most of clinical teaching and research took place outside the hospital. One of the reasons was that hospital populations were made up in the past almost exclusively of poor people and, to a large extent, of social outcasts. Such a population was regarded as unrepresentative for teaching, because it was profoundly different from the patients seen by physicians in their regular practice. The aspect of clinical experience then most useful as well as most remunerative dealt with the ailments seen among the well-to-do. The physician could gain experience in this field only by observing patients in their normal bourgeois surroundings. Hence the importance of apprenticeship, during which the student learned so much by helping an experienced physician in his practice, acquiring not only medical skills, but also the difficult art of relating the problems of disease to the actual circumstances of the patient's daily life. Today hospital training must substitute for the technical and human training the experienced practitioner was expected to impart to his apprentice. To achieve this end, however, the hospital must concern itself not only with the pathological conditions found on its wards but also with the medical problems of the whole community that it serves.

In the past, the role of the medical school was merely to provide a systematic background, in the form of dogmatic teaching, for the practical experience gained by the apprentice at the bedside in the home of patients. Today its role is still to teach the knowledge on which scientific medicine is based: the theoretical aspects of human functions, physical and mental; the ecology of human relations; the rational bases of prevention and therapy of disease. In contrast, hospital teaching in the medical center should reflect the changing aspects of disease as they manifest themselves in a given community at a given time. In other words, the hospital should supplement the universal aspects of medicine, which are the main concern of the medical school, with emphasis on the problems of health and disease that are peculiar to its community at a given time, with which it must deal in order to fulfill its social function.

4. Individual factors in medical care

If the future of man as a biological species were the only issue at stake, personalized medicine, as practiced for the benefit of individual patients, would have little practical justification. Disease control through social and environmental techniques is far more effective than personalized medicine, because it benefits large numbers of people, at lower cost of money and effort. But the distinguishing characteristic of human medicine is precisely that it is so much concerned with the individual person. Statistical morality has no place at present in the medical ethics of Western civilization because the individual patient still constitutes the unit of value, irrespective of biological and economic consequences.

The importance of the healer and his attitude toward the patient to whom he ministers is most obvious, of course, in psychiatric diseases. This was clearly recognized long before the modern psychoanalytic era, as illustrated by the excellent therapeutic results achieved during the first decades of the nineteenth century through the humanitarian and compassionate management initiated in France by Philippe Pinel (1745–1826), in England by William Tuke (1732–1822), and in the United States by Dorothea Dix (1802–87). Wherever kindness and understanding prevailed in the management of mental patients, the results could bear comparison with the best obtained today. During the years 1833 to 1846, for example, 70 per cent of the patients admitted at the Worcester State Hospital within one year of the onset of their illnesses were discharged as cured or much improved. The discharged patients were followed up until 1893, and it was ascertained that half of them had experienced no recurrence. These figures are the more remarkable when one considers that a substantial portion of the patients must have had illnesses such as cerebral syphilis, for which there was no effective treatment.

Eventually, however, the humane attitude initiated during the early nineteenth century began to wane. General care deteriorated in mental hospitals and probably reached its lowest level during the first decades of the present century. In large part, this deterioration in institutional care was due to a gradual increase in the size of the mental hospitals and a consequent disruption of the close relationship between the staff and their patients.

The therapeutic effect that the physician exerts by his very presence, irrespective of any medication he uses, constitutes one of the most ancient components of the medical art. Countless old paintings illustrate the well-known fact that the spiritual and physical activities of the healer were closely interrelated in ancient civilizations; for example, Christ is often depicted holding a urine bottle, the medieval symbol of the physician. Even more often perhaps, Christ the Healer and his apostles are shown curing the sick without any help from the instrumentalities of medicine. Evidence of healing through the power of faith has continued to accumulate. The votives in the Asclepiades and in the shrines of all religions have their well-documented counterparts in the cures performed every year at Lourdes and by the various forms of faith-healing. Placebos, as we have seen, often produce effects similar to those of the most powerful drugs (Chapter X.1).

In all societies and at all times, physicians have exploited, consciously or unconsciously, the fact that body ailments can be mollified by influencing the mind. The costumes, chants, and dances of medicine men among primitive people constitute an essential part of their therapeutic abilities. The blue frock of the physician a few generations ago, his gold-headed cane, and his dignified carriage were the European symbols of the same attitude. The modern physician has given up these props, but he still cultivates a certain kind of bedside manner that contributes to his therapeutic effectiveness. He still believes with Alan Gregg that "a miraculous moment comes when the doctor becomes the treatment."

Even more important, every generation of clinicians has experienced and taught what Peabody expressed in his famous remark, "The secret of the care of the patient is in caring for the patient." Whatever the system of medicine in vogue, patients want and often actually need to be treated as individual human beings; they want their physician to deal with them as if they required special attention and were entitled to it. Today as in the past, the French motto carved on E. L. Trudeau's statue in Saranac Lake symbolizes the attitude of the true physician:

Guérir, quelquefois	To cure, sometimes
Aider, souvent	To help, often
Consoler, toujours	To console, always

While fully aware of the limitations of his science and therapeutic armamentarium, the dedicated physician knows that he can succor suffering and contribute to his patient's recovery by helping him to regain confidence. This highly personal aspect of the physician's therapeutic action is widely recognized among the general public and is frequently discussed in lay writings. The following quotation from a popular magazine illustrates the public awareness of the role played by personal relationships in the care of the sick:

> Of all human acts, few can match the quiet splendor of the moment when the pale and tremulous fingers of a sick person are grasped in the firm reassuring hands of a compassionate physician. This simple act mutely promising that all the powers of modern science and human thought will be unsparingly invoked to restore health is among the finest deeds of humankind. It is more than ritual. When pain and fear make a sick person feel that all is lost, the laying on of healing hands brings solace and hope. Its strength can even turn the tide of illness and amplify the curative effect of the strongest wonder drug. It remains today as it always has been—man's oldest medical miracle [*Life,* October, 1959].

The personal approach to medical care is extremely costly in terms of physician time and, as already mentioned, is of limited value to the community as a whole. There is indeed something economically extravagant in the fact that such a large percentage of a highly trained physician's time and efforts should be devoted to a kind of laying on of hands, almost completely unrelated to his prolonged sophisticated and costly scientific training. Yet it is also true, as already mentioned, that most patients want their physician to become personally involved. For lack of this personal involvement, they will appeal to a witch doctor or other extramedical source of spiritual support. Hence, the proliferation of health cults and fads even in the most sophisticated social groups.

The physician derives much of his personal influence from an awareness, conscious or unconscious, of his patient's somatic and psychologic idiosyncrasies. Such knowledge, even though it may be highly accurate, pertains more to artistic apprehension and insight than to science, since it is concerned with unique situations. The importance of apprehending the unique aspects of a given medical situation and of evaluating it within the framework of existing scientific knowledge has been so ably expressed by Paul Weiss that it seems best to quote at length his own statement of the problem:

> We need not labor the role of the study of biology as a basis for medicine, for it is almost universally taken for granted. It is to be made clear at the outset, however, that this does not reduce medicine to just another branch of the biological sciences. Medicine is rather a hybrid of two equal parental lines, only one of which lies in the science of objective measurement and logical deduction, while the other stems from the subtle, but subjective, powers of evaluation and judgment of the human mind to be tapped whenever the doctor faces the individual patient, the single case in its uniqueness. Science does not deal with unique events; it only encompasses them. It deals with the general rules, usually statistical, common to large numbers of cases—the average behavior of categories. Physics does not deal with a particular electron or a given atom, nor does meteorology deal with a specific cloud in its uniqueness. No two cells will ever behave exactly alike and no two disease courses duplicate each other down to the last detail. Predictive science takes an interest in individual cases only for what they may hold in supporting old or directing new generalizations and theories. It then discards them as nonrecurring items. Scientific prediction can be positive about what cannot happen and what might happen, but it can only approximate reality in anticipating just precisely what will happen in the individual event. This is where the physician cannot afford to be a scientist. His interest must sharpen as it focuses on the single, unique, and nonrecurring specimen, the human individual whom he aims to help.
>
> There are thus two strains running through medicine, one of impersonal scientific method, which furnishes the rules and tools for guiding judgment, and another of personal responsibility for judicious decisions to deal with each specific and unique case and situation at hand. The trend over this past century for the scientific line to grow in vigor and in volume should be no cause for alarm, since the two lines are not com-

> petitive, but strictly complementary. In the expanding universe of human welfare, there must be room for both happily to expand in unison.
>
> As science, with its method, tests, and critical checks, provides personal judgment with ever safer guides and narrower margins for error, it expands legitimately into what the ministers of the healing art of old might have claimed to be their exclusive domain. But as it does so, it also frees time and energy for the cultivation of those human values and powers of application for which science has furnished no proper substitute. Let us not overlook the fact that the human body contains some of the subtlest precision devices of discrimination and evaluation. Our eye can detect a few quanta of light, our nose a few aromatic molecules. True, we can design technical apparatus of equal sensitivity to register the elementary stimuli involved, but for the perception and evaluation of the kaleidoscopic variety of patterns and combinations in which they meaningfully appear in nature there is no substitute for that discerning power of the human mind which we call judgment. The time and effort that science saves us could, therefore, be profitably spent on sharpening by practice those peculiarly human faculties [Weiss, 1964].

Many great clinicians have clearly recognized the parascientific aspect of their skill and its similarity to the perceptiveness of great novelists. Sydenham advised his students that reading *Don Quixote* would help them qualify for practice, and Osler recommended George Eliot's *Middlemarch* for the same purpose. Granted that the art of interpersonal relations still depends on intuitive insight, some facts suggest nevertheless that it would be possible to introduce scientific methods in its study. Controlled tests have revealed, for example, that one psychiatrist may be highly successful with certain patients and yet fail with others who respond well to the care of another psychiatrist. And similar specific affinities between doctors and patients are observed in all other fields of medicine. The mere classification of such affinities might be a first approach to their scientific analysis.

In any case, their existence points to the fact that doctor-patient relationships involve more than good public relations on the part of the doctor, and more than his desire to do good. The interplay between the patient and his doctor is an expression of psychobiological processes that certainly come within the realm of medical science. The practice of medicine has not become too scientific, as is so often claimed. Rather, medical science must broaden its outlook so as to include within its scope not only the physicochemical aspects of the body machine but also the psychobiological determinants of human relationships.

5. Social determinants of disease control

In 1902 William Osler painted a picture of medical progress that cannot but appear disconcerting to those who believe that most of the important medical advances have occurred during the past few decades. In Osler's words:

> The average sum of human suffering has been reduced in a way to make the angels rejoice. Diseases familiar to our fathers and grandfathers have disappeared, the death rate from others is falling to the vanishing point, and public health measures have lessened the sorrows and brightened the lives of millions.

Unbelievable as they may appear today. Osler's statements were certainly correct. By 1902 many of the worst nutritional and microbial diseases of the nineteenth century were in full recession, even though vitamins and antimicrobial drugs had not yet been discovered. As we have seen, several independent biological forces have contributed to the change in the pattern of disease; but the most important during the late nineteenth century were certainly the social factors that resulted in a general improvement of living standards and thereby profoundly changed the ways of life. Since the diseases most characteristic of modern societies also have social determinants, it is very probable that advances similar to those noted by Osler in 1902 with regard to nutritional and microbial diseases could be brought about by appropriate changes in the ways of life. But the fact is that we know very little about the environmental determinants of the diseases characteristic of our societies, and behave as if we cared even less.

Laymen, as well as many physicians, seem to regard as truly preventable only those diseases which can be controlled without individual effort. Typhoid fever is called preventable because the chlorination of water can be achieved through anonymous and unnoticed administrative measures; poliomyelitis is preventable because vaccination is an effortless procedure; nutritional deficiencies are preventable because vitamins can be bought at low cost. In contrast, prevention of the chronic diseases and the behavioral disorders will probably demand an entirely different approach, much more exacting because it will depend on individual discipline. Environmental control carried out by government agencies and prophylaxis obtained by a simple visit to a physician should be supplemented by individual control of behavior and by continued exercise of personal responsibility. Almost everyone agrees that overeating, lack of physical exercise, excessive consumption of alcohol, chain cigarette smoking, too great reliance on drugs, etc. contribute to the disease load in our communities. Yet few are those who seriously attempt to reform their ways with regard to these threats to health, except under special conditions and for a short time during periods of scare propaganda.

At present, the atmosphere in our social environment seems unfavorable to disease control through the exercise of individual responsibility. The general skepticism on this score comes out clearly in the following statement by a very wise physician, who is also a well-known professor of medicine in the United States but prefers to remain anonymous on this occasion: "It is part of the doctor's function to make it possible for his patients to go on doing pleasant things that are bad for them—smoking too much, eating too much, drinking too much—without killing themselves any sooner than is necessary." [2]

Thus it is certain that many social forces will in the future

2. As seen in the following quotation from a medical journal, even physicians are wont to make fun of attempts to prevent disease through control of one's individual life:

> Thumbnail Sketch of the Man Least Likely to Have Coronary Heart Disease:
> An effeminate municipal worker or embalmer,
> Completely lacking in physical and mental alertness and without drive, ambition or competitive spirit who has never attempted to meet a deadline of any kind.
> A man with poor appetite, subsisting on fruit and vegetables laced with corn and whale oils,
> Detesting tobacco.
> Spurning ownership of radio, TV, or motor car,
> With full head of hair and
> Scrawny and unathletic in appearance,
> Yet constantly straining his puny muscles by exercise;
> Low in income, B.P., blood sugar, uric acid, and cholesterol,
> Who has been taking nicotinic acid, pyridoxine, and long term anticoagulant therapy
> Ever since his prophylactic castration.

limit the extent to which knowledge can be applied to the control of disease. Some of the limitations have their roots in human behavior and therefore are unpredictable at the present time. Others appear inescapable, in particular those arising from increasing costs and especially from shortages of personnel trained for the performance of ever more complex techniques. Still other limitations have a less honorable basis and hopefully could be eliminated, namely those derived from economic reasons. In this respect, it is rather shocking to note the weaknesses with which governmental agencies and medical institutions approach problems such as environmental pollution, or excessive consumption of alcohol and tobacco, on account of economic pressures. Support of medical research and action in public health matters are most vigorous in those situations which do not conflict with economic interests. Everybody is willing to devote large funds and great scientific talent to the study of intermediate metabolism, chemical genetics, electron microscopy, the structure of viruses, poliomyelitis, or muscular dystrophy. But support is much less evident if there is any indication that the scientific studies could threaten vested interests, whether it be those of industries poisoning water supplies or of advertisers urging young people to develop habits known to be noxious.

6. Safety Regulations and Prospective Epidemiology

Rachel Carson in *Silent Spring,* Lewis Herber in *Our Synthetic Environment,* and other authors who have discussed the threats to health posed by environmental pollution curiously end their indictment on a rather hopeful note. They seem to believe that ways can be found to retain the advantages derived from the use of chemicals and other technological innovations, and yet to limit and control their use so effectively that all dangers can be avoided. In my opinion, this Panglossian attitude is like whistling in the dark. On the one hand, it would be just as difficult to decrease significantly the use of chemicals as it would be to cut down on the number of automobiles. On the other hand, it is as impossible to predict at present the delayed toxic effects of most technological innovations as it was half a century ago to predict that exposure to small doses of radiations would cause cancers 20 years later.

Many examples could be quoted to illustrate the dependence of our economy and ways of life on synthetic chemicals. The most obvious, probably, concern food production and food technology. In theory it is possible by a skillful practice of organic agriculture and biological pest control to achieve large farm yields without using synthetic chemicals. But these practices no longer fit the pattern of life in our societies. They would soon be rejected because they involve a kind of physical labor men resent more and more. The elimination of chemicals from modern agriculture and pest control would upset the present balance in human life even more profoundly than their use upsets plant and animal life. High yields of foodstuffs, capital and labor input, and a certain form of social structure have become essential aspects of the order of nature as far as mankind is concerned. For a simpler example, one need only try to imagine the practical problems that would have to be solved before laundering equipment could be redesigned to avoid the use of detergents.

All technological innovations, whether concerned with industrial, agricultural, or medical practices, are bound to upset the balance of nature. In fact, to master nature is synonymous with disturbing the natural order. While it is desirable in principle to maintain the "balance of nature," it is not easy to define the operational meaning of this idea. Nature is never in a static equilibrium because the interrelationships between its physical and biological components are endlessly changing. Furthermore, man placed himself part from the rest of nature when he began to farm the land and even more when he became urbanized. The survival, let alone growth, of his complex societies implies that he will continue to exploit and therefore upset nature. The real problem, therefore, is not how to maintain the balance of nature, but rather how to change it in such a manner that the overall result is favorable for the human species.

Every effort should be made of course to determine beforehand the potential dangers of technological innovations. But it would be unrealistic to believe that *all* these dangers can be recognized and avoided. New techniques and new substances are introduced so rapidly that adequate testing of all of them would paralyze progress. The recent experiences with drug control will suffice to illustrate the magnitude of the difficulties arising from any attempt at strict enforcement of safety regulations. In October 1962, the President signed into law a new Food and Drug Act, which greatly increased the regulatory powers of the Food and Drug Administration. In January 1963, the FDA issued a set of regulations instructing pharmaceutical manufacturers to submit detailed reports of all drugs currently in the course of being tested on human beings in the United States. By June of the same year, approximately 2,500 reports were on hand, each containing many hundreds or thousands of pages of data! Needless to say, every individual report required several months of study, and in many cases its analysis pointed to the need for further experimental and clinical tests.

Dr. Frances O. Kelsey, then chief of the drug investigation branch of the FDA, was disturbed by the lack of time, facilities, and office space for the study of so many drug applications. She stated in a newspaper interview:

> Unless we can work efficiently, the belief of the American people that the Government is now providing greater security over drugs will be a false one. We must go through the reports as fast as possible to see if there is any danger in the drugs being tested. If the reports are lying around in packing cases, unopened and unexamined because we don't have the room, what good are they? [New York *Times,* May 7, 1963].

At the time of this interview, 60 physicians had been assigned to the study of the 2,500 reports, each being therefore responsible for evaluating more than 40 drugs! Dr. Kelsey's own experience with thalidomide illustrates the discouraging complexity and magnitude of the task. Furthermore, it takes more than room, more than time, and even more than people to determine the potential dangers of technological innovations. It takes a kind of theoretical knowledge that does not exist in most cases and, unfortunately, is not being vigorously pursued. However strict the safety regulations, and however diligent the work of those who try to apply them, accidents will happen because the theoretical basis of toxicology is not sufficiently developed to permit a prediction of dangers to health.

Since there is no progress without risks, it might be feared that the rapid development of new techniques and new products implies the immolation of mankind to a technology gone wild. But the future need not be as dark as that. We must abandon, it is true, the utopian concept that safety regulations can protect us completely from the health dangers of the modern world, but this does not mean accepting passively the consequences of these dangers. As mentioned earlier, there is much that could be done to gain better knowledge of the biological effects of new substances and new techniques (Chapter VIII.4). In addition, it is urgent to develop epidemiological techniques to recognize as soon as possible early manifestations of toxicity in the population at large and to trace their origins to technological and social changes. This approach, which might be termed prospective epidemiology, is one of the steps in the process of continuous self-correction, without which scientific knowledge cannot make a lasting contribution to health improvement.

Health surveys such as the one presently conducted on a national scale by the United States Public Health Service will provide useful data for the recognition of new pathological states associated with the modern ways of life. But such surveys are far too ponderous to detect the early manifestations of new dangers to public health. They should be supplemented by an elaborate system of listening posts designed for recognizing any sign of abnormality in the population so that more pointed epidemiological surveys can be rapidly organized. To a limited extent this is being done for detecting resistance of pathogens or their vectors to antimicrobial drugs or to insecticides. Largely as a result of the impetus provided by the thalidomide episode, the World Health Organization has initiated a campaign for the development of a worldwide warning system for toxic effects of drugs (resolution passed in May 1963, at the Sixteenth World Health Assembly). The FDA is supporting this program as a first step toward an international drug safeguard plan.

There is no doubt that physiological, chemical, and hematological tests could be devised to detect abnormalities in human populations and thus broaden the coverage of potential threats to health. Needless to say, such programs are constantly being carried out at the present time, but usually in an episodic manner on a limited scale, chiefly to deal with special situations. What is proposed here is that this form of prospective epidemiology become an integral part of social management in all industrialized societies. Its aim would be early containment of new threats to health, rather than the utopian goal of complete prevention or eradication of disease.

The development of a social sense of alertness to the inevitable dangers of technological innovations is part of the social adaptive process to the conditions of the modern world. Prospective epidemiology could be regarded as a kind of social protective organ, at least as essential to disease control as are the safety regulations by which official agencies attempt, vainly, to protect the general public against unforeseeable dangers.

XVI. Medicine Adapting [1]

1. Medical research and social goals

Biomedical technology is rapidly approaching a state of development that will make it possible to modify several aspects of man's nature. The transplantation of organs and the use of mechanical prostheses may soon become commonplace. Personality will be increasingly modifiable through physiological and embryological manipulations. Even the genetic make-up of man may become amenable to willful alteration. Since most of what we try to do will probably come to pass, we must give thought to the long-range implications of our scientific programs; we must try to evaluate prospectively the potential effects of manipulating the human body and mind, lest there be analogues of the nuclear bomb in the medical future of mankind! Unfortunately, while the scientific method is immensely effective for dealing with the technical aspects of biomedical problems, it provides no philosophical basis or ethical guidance for relating technical solutions to the fundamental needs or aspirations of man.

The word progress is extremely ambiguous. It means, of course, moving forward on a certain road, but not necessarily the right road. It denotes a continuous expansion of knowledge, but is almost meaningless with regard to the betterment of the human condition, unless the goals to be reached are stated. The applications of scientific techniques are commonly based on social and ethical values that are ill defined and often contradictory. Preserving human reproductive cells is a technical advance, but even if it could be used effectively, selective reproduction would be of little avail to the welfare of man because there is no agreement as to the kind of human traits that are most desirable. Controlling man's desires and thoughts by drugs is an exciting scientific

1. Many of the views expressed in this chapter are the outgrowth of extensive discussions with Dr. Walsh McDermott, of Cornell University Medical College. The articles listed under Dr. McDermott's name in the bibliography present but a small part of the ideas he has contributed to the formulation of my own attitude.

achievement, but with trivial or even undesirable consequences from the point of view of man's future if control is used only to foster adjustment to a bland life of togetherness.

Even the words health and disease, as we have seen, do not provide clear guides for medical action. They are meaningful only in the light of social criteria, which in turn are determined by beliefs and mores that differ from one group to another and even from one person to another (Chapter XIII.2). At the end of a symposium on *Man and His Future* held in London in 1962, one of the participants emphasized in the following words the wide diversity of opinion regarding what would constitute desirable goals, and the proper scientific approach for reaching them.

> I really do not know, even if we took a census of opinion, what principles we would teach or what beliefs we would try to inculcate. This is the thing that has impressed me most about this meeting—the sheer diversity of our opinions. . . . This diversity of opinion is both the cause and the justification of our being obliged to do good in minute particulars.[2] It is the justification of what Karl Popper called "piecemeal social engineering." One thing we might agree upon is that all heroic solutions of social problems are thoroughly undesirable and that we should proceed in society as we do in science. In science we do not leap from hilltop to hilltop, from triumph to triumph, or from discovery to discovery; we proceed by a process of *exploration* from which we sometimes learn to do better, and this is what we ought to do in social affairs [Medawar, 1963].

Granted the wisdom of these words, it is also true that human institutions cannot merely drift if they are to survive. Like individual human beings, they must have a direction or they flounder. Each civilization is characterized by the special kind of "minute particulars" it elects to emphasize, and by the patterns it creates with these particulars. Societies operate on certain assumptions that influence the way they practice science and use the technologies derived from it. Despite our pathetic attempt at objectivity and the assertion that practical wisdom and intellectual honesty

2. The expression "minute particulars," as quoted here by Professor Medawar, is borrowed from William Blake, who used it repeatedly in his criticism of the painter Joshua Reynold's *Discourses*. "He who would do good in anything," William Blake claimed, "must do it in minute particulars."

demand that we concern ourselves with "minute particulars," we as scientists are in fact highly subjective in the selection of our activities, and we have goals in mind when we plan our work. We make a priori decisions concerning the kind of facts worth looking for; we arrange these facts according to certain patterns of thought we find congenial; and we develop them in such a manner as to promote social purposes we deem important.

Admittedly, there is no way to state with precision what are the most desirable attributes of man. Nevertheless, history shows that the course of medical science is not determined by blind forces or by the unplanned accretion of "minute particulars." Science, especially medical science, is not random behavior or idle curiosity without purpose. It implies responsibility to certain basic needs and goals of human life. At any given period, the fields of research selected for emphasis reflect large social decisions, conscious or unconscious. These decisions, in turn, are influenced by prevailing beliefs as to the way the physician should ideally deal with man—whether he should concern himself primarily with the body, regarded as a machine to be mended, or whether he should base his action on the doctrine that health cannot be achieved without regard for the patient's memories, fears, and aspirations.

The presuppositions on which medicine operates are thus conditioned by the general philosophy of the social group as a whole. In the words of Oliver Wendell Holmes (1860), "The truth is that medicine, professedly founded on observation, is as sensitive to outside influences, political, religious, philosophical, imaginative, as is the barometer to the changes of atmospheric density." On the other hand, physicians play a large part in shaping social convictions and expectancies by creating an intellectual and emotional atmosphere consonant with their scientific faith. In 1850 scientific orthodoxy taught that man must return to the ways of nature. In 1950 the official dogma was that a specific therapy must be discovered for each type of ailment. The dream of the philosopher's stone lasted one thousand years; then it was replaced by the faith that medical salvation would be found in specific agents which do only one thing; now the belief is slowly growing that "we can never do merely one thing, therefore we must do several in order that we may bring into being a new stable system" (Hardin, 1963). History is woven out of the fibers and the pigments available in a given place at a given time, but the pattern of the fabric is determined by the weavers. In general, the weavers follow a traditional design. Now and then, however, a few of them depart from tradition and thus give to the fabric entirely new and distinctive patterns.

A general assumption in our communities at the present time is that health and the good life will automatically emerge if we focus our scientific efforts on the production of things and on improving the body machine. Such an attitude is probably responsible for much incoherence in our technological civilization, but unfortunately the formulation of worthwhile social goals is now far more difficult than it was in the past. During the eighteenth and nineteenth centuries, for example, the production of food, shelter, and clothing, the limitation of physical effort, the relief of obvious forms of suffering, constituted well-defined objectives that could be reached by technological improvements. The pursuit of knowledge for its own sake was also simpler. Scholars worked individually on problems of their own choice. Receiving little recognition or help from the community, they had much intellectual freedom and could select their fields of endeavor. The situation is now different because the purely scientific aspects of survival have largely been solved, and also because most scientists are directly or indirectly paid by the state. As a result, the question of what studies to prosecute, which used to concern almost exclusively the individual scientist, now must be considered by the social group.

The most extensive and sophisticated research programs of Western medicine are naturally focused on certain disease states that are causing public apprehension in prosperous countries. But there are many other medical issues of equal or greater importance that are grossly neglected. Some concern problems of disease very prevalent in the underdeveloped countries, i.e. in four-fifths of the world! Others are of relevance to mankind at large.

Comparison of disease patterns in prosperous countries 50 years ago with what they are today brings to light the puzzling fact that several medical problems have all but disappeared without benefit of scientific understanding, through a process which has been called "historic bypass" (McDermott, 1964). For example, infantile diarrheas are now scare in Western countries despite the fact that their etiologies are not at all understood; likewise several respiratory infections became less frequent or at least less severe long before specific methods of prevention and therapy became available. These various conditions, however, remain highly destructive in the underprivileged parts of the world; hence the urgent need to study their social as well as biological determinants.

The factors involved in adaptation constitute another large gap in medical knowledge, one which has a universal bearing on health. We devote an enormous amount of skill and foresight to make food available throughout the year and at all hours of the day, to protect our bodies from heat and cold, to avoid contact with pathogens, to improve and multiply labor-saving devices, even to minimize the effort of learning. Scientific medicine, furthermore, provides a seemingly endless supply of drugs to correct physiological disturbances, to control infection, to relieve pain, to increase activity, to induce sleep or tranquility. In brief, modern technology and scientific medicine provide for man a sort of hothouse environment, in which he is sheltered from efforts, trauma, and stresses.

Technological and social advances have of course made life easier and decreased the incidence of certain diseases. Unfortunately, their very success has rendered man less and less able to face the world unaided, and has made him more and more dependent on medical assistance. Moreover, this success has led to the complete neglect of another approach for helping man to deal with the external world, namely, the cultivation of his adaptive resources. Civilization has attempted to solve problems of human life by eliminating effort and suffering, but a completely sheltered life entails some indirect dangers, even for the person who enjoys it. On the one hand, it weakens his potential abilities for adaptive responses through disuse atrophy; furthermore, it causes him to lose come of his physical and mental pleasures, because these are the expressions of responses to environmental challenges.

Problems of adaptability are encountered of course in all aspects of ordinary life, but the demands made on man in the modern world change so rapidly in kind that it is difficult if not impossible to define what adaptability should be for. Today's bank teller has to meet challenges very different from those encountered by yesterday's accountant, and soon he will have to adjust to a completely new kind of demand created by automated work. In September 1959, the United States Air Force Academy conducted in Denver a conference on "Fitness of American Youth." Fitness for combat seems to have little to do with the qualities desirable in ordinary human beings living in a world at peace, yet some of the facts brought out during the discussions are relevant here because they illustrate so vividly the problems that have to be considered in discussing the challenges of tomorrow.

It was emphasized at the Denver conference that there is no such thing as fitness per se with regard to military service, because fitness must always be defined in terms of a particular combat situation. In consequence, the armed forces find it necessary to revise the physical standards of health at frequent intervals, in order to keep them in tune with the changing requirements of military service. With propeller-driven aircraft, for example, there were many situations in which survival depended on strength of arm and limb. Moreover, the pilot of a fighter airplane in World War II had to watch for enemies in the sky by direct visual perception. His head swiveled from side to side looking to the rear, and for this reason calisthenics to develop neck muscles were part of training.

Today, power controls have lessened physical requirements of the aircraft operator, and with electronic vision the fighter pilot never needs to look to the rear. In any case, direct vision would be of little help in modern air combat because of the terrific speeds at which aircraft approach each other. At 600 miles per hour, and this is now moderate speed, half a mile means little more than a second, clearly not enough time for the pilot to see, to react, and to change the direction of his aircraft. As a result, keenness of distant vision no longer means the difference between life and death for the fighter pilot; this attribute has been superseded by keenness in ability to detect slight changes on electronic dials and gauges. More generally, strenuous physical conditioning programs are no longer as directly relevant to performance in the armed forces as they used to be. And in fact, recent tests indicate that pilots at the peak of physical form do not score any better in difficult operations than do those of comparable groups who are less well endowed physically.

The changes in relevance of physical prowess to military performance have many counterparts in civilian life. Effectiveness in modern technology depends to a large extent on dial-watching and on reading printed matter. Whereas physical stamina and distant vision were once extremely important, muscles are now called into play chiefly during leisure time, and nearsightedness has become almost an asset in several professions. The present trends of life seem to provide justification for the child who does not want to walk because he considers it old-fashioned and for his mother who dissuades him from engaging in physical exertion or exposing himself to inclemencies because modern existence is and will increasingly become air-conditioned and effortless. And yet this attitude may have unfavorable consequences in the long run (Chapter X.2). A state of adaptedness to the conditions of today is no guarantee of adaptability to the challenges of tomorrow.

Many of the challenges man encounters in the modern world affect the mind rather than the body. For this reason adaptive mechanisms having their seat in the mind are as important as those affecting the body machine. Man can learn by self-discipline and experience to become more proficient in intellectual tasks, and even to increase his resistance to suffering. A society that depends on sedatives or stimulants cannot achieve the resilience necessary for survival, let alone for growth.

The view that adaptation is a creative endeavor which demands continuous effort clearly applies to the process of learning. In a world where everything changes rapidly, "practical" information soon becomes obsolete. The techniques and equipment that are the most up-to-date expressions of knowledge during the school years are usually outmoded by the time the student becomes an adult. The kind of knowledge most likely to have permanent value and to be useful in practice is theoretical knowledge. Even though it appears of no practical use at the time it is acquired, theoretical knowledge is the most useful for the future because it is applicable to a wide range of conditions. In any given field, the leaders are rarely those who have entered professional life with the largest amount of practical information, but rather those who have breadth of understanding, critical judgment, and especially discipline of learning. The intellectual equipment most needed is that which makes it possible to adapt rapidly to new situations, as they constantly arise in the ever-changing world.

These qualifications are not acquired without effort, and may even demand painful effort. In fact, it may well be dangerous to make learning so easy that it becomes passive and effortless, because one of the most important aspects of education should be to instill the willingness to engage in difficult tasks. Like health, learning cannot be acquired passively. It is an active process, and the measure of its success is the extent to which it increases adaptability.

The conviction that survival and growth demand strenuous and constant adaptive effort does not imply a tragic view of life. Learning, self-mastery, and the willingness to respond creatively to the environment constitute the very experiences which give structure and meaning to life. In final analysis, the potentialities of human beings are realized only by meeting challenges. Man cannot progress without effort; man without effort is sure to deteriorate and cannot be happy or even contented.

It might be assumed that such complex problems of motivation are outside the realm of medicine. However, the very fact that man is a goal-directed animal makes it unrealistic to limit medical science and action to a concern with the pathological derangements of the body. For reasons that are still obscure, lack of fulfillment is at the root of much organic as well as mental disease everywhere in the world. In contrast, the satisfactions derived from vigorous responses and from fulfillment of purpose constitute powerful adjuncts to health; they help man to function effectively in spite of disease, and indeed they often help him to overcome disease.

2. Dilemmas of modern medicine

Some of the therapeutic and preventive techniques now in use are so powerful and so indiscriminate in their effects that they influence not only the person to whom they are applied, but also the welfare of the whole community, indeed the very future of the human race. The possibility of postponing death in every age group first comes to mind as an example of a situation that is creating difficult dilemmas for medical conscience. To save the life of a child suffering from some hereditary defect is a humane act and affords great professional gratification; but its long-range consequence is commonly to magnify the medical problems of the future. Likewise, prolonging the life of an aged and ailing person must be weighed against the possible hardships that medicated survival entails for this very person and also for the community. Such ethical problems are not new, but they were rarely encountered in the past because the power of action of the physician was so limited. They are bound to become more frequent and more disturbing as the physician becomes more able to prolong biological existence in human beings who cannot derive either profit or pleasure from life, and whose survival creates heavy social burdens.

By postponing deaths due to irreversible physical and mental defects, we allow the accumulation of persons who need continuous and exacting medical supervision. By making physical life almost effortless and by minimizing exposure to infection and inclemencies, we reduce the opportunity for the adaptive reactions that would otherwise increase the resistance of the body and the mind to the accidents and challenges unavoidable in life. Medical science now makes it possible for almost every newborn infant to survive, however defective he may be, whatever the inadequacies of his hereditary endowment. Admittedly, genetic defects and physiological deficiencies need not be a serious handicap in a society equipped to correct their manifestations. But a time may come when the accumulation of the weak and the sick will constitute an economic and medical burden difficult to manage even by a prosperous and dedicated society. As health becomes more and more dependent on specialized and onerous care, some aspects of medical ethics will have to be reconsidered in the harsh light of economics.

Modern medicine faces therefore a paradoxical situation which has no precedent in history. On the one hand, science can eventually solve the technical aspects of almost any medical problem. On the other hand, the application of medical knowledge to the prevention and treatment of disease will be necessarily limited by economic and other social factors. Choices have to be made among all the possibilities for medical care and disease prevention, but there is no agreement as to the social or ethical bases on which to make choices. Public attitudes on this score reveal indeed some strange paradoxes. Thus, research funds are lavished on the development of techniques to remedy the infertility of the infertile, and also of techniques to limit the fertility of the fertile! The importance given to a particular disease, or group of diseases, differs from one culture or one social group to another, and furthermore changes with time. At present the death of a few children from poliomyelitis is regarded as a national calamity. But slaughter on the highways is accepted almost as a matter of course, even though this disgraceful tragedy is unquestionably the expression of a social disease.

In addition to the difficulties arising from judgments of value, there are other dilemmas which have their origin in scientific uncertainties. As emphasized in preceding chapters, technological innovations involve dangers that cannot always be foreseen. Thoughtful men are of course concerned with safety, but they must accept risks lest social action be paralyzed. It is almost certain, for example, that any drug or other substance possessing biological activity will prove to have some toxicity. Each one of the drugs introduced into the practice of medicine during the past 20 years —from penicillin to cortisone, and to the tranquilizers—is now known to cause severe toxic reactions under certain circumstances; this is true even of aspirin. Thus, the problem is far more complex than ruling out of use potentially toxic substances. What is involved is to weigh advantages against dangers, so as to evaluate the justification of taking risks with human lives.

The case of isoniazid illustrates well how medical ethics must be based on educated common sense rather than on hard and fast regulations. It has been known for many years that isoniazid causes neurological symptoms in a certain percentage of human beings; and recent publications indicate that it can elicit cancers in several strains of mice. On the other hand, it is the most effective drug for the treatment of tuberculosis. Clearly, therefore, isoniazid must be used despite its potential dangers. And a similar case could be made for many other substances that have become essential in medicine, industry, or agriculture, even though they present a real danger for human health.

The report of the Committee on Tissue Culture Viruses and Vaccines, appointed by the director of the National Institutes of Health, provides a striking illustration of the technical difficulties involved in predicting the dangers of a technological innovation, and of the dilemmas they create for physicians and medical scientists.

Many, if not all, of the animal tissues used in the preparation of viral vaccines harbor in a latent form certain viruses potentially capable of causing disease, indeed of acting as carcinogenic agents. Needless to say, stringent regulations have been formulated in an attempt to rule out accidents that could arise from these latent viruses. But it must be acknowledged that dangers persist because not all latent viruses have yet been recognized, and because many of them exist in the tissue cultures in a form not detectable by presently available techniques. On the other hand, the public health benefits derived from the use of viral vaccines are so considerable that a certain amount of risk is justified. A paragraph in the report, published under National Institutes of Health sponsorship, is worth quoting here because it constitutes such a clear and reasonable statement of a point of view applicable to all other innovations in medicine and technology:

> We agreed readily that continuously cultured tissue cells afford numerous advantages in the propagation of viruses for vaccines. However, because such cells tend to develop characteristics suggestive of malignant change, and theoretically oncogenic activity might be associated with viruses propagated in them, we recognized that present knowledge permits only carefully qualified approval of their application to vaccine production. On the other hand, we were mindful that too cautious an opinion would discourage the continued research needed for better definition of the permissible limits of such application.

Another official statement concerning a similar problem is found in a recent report of the Federal Radiation Council, the agency now responsible for setting radiation standards in the United States:

> If . . . beneficial uses were fully exploited without regard to radiation protection, the resulting biological risk might well be considered too great. Reducing the risk to zero would virtually eliminate any radiation use, and result in the loss of all possible benefits. It is therefore necessary to strike some balance between maximum use and zero risk. In establishing radiation protection standards, the balancing of risk and benefit is a decision involving medical, social, economic, political and other factors. Such a balance cannot be made on the basis of a precise mathematical formula but must be a matter of informed judgment.

As mentioned earlier, the background of theoretical knowledge is not sufficient for the adequate testing of all technological innovations with regard to their potential dangers, especially those of delayed character. Furthermore, there will never be enough scientific personnel or laboratory facilities to carry out all the tests needed, over a long enough period of time. Indeed, there cannot be any hope of fulfilling these conditions during the foreseeable future. Hence, decisions must be taken in the face of scientific uncertainties.

The problems of decision created by the dilemmas of modern medicine demand a new kind of sociomedical statesmanship involving not only physicians and medical scientists but the citizenry at large. On the other hand, the citizen can play a useful role in such complex sociomedical decisions only if he has been sufficiently educated. And this brings up one of the most difficult problems of social and medical ethics, namely the extent to which it is justified to condition the public mind by any means whatever.

There are, of course, cases which present no difficulty. Everyone is against the use of drugs or of brainwashing for taking advantage of an individual person. But in most situations the issues are far more complex. Where does education stop and propaganda begin? At what stage of evidence does it become justified for the scientific community to claim that it knows the truth and to advocate a program of action? How far should concern for the public at large, or for a social cause, affect the advisability of using a drug, or any other medical procedure, to control the health or behavior of an individual patient or of a limited human group? Two examples will serve to illustrate the wide range of ethical difficulties created by decision-making in medical matters.

A bill to make immunization against poliomyelitis compulsory was recently considered in South Carolina. According to its supporters, the voluntary program of vaccination "had been rather a fiasco" and "when the public refuses what is best for it . . . the more resistant should undergo some legal coercion for the good of all" (*J. South Carolina M. A.*, 1961). The question at issue here is not the value of a particular kind of antipoliomyelitis vaccine for an individual person, but whether knowledge was sufficient in 1961 to make immunization compulsory "for the good of all." Even during recent years there have been many fatal accidents resulting from vaccination, not only against poliomyelitis. Some 20 years ago, many cases of postvaccinal encephalitis followed vaccination against smallpox in Holland and in New York City. At the 1962 meeting of the British Medical Association in Belfast, it was reported that 27 babies were known to have died from the sequelae of smallpox vaccination out of 2,600,000 primary vaccinations in a period of 10 years (the number of deaths may well have been larger since reporting is not compulsory). The experience is much the same in the United States. Yet, the smallpox vaccine is unquestionably one of the best understood, and constitutes the most effective of all the products used in preventive medicine.

A more drastic interference with individual freedom was recently enacted by the administration of the Territory of New Guinea in an attempt to curb the spread of a grave neuropathological disturbance known as kuru. This disease is characterized by motor incoordination of cerebellar type, inability to walk, aphoria, dysphagia; it usually terminates with death from marasamus or bronchopneumonia within a few weeks after appearance of the first symptoms. Kuru was first recognized among the Fore people, a tribe some 35,000 strong, inhabiting the eastern highlands of New Guinea and still living according to a Stone Age civilization, with ceremonial cannibalism. As kuru seems to be confined to this single tribe, it was first thought to constitute a purely genetic disease. The assumption was that it is transmitted by a gene that behaves apparently as a Mendelian dominant in females and a recessive in males. The presumed homozygotes of both sexes die in childhood, usually before adolescence; heterozygous females die later, after most of them have produced children. Since many of the Fore men were assumed to be heterozygous carriers of the gene for kuru, it was feared that the disease would spread and afflict the populations of other districts. To counteract this danger, the administration's plan was to quarantine the Fore tribe; its members are not allowed to emigrate from the tribal area.

The fact that kuru has existed for less than 100 years and is increasing in incidence as well as appearing in other districts now makes it seem probable that nongenetic factors are involved in its etiology. Its control, therefore, might be achieved by techniques different from the eugenic approach taken by the administration of the Territory of New Guinea, which was based on insufficient scientific information. Disease control through an official eugenic policy is without scientific precedent, a fact which makes the quarantine of the Fore people a landmark in the history of medical ethics.

3. The manipulation of man's nature

At least one in every five persons is said to be encumbered with a mutant gene that arose in a reproductive cell of one of his parents. It is likely that the mutation rate is increased at present by certain new factors in modern life, such as exposure to radiation, mutagenic drugs, cosmetics, contraceptives, food additives, industrial wastes, and environmental pollutants. Furthermore, the new techniques of prophylaxis and therapy may be interfering somewhat with the processes of selection that upheld in the past the genetic quality of the human race.

The fear that genetic self-correction may no longer operate is creating a renewal of scientific and popular interest in the problems of eugenics. In fact, some geneticists claim that if man is to avoid genetic deterioration "about 20 per cent of the population who are more heavily laden with genetic defects than the average must in each generation fail to live until maturity, or if they do live, must fail to reproduce" (Muller, 1963). The eugenic revival, however, goes far beyond an attempt to correct genetic deterioration. Man is so immensely complex genetically that the temptation is great to try to improve him by selecting from among the many different human genotypes those whose phenotypes exhibit desirable qualities.

The dream of voluntary reproductive selection, first formulated by Galton and the early eugenists, is now coming within the range of technical possibility through the development of techniques for preserving frozen spermatozoa. Frozen germ cells can remain viable and available in storage banks for indefinite periods of time, and can be utilized for artificial insemination long after the death of the donor. Preservation of the female ovum also looms as a possibility in the foreseeable future. There is no doubt, therefore, that children can be produced from the reproductive cells of parents selected for certain desired genetic traits. But the technical problems of artificial insemination are only a minor aspect of those posed by the desire to improve the genetic endowment of the human race.

The proponents of eugenics claim that germinal choice through the use of semen derived from men whose lives have proved that they possess the desired qualities constitutes "the most practical, effective, and satisfying means of genetic therapy" (Muller, 1963). According to them, controlled reproduction provides a chance not only to reverse the downward genetic trend by compensating for ailments and weaknesses, but also to help mankind progress toward a higher estate under scientific guidance.

Granted that artificial insemination has proved a very useful practice in animal husbandry, there are many purely scientific problems that must be solved before the science of genetics can be usefully applied to the improvement of human populations. One of them comes from the likelihood that heterozygosity and diversity constitute positive values because they increase the chance of the species, or of the social group, to achieve fitness as the environmental conditions change. As stated by one of the upholders of this point of view,

> heterozygosity represents a compromise between getting the best out of individuals while maintaining a population versatime enough to cope with hazards that change from time to time and from place to place. A case can be made for saying that a genetical system that attaches great weight to genetic diversity is part of our heritage, and part of the heritage of most other free-living and outbreeding organisms [Medawar, 1960].

Many different possible adaptive peaks exist in nature, but an overspecialized population can occupy only one of these. If circumstances change drastically, its chance of survival is small because it may not be able to adapt rapidly enough to the new environment. Indeed, scientific control of man's nature may turn out to be a suicidal policy, since evolution and progress depend upon the variety of the material subject to selection. If holistic control should ever lead to the equalization of human bodies and human minds, such uniformity might spell the end of progress.

The resistance to malaria associated with the sickle-cell anemia gene constitutes the now classical case to illustrate that genetic selection has meaning only in terms of well-defined environmental conditions. Unfortunately, the genetic factors of human population are so poorly documented that it is not possible at present to define their relation to eugenics except in very general terms. The wide range of possibilities with regard to important human affairs becomes apparent, however, from the mere mention of some of the tentative hypotheses that have been formulated. One of these concerns the possibility that the genetic proneness to obesity had survival value in the past when periods of food shortage were frequent; now in contrast, this genetic constitution is associated with shorter life expectancy wherever food is always abundant. Some forms of obesity would thus appear to be the expression of a polymorphic system for which modern man is now paying an ancestral debt. Similarly, diabetes has been postulated to be the product of a genotype that is "thrifty" of sugar, for a time at least, and thus might have been of advantage when famine was more frequent than feasts. It has even been suggested, perhaps facetiously, that the genetic constitution favoring schizophrenia might under the proper social conditions express itself in a type of behavior making for greater success in power politics or in big business, and therefore be advantageous in our competitive societies.

Even assuming that the genetic problems of "voluntarily conducted germinal choice" can be solved, the criteria for choice will long remain a source of difficulty. It is true, of course, that "practically all people venerate creativity, wisdom, brotherliness, loving-kindness, perceptivity, expressivity, joy of life, fortitude, vigour, longevity" (Muller, 1963). But the choice of a suitable semen donor is not greatly facilitated by the statement of these ideals. Opinions about the world and men change and so will necessarily the selection of semen donors. Everyone agrees, of course, that it would be desirable to eliminate gross physical and mental defects, though even this limited program poses problems of judgment and execution far more complex than usually realized. But the problem of positive qualities raises questions of a more subtle nature.

It is relatively easy to formulate a genetic program aimed at producing larger pigs, faster horses, better hunting dogs, or more friendly cats, but what is the ideal human being? Our present ways of life will soon be antiquated, and the future may demand qualities undreamt of at the present time. For all we know, resistance to radiation, noise, intense light, crowding, and the repetition of boring activities may be essential for biological success in future civilizations. Who knows, furthermore, whether mankind is better served by the gentleness of Saint Francis of Assisi and Fra Angelico or by the dynamism of space explorations and modern art? Is the higher type of society one that prizes, above all, individuality and self-development, or one that regards devotion to the common welfare as the highest standard of morality? Is it not dangerous to reproduce a trait that once had appeal because of its uniqueness? How many Beethovens would it take to make his genius commonplace? Furthermore, would a human being endowed with Beethoven's genes have the kind of musical genius suited for the automation age?

The fundamental difficulty in formulating a program for the genetic improvement of man is that we do not know what we want to become or where we want to go. To a large extent we still hold a static view of mankind and have difficulty in conceiving of life as a continuous act of creation, in which man has become the most important actor.

The practice of a eugenic policy would thus require great wisdom; but there is little indication that such wisdom is widespread today, or that we know how to recognize it where it exists. The likelihood is that if people really had the chance to choose the fathers of their children, they would be likely to choose more pronounced projections of their self-images. The disturbing diversity of opinion as to desirable human goals that came to light among a group of illustrious scientists, recently gathered to discuss the "future of man," made it clear that even the most learned and sophisticated human beings do not have the wisdom required to formulate long-term eugenic objectives (Wolstenholme, 1963).

Many biologists believe that results even better than those which might follow selection from existing genotypes could be accomplished more rapidly by direct mutagenic operations on the genetic material. Others advocate that man could be rapidly improved by modifying the course of development and physiological processes. Still others are in favor of creating artificial contrivances capable of replacing or supplementing normal bodily structures. The following quotations illustrate statements recently made with regard to these different possibilities.

> Perhaps, following the current use of plastic heart valves and arterial walls, it will be possible to construct plastic prostheses of heart or kidney which would be accepted by the human body and remain functional in it [Koprowski, 1963].
> It would be incredible if we did not soon have the basis of developmental engineering technique to regulate, for example, the size of the human brain by prenatal or early postnatal intervention. In fact, it is astonishing how little experimental work has been done to test some elementary questions on the hormonal regulation of brain size in laboratory animals or the functional interconnexion of supernumerary brains. Needless to say, "brain size" and "intelligence" should be read as euphemisms for whatever each of us projects as the ideal of human personality. . . . Only preliminary suggestions are possible, but even imperfect ones may help to illuminate the possibilities:
>
> (1) Accelerated engineering development of artificial organs, e.g. hearts, which may relieve intolerable economic pressures on transplant sources.
>
> (2) Development of industrial methodology for synthesis

of specific proteins: hormones, enzymes, antigens, structural proteins. For example, large amounts of tissue antigens would furnish the most likely present answer to the homotransplantation problem and its possible extension to heterotransplantation from other species. Structural proteins may also play an important role in prosthetic organs.

(3) A vigorous eugenic programme, not on man, but on some non-human species, to produce genetically homogeneous material as sources for spare parts [Lederberg, 1963].

Thus medical science is now contemplating the supplementation and modification of man by biotechnological procedures and mechanical contrivances that alter his very personality. The requirements for space travel may encourage even more drastic changes. Under these conditions, it has been suggested:

> The human legs and much of the pelvis are not wanted. Men who had lost their legs by accident or mutation would be specially qualified as astronauts. If a drug is discovered with an action like that of thalidomide, but on the leg rudiments only, not the arms, it may be useful to prepare the crew of the first spaceship to the Alpha Centauri system, thus reducing not only their weight, but their food and oxygen requirements. A regressive mutation to the condition of our ancestors in the mid-pliocene, with prehensile feet, no appreciable heels and an ape-like pelvis, would be still better. There is no immediate prospect of men encountering high gravitational fields, as they will when they reach the solid or liquid surface of Jupiter. Presumably they should be short-legged or quadrupedal [Haldane, 1963].

Ultimately the question will arise as to the identification of the person modified by biotechnology. What is the moral, legal, or psychiatric identity of a human being so modified by medical manipulation that he has almost become an artificial chimera? Poets have long been aware of the loss of human quality symbolized by the "hollow man." Yet at a recent meeting a physician concerned with space medicine was willing to advocate the creation of an "optiman," in whom essential organs had been replaced by different mechanical contrivances more efficient for operation within the environment of a space ship than those provided by nature. The present interest in the creation of an "optiman" may have encouraged the use of "instant men," consisting of large bodies of plastic having the same sound-absorptive powers as a human audience, to test the acoustic qualities of the new Philharmonic Hall in Lincoln Center! The poverty of the results shows that musically as well as biologically "instant men" and "optimen" are poor substitutes for real men of flesh and bone, despite all their imperfections.

The intensity of efforts devoted to problems of human development makes it likely that techniques will soon become available to affect the human brain by prenatal or early postnatal intervention. Even at present, mental states can be influenced by many different techniques, from yoga to hypnosis and drugs. In man as well as in animals, electric stimulation of a particular area in the brain can produce a sense of well-being in the whole organism. Similar effects can be produced by drugs such as mescalin, lysergic acid, and psilocybin. This kind of knowledge is immensely exciting because it enlarges the understanding of the human mind, but by the same token, it is also frightening because, almost inevitably, knowledge is used for control. Electrodes have already been placed in the pleasure centers of the brain. Presumably they were the brains of hopeless psychotics, but it takes little imagination to realize how the power to manipulate human behavior could become an instrument of tyranny.

The moral aspect of the biomedical techniques that alter man's nature must be emphasized because it will often be difficult to recognize the real motivation of those who use them. At the least, a sharp distinction must be made between changing people's ideas by dialogue and by manipulation. The manipulator regards the other member of the interplay almost as an object; whereas in dialogue, there is a reciprocity of interaction and therefore a greater chance that human freedom and rights are respected.

Fifty years ago, a physician, like an engineer, could approach his tasks with the confidence that he was acting as a benefactor of humanity. In contrast, medical technology is now so powerful and often so indiscriminate that it can damage human personality even as it improves the functions of the body. The most cruel dilemma of modern medicine is to decide which aspects of man's nature can be ethically tampered with and which ones should be respected at all cost. There is no guide to resolve this dilemma, beyond Montaigne's admonition: "Science without conscience is but death of the soul."

4. Stability vs. Adaptability of Scientific Institutions

a. Logical vs. willed future

Science deals with concrete facts, but individual scientists like artists use facts to different ends and even for conflicting purposes. One painter employs colors and shadows to convey the play of light on a vase and another to enliven a philosophical allegory. One poet converts the spectacle of human life into images of sensuous pleasures and another into stanzas to the glory of God. Similarly, scientific facts can be used to describe different aspects of reality. Some biologists emphasize that the human body functions as a chemical machine, others that the most characteristic attribute of man is his ability to symbolize events. For some physicians the ultimate goal of medical knowledge is to build more efficient semi-mechanical "optimen," and for others to create an environment suited to the ordinary man of flesh and bones, *l'homme moyen sensuel.*

Scientists make a priori judgments as to the facts worth looking for and thereby impose a pattern on the social applications of science. There is no doubt, of course, that "he who would do good in anything must do it in minute particulars" (Chapter XVI.1), as William Blake wrote. But it is equally true, again in his words, that "what is now proved was once only imagined." Thus, while science is obviously based on facts, the scientist can not be entirely objective or neutral. As a result, science and technology can be illiberal and can even become immoral when pursued without concern for their human and social consequences.

The importance of value judgments is particularly great in medical science and the public health practices derived from it, because they can have such a profound influence on human life. This problem may be ignored by the physician in medical practice because his code of ethics instructs him to do whatever he can for the patient to whom he ministers. But public and private medical institutions cannot help playing God to some extent. When health organizations decide to spend a large percentage of their resources on certain disease problems, they are almost compelled thereby to neglect others that are just as important. And so is it for the scientist when he elects to devote his energies to one field of research instead of another.

Admittedly, there is an element of intellectual conceit, and also of naiveté, in even attempting to postulate what would be a proper scientific course of action for the future welfare of mankind, because the future is so uncertain. But a defeatist attitude on this score would be justified only if human life were completely determined by blind forces. In practice, the future is the creation of man as much as it is the result of circumstances. True enough, the "logical" future is the expression of natural forces and antecedent events. But there is a "willed" aspect of the future, which comes into being to the extent that men are willing to imagine it and to build it. As H. G. Wells stated in *A Modern Utopia:* "Will is stronger than Fact; it can mold and overcome Fact. But this world has still to discover its Will." Wells' remarks apply particularly well to the medical future. Modern medicine might be compared to a mighty and glamorous ocean liner with powerful engines and luxurious appointments, but with no compass and an absurdly small rudder. It moves fast but its course has not been charted; its ports of call are uncertain and its destination unknown.

Each particular field of medical science has its own built-in logic, which determines in large measure the course of its development. For example, the study of tissue structure will certainly become more and more concerned with the description of macromolecules and the forces that hold them together in a certain orientation. Anatomy, histology, and elementary structure, as revealed by electron microscopy and by chemistry, will eventually be unified into a science of molecular morphology. Pharmacology will similarly continue to evolve from observations of the gross effects of individual drugs on the organism, to the determination of active molecular groups and the manner of their reaction with susceptible bodily structures; eventually this knowledge will guide the synthesis of tailor-made molecules having the desired activities and selectivities. As a last example, one can envisage that the study of mental processes will move in several independent lines, focused on the transmission of nerve impulse, the storage mechanisms that constitute memory, the manifestations of the subconscious, the various forms of conditioning, and the effect of drugs on the mind.

While it is not possible to predict the evolution of any particular field in its details, there exists among specialists a broad area of agreement as to the techniques to be developed and the fundamental knowledge required for certain kinds of studies. The built-in logic of the individual sciences therefore makes it possible to prepare a blueprint of the logical future of scientific developments and gives a striking uniformity to the teaching centers and research institutes the world over.

Magnificent as they are in logical structure and in achievements, today's medical sciences almost completely leave out of consideration certain crucial aspects of man's nature and of human life. For example, they scornfully delegate to psychologists and sociologists the responsibility of dealing with the responses of living man to his total environment. While it is true that the problems of organismic and environmental medicine are less appealing to the laboratory investigator than are the phenomena of intermediate metabolism or molecular biology, this does not make them less important. Their discouraging complexity comes from the fact that the phenomena of human life always involve multiple variables that are changing in their relationships to each other, moment by moment. Yet they must be studied lest some of the most important areas of medicine remain as neglected as they are today. The logical aspects of the medical future are easy to recognize and they demand only earnest work and competent knowledge of the orthodox biological sciences, but the willed aspects require the spirit of the pioneer willing to imagine the possible existence of other lands and to venture on uncharted seas.

b. The study of multifactorial systems

There is nothing new of course in the statement that physicians must concern themselves with the whole man in his total environment. Hippocrates himself taught that "to heal even an eye one must heal the head and even the whole body." Plato repeatedly stated the same conviction in his dialogues, especially in *Charmides:*

> The reasons why the cure of so many diseases is unknown to the physicians of Hellas is because they are ignorant of the whole which ought to be studied also; for the part can never be well unless the whole is well . . . for this is the great error of our day that the physicians separate soul from body.

In *The Scarlet Letter* Nathaniel Hawthorne expressed much the same thoughts in writing of the physician Roger Chillingworth:

> He deemed it essential, it would seem, to know the man before attempting to do him good. Whenever there is a heart and an intellect the diseases of the physical frame are tinged with the peculiarities of these. . . . The man of skill, the kind and friendly physician strove to go deep into his patient's bosom, delving among his principles, prying into his recollections, and probing everything with a cautious touch like a treasure seeker in a dark cavern.

Successful physicians of all times have known that no disease state can be understood without considering the patient as a whole. But while this general truth is given lip service in medical schools, it is not taught with the thoroughness that comes from conviction, and it has not generated many pointed research programs. Yet the observations made on higher animals that have been reported throughout this book leave no doubt that useful experimental models could be designed to study most if not all aspects of the response of the whole man to his total environment.

For the experimenter, the interrelationships between man and his environment present some formal analogies with the complex problems encountered in airplane construction, weapons research, economic planning, and other multifactorial situations. In all these problems of complexity, the number of possible alternative choices is far too great to permit building univariable models of them all. Hopefully, the experimental approach known as systems analysis might help in making a choice among the multiple alternatives. Through this approach, a number of factors closely interrelated yet constantly changing in a somewhat predictable manner are identified. If appropriate mathematical models can be found, or developed, their fluctuations can be reduced to mathematical formulation and can thus be subjected to simultaneous study. With computer aid, the change produced by one variable present in the system or introduced at will can be compared with the change resulting from another variable.

There is as yet no evidence that systems analysis or other related approaches, such as games theory, will be helpful in the study of the multivariable situations encountered in medical problems, but it is certain in any case that such problems will have to be faced, because they will increasingly dominate the medical picture of urban and industrial societies. If schools of medicine continue to keep aloof from the experimental study of situations too complex for analysis by the conventional laboratory sciences, then other institutions will move into this vacuum, because society cannot afford to ignore the biological consequences of man's responses to the forces of the modern world. The danger, as already mentioned, is that large areas of health and disease might become the province of scientists and technicians inadequately prepared to deal with the human aspects of biomedical sciences. Whatever their technical limitations, most physicians are conditioned by their training and clinical experience to approach the study of man in health and disease with a sense of respect for the complexity and plasticity of his nature. Medical training develops an awareness of the fact that man has the ability to respond in a creative way to a changing environment, if he can manage not to be overwhelmed by it.

Population problems illustrate in a telling manner the extent to which new methods are needed in order to determine how multiple biologic units interact with each other and with the many variables in their environment. Population genetics, dealing with the comparative prevalence of physical or physiological traits, has already reached a state where a method akin to systems analysis can prove of advantage. The implications for man's genetic future of the rapid changes now taking place in the world are of course incredibly complex, but they are not necessarily beyond scientific analysis.

> Given the basic data, even the simplest predictions involve equations with ten or twenty variables. No one is more delighted at the advent of the high speed, electronic computers than the geneticist; now for the first time can some of our problems in human genetics be explored in the requisite degree of complexity [Neel, 1963].

Whether the study of the psychological characteristics of individual persons or populations will ever be amenable to experimental approach is still *sub judice,* but in this case again a few questions can at least be formulated in useful terms. Experience indicates, for example, that whereas a passive conformist temperament constitutes a weakness during the emerging phase of a society, it makes life easier during a stable phase. Furthermore, the traits most useful and therefore socially acceptable or desirable in a group probably tend to accumulate in it through sexual selection. Thus, subtle interrelationships exist between genetic heredity and the social environment. On the other hand, selection for socially desirable traits may have biological consequences not necessarily favorable under all conditions. Many fascinating observations have been made in this regard by contrasting the behavior of Norwegian rats obtained from wild populations and from colonies that have long been domesticated. The wild rats are better fighters, but they are also less resistant to many of the artificial stresses imposed upon them in the laboratory.

It is unwarranted, of course, to apply directly to man the knowledge derived from observations on rodents. But there is no doubt that recognition of the fact that profound genetic and physiological changes can rapidly occur in a population calls for a broadening of the concepts of medical responsibility. Medical education should be concerned not only with the care of sick individuals but also with the fate of human populations on a time span reaching far into the future. Medical research must develop techniques to study the interplay between genetic and environmental variables.

The urgency of certain practical problems has stimulated during recent years some interest in the study of natural ecosystems. Attempts are being made to use systems analysis in the study of complex biological problems involving the interplay between organisms and the multiple factors of their environment, for example the delayed effects of ionizing radiations on various ecosystems in nature, and on characteristics such as longevity under various environmental conditions. The new science of bioastronautics is also making use of multivariable analysis for studying the kind of closed ecosystems that will be needed in space vehicles.

The training of men for operations in the arctic, in the tropics, and in space ships, and the effects of brain washing, sensory deprivation, and other unusual emotional difficulties are among the very practical problems that urgently demand the development of new techniques for the study of complex human ecosystems. The responses of human populations to the environment in which they live also present urgent problems to organismic and environmental medicine. A few enlightened architects and city planners are emphasizing the need to reconsider the design of dwellings and urban developments, in order to make them better suited to the physiological and psychological requirements of human beings.

c. Research facilities for organismic and environmental medicine

Experimental models in animals can be designed for the study of the responses man makes to his total environment. But it is certain that such study will demand profound changes in the facilities now available for biological experimentation, and even more in the attitude of experimenters. Almost any kind of tissue will do for the electron microscopy of subcellular elements or for the analysis of intermediate metabolism. But highly evolved organisms, in a known physiological state, are required for studying responses to environmental stimuli and insults, especially if the results are to be extrapolated to man.

The most obvious need, yet so strangely neglected, is for experimental animals of known genetic constitution and proper physiological state—also for adequate facilities to maintain them and study them under a wide range of conditions. Needless to say, studies involving social interrelationships and population problems require groups of different sizes, often quite large. Even more important perhaps is the necessity to have complete knowledge of the history of the organisms under investigation, since early influences, both prenatal and postnatal, affect so profoundly all physiological and behavioral characteristics. In view of these needs, it is shocking to observe the poor quality of experimental animals used in biological investigations and the primitive character of the facilities in which they are housed, even in our most famous medical schools and research institutes. It is humiliating for biologists to compare the lack of interest in the quality of the biological material with the luxury and modernity of the physical and chemical equipment!

Many problems will demand highly organized long-range programs, different from the individual type of research that prevails today. The greatest stumbling block in the development of such programs may be created by psychological difficulties. Academic scientists will find it uncongenial to dedicate themselves, often almost anonymously, to collective tasks of which the outcome may not become apparent during their own lifetime. The highly individualized structure of modern academic science may have to be supplemented by a collective form of intellectual life akin to that which made possible the acquisition and transmission of learning in the medieval monastic orders, and which is now flourishing in industrial research.

For historical and administrative reasons, the existing medical schools, federal laboratories, and other research institutes will find it difficult to develop the kind of programs envisioned here. The most spectacular scientific achievements of medical institutions during recent decades have been in the study of acute pathological processes and molecular biology. The approach to these problems has determined in a large measure the organization of medical research and, what is more important, its scientific ideals. The emphasis on simplified one-variable systems has paid such large dividends during the past 300 years that medical scientists have rejected instinctively and often unknowingly any question that could not be so studied. Unfortunately, as repeatedly stated in these pages, many of the most important problems of mankind cannot be studied by the conventional methods developed for simplified laboratory models.

In theory, of course, any young investigator wishing to work on the complex problems of organismic or environmental medicine is free to do it and needs only to get financial support. In practice, however, the institutional structure of medical schools and research institutes does not foster such interests. Indeed, the intellectual atmosphere of medical institutions is so indifferent to them as to constitute a real deterrent. Furthermore, financial support for research of this nature can be obtained only when there is a precedent of success. Administrators responsible for the development of teaching and research programs naturally favor questions that are well defined and within an accepted field. They shy away from topics that are less safe, however important they may be for human welfare.

The all but universal tendency among medical scientists and administrators of research is to limit the qualification of "fundamental" to the problems that are reducible to simple laboratory models and preferably to molecular biology. As a result, the intellectual atmosphere in the scientific establishment is becoming increasingly incompatible with the study of biomedical problems that involve more complex living systems. In many ways, the present situation of organismic biology and especially of environmental medicine is very similar to that of the physicochemical sciences related to medicine around 1900. At that time there was no place in the United States dedicated to the prosecution of physicochemical biology, and the scholars who were interested in this field were treated as second-class citizens in the medical community. Fortunately, a few philanthropists were made aware of this situation, and they endowed new kinds of research facilities to change the trend. The Rockefeller Institute is probably the most typical example of a conscious and successful attempt to provide a basis of physicochemical knowledge for the art of medicine.

As a result of such initiative, physicochemical biology acquired such glamor that its specialists have become the most honored citizens of the medical establishment. Any activity, however trivial, that deals with submicroscopic particles or subcellular chemical phenomena is labeled "fundamental," whereas efforts to formulate, investigate, or teach the phenomena of life as experienced by the whole organism are considered scientifically unsophisticated. Yet, the study of these phenomena demands great ingenuity, originality, and initiative precisely because they are so complex and so neglected. Organismic and especially environmental medicine constitute today virgin territories even less developed than was physicochemical biology 50 years ago. They will remain undeveloped unless a systematic effort is made to give them academic recognition and to provide adequate facilities for their exploration.

d. Medical science and social needs

Institutions devoted to medical problems, like all human institutions, fulfill two different but complementary roles. One is to provide stability of purpose against the pressure of daily events; the other is to provide a mechanism for prompt adaptive responses to changes in the environment and in the problems of the social group. Medical schools and research institutes have been remarkably successful in assuring the stability required for the long-term intellectual effort from which is progressively emerging the scientific structure first envisioned during the seventeenth century. In contrast, they have not met effectively the new problems created by technological civilization, nor have they prepared themselves to explore the consequences of biotechnological changes for man's future. Yet the most important problems of human biology today are those created by the fact that ours is an adaptive civilization.

The tenacity with which orthodox medical institutions are pursuing the dream of defining all the "minute particulars" of life constitutes proof of their admirable stability. But stability without adaptability to the new problems posed by modern life is socially dangerous and perhaps suicidal. An ever-increasing percentage of resources, as well as of professional and student time, is devoted to the end results and terminal aspects of disease states that have their origin in the interplay between the genetic make-up of man and his environment. But concern with the origin of these problems is at a very low ebb in medical research, teaching, and consequently practice.

Investigators in the biochemical sciences might find it useful now and then to evaluate their scientific activities in the light of Kant's admonition in *Dreams of a Ghost Seer*:

> To yield to every whim of curiosity, and to allow our passion for inquiry to be restrained by nothing but the limits of our ability, this shows an eagerness of mind not unbecoming to scholarship. But it is wisdom that has the merit of selecting from among the innumerable problems which present themselves, those whose solution is important to mankind.

The luxurious hospitals, lavish research facilities, and scientific programs of training, of which we are justifiably so proud, are likely to appear within a few decades as magnificent cenotaphs to concepts which once were vital forces but which are no longer generating truly new scientific departures. The modern medical establishment brings to mind the tuberculosis sanatoria that reached a state of technical perfection just before they had to be converted into sports hotels because of the revolution in therapeutic procedures. As Thomas Mann himself wrote of his famous book on sanatorium life: "*The Magic Mountain* became the swan song of that form of existence. Perhaps it is a general rule that epics descriptive of some particular phase of life tend to appear as it nears its end." The nauseating and misleading articles on medical miracles and on the synthesis of life in vitro that crowd today's newspapers and magazines cannot be compared of course with *The Magic Mountain,* but they are its degenerate descendants. In most cases what is called "modern" medicine is the present fruition of nineteenth-century science rather than the beginning of a new medical era.

History shows that philosophies and institutions which have run out their truly creative course continue to flourish after they have entered a stable phase in which the orderliness of classicism is mistaken for strength. Indeed, classical philosophies and institutions commonly appear most successful and permanent just before the beginning of their decay. The orthodox disciplines of contemporary scientific medicine will of course continue to be productive of their own kind of products and long remain prosperous, as do all orthodoxies. But they will eventually founder in a sea of the irrelevant if they become hypnotized by their own self-created problems, and continue to neglect those presented by real life.

Medical science must reconsider some of the assumptions on which it decides the relative importance of various subjects, and shift some of its emphasis from elementary units and reactions to the manner in which man responds to the threats he encounters in the world. To deal with these problems, medical institutions will have to escape from their bondage to a scientific tradition that has its origins in the seventeenth century. They must forego the security of the future built on the logical outgrowth of the past for the uncertainties of a future so willed that organismic and environmental medicine finally receive their share of attention. Such a course naturally entails intellectual discomforts and risks. "To venture causes anxiety," Kierkegaard wrote, but "not to venture is to lose oneself." What is in danger of being lost in this case is the identification of medicine with the health of the people.

5. The health of the people

Ideally, the approach to disease control should be the same in all countries of the world, whether prosperous or underdeveloped. But in practice, social and economic factors condition both the incidence and manifestations of the various types of disease and the application of medical knowledge (Chapters IX, XIV.4 and 5, and XV). Each society must therefore have its own system of medicine and public health, suited to its particular needs and to its resources. In order to make the best of what modern science has to offer, the developing countries must not imitate the types of hospitals, medical schools, research institutes, and investigations fashionable today in the prosperous areas of the world. They must adapt rather than adopt the techniques and products of Western medicine. In many cases they will have to devise new approaches not only to solve the medical problems peculiar to their present stage of development but also to avoid repeating the mistakes made in the Western countries.

Malnutrition and infection constitute today the two most important causes of disease and early deaths in many parts of the world. The failure to control these problems is not due to lack of scientific knowledge, but to local factors that handicap its application (Chapter XV). Thus, in a disheartening but very real way, the most urgent problems to be solved before health can be improved in the emerging countries fall in the domain of social planning rather than of experimental medicine. By the same token, certain aspects of medical policy must differ in the underprivileged countries from what they are in the prosperous parts of the world.

It might be assumed that the most effective way to improve rapidly the medical situation of an emerging country would be to introduce trained personnel from the outside—physicians, scientists, teachers, and technicians. In reality, however, the world's capital of such specialists is far too small to permit their export in numbers sufficient to have a significant effect. A single country such as Afghanistan or Ethiopia would need a transfusion of 1,500 to 2,000 physicians to achieve a ratio of one physician per 10,000 population; and yet this would be inadequate since the optimum in Western countries is assumed to be approximately one per 1,000! Clearly, the requirements for the world at large far exceed the numbers of Western-trained physicians, scientists, and technicians available anywhere.

Even the local training of adequate numbers of physicians present difficulties for which there is no answer in the immediate future. On the one hand, the very process of teaching demands in most cases the actual and continued presence of highly trained specialists because a great deal of expert medical knowledge can be transferred only by direct sharing of experience. Furthermore, the communication of modern medical knowledge requires recipients who are at least partially educated in Western science and modes of thought. Finally, the mere organization of new medical schools on a large scale constitutes a formidable economic and administrative enterprise, even in prosperous countries with a highly developed educational system.

Since trained physicians or technical knowledge cannot possibly be transferred on a scale sufficient to meet the requirements of the emerging countries, the chance for rapidly improving their health status seems at first sight very slim. The problem, however, becomes more manageable when medical needs are formulated not in the abstract but with regard to dominant disease patterns. As was stated earlier, the social mores and medical ethics of Western countries make it inevitable that the largest percentage of medical effort be devoted to a form of personalized medical care, which is very expensive yet not highly productive socially or economically (Chapter XV). But, in contrast, the most urgent need of the emerging countries is the control of nutritional and microbial diseases. This problem is best handled through the methods of preventive medicine that can achieve improvement of collective health at fairly low cost.

Some conclusions published in a World Health Organization monograph devoted to the state of hygiene and public health in Europe are worth quoting in this regard:

> The improved health owes less to advances in medical science than to changes in the external environment, and to a favourable trend in the standard of living. We are healthier than our ancestors not because of what happens when we are ill but because we do not become ill; and we do not become ill not because of specific protective therapy but because we live in a healthier environment. In its preoccupation with the minutiae of diagnosis and the pathogenesis of disease, medicine is in danger of neglecting what has hitherto proved its most powerful resource—the manipulation of the external environment [Grundy and Mackintosh, 1957].

While the type of internal medicine and surgery practiced in the prosperous countries of Western civilization contributes little to the creation of wealth, collective disease control is an essential ingredient of economic growth in much of Africa, Asia, and Latin America, and therefore must be given high priority in any form of social planning. Needless to say, the individual human being has the same unique value and dignity everywhere, but the economic and especially personnel resources of poor countries are so limited that it would seem wise for them to focus their medical effort at first on the collective aspects of preventive medicine. In this light, the evolution of medical services in the emerging countries might present some analogy to their economic evolution.

The first stage in the change away from primitive patterns of disease control in a country undergoing industrialization (or more usually from total absence of effective disease control) might be the introduction of measures not customarily regarded as pertaining to health—such as the development of roads, bridges, telephones, dams, and irrigation of farm land. Increased food production is the most essential need for health improvement. Furthermore, the possibility of rapidly reaching a remote area over good roads or of spanning large distances by telephone might be critical for stamping out an epidemic or overcoming famine in a local area.

The second stage of development would be as nonpersonal as the first stage. It would consist in measures designed to bring about environmental changes beneficial to health, such as draining of swamps and large-scale spraying of residual insecticides.

The third stage would introduce a more personal relationship, but one that is noncontinuing. Examples would be immunization programs implemented every two or three years in remote villages, or elementary training of selected members of a tribe in sanitary and nutritional practices. In these cases, the deliverer of the scientific biomedical technology establishes some personal contact with the recipient population, but he is not expected to be on call or to return soon.

These three first stages depend upon the use of indirect methods for the application of the biomedical sciences, rather than upon personal physician-patient relationship. The latter approach to disease control represents an entirely different kind of medical action and creates in the recipient public new hopes and eventually certain demands that can be met only through a far more costly type of medical organization.

In the fourth stage, the community would assume that individual persons are entitled to medical care on a continuing basis, whether this care meets a real biological need or is merely a response to a personal demand. Such personal service may be paid for on a cash basis, through insurance programs, or by government agencies, but the purely financial aspects are irrelevant here. What matters is that moving into this fourth stage constitutes a crucial and probably irreversible step. It is crucial because, as already mentioned, personal medicine has a very low economic return when compared with collective preventive measures. It constitutes therefore a heavy financial burden for any society trying to lift itself by its bootstraps. It is irreversible because of the new emotional attitudes it creates. When a baby dies in a primitive village, the mother naturally grieves the loss, but she accepts it as one of life's inescapable tragedies. However, once she has discovered that a physician can save babies from death she will regard society's failure to make the resources of modern medicine available to her as an act of social injustice or even as deliberate cruelty.

The fifth stage refers to medicine as practiced by physicians belonging to or in direct contact with a modern medical center. Ideally in this situation, the physician assimilates as fast as possible all the new contributions of biomedical sciences and applies them to the health problems of individual persons.

It goes without saying that the various stages outlined above do not necessarily occur in sequence. They usually overlap and furthermore their concrete expression retains traits peculiar to the civilization in which they occur. The Navajo Indian eagerly accepts isoniazid treatment for tuberculosis yet continues to depend on a tribal chant for other kinds of ailments. Primitive people in tropical forests will participate in the spraying of residual pesticides for the control of malaria, and ask the medicine man on the same day to help them overcome the malevolent effects of evil spirits. In view of the shortage of modern physicians, it is probably desirable that indigenous medicine men be allowed as long as possible to render personal services to patients whose psychological needs they understand intuitively. This would leave more time to scientifically trained personnel for the tasks of disease prevention based on modern biotechnology.

Culturally disparate systems of medicine do in fact coexist everywhere in the world. Countless Americans whose survival depends on the federal, state, and city systems of preventive medicine resort to medical healers in whom they have faith, religious or otherwise, for most of the ailments from which they suffer. The medicine men in the Indian reservations have their counterpart on Main Street, Broadway, and Park Avenue all over the United States, as well as in all the most sophisticated cities of Europe. In Thoreau's words, "The medicine man is indispensable to work upon the imagination of the Indian by his jugglery—and of like value for the most part is the physician to the civilized man."

To a very large extent, the traditional structure of medical services is determined by the fact that medicine started as a healing art. Most administrative and educational programs are arranged as if the primary and almost the only role of the physician were to treat individual persons who are sick. Yet, one of the most important aspects of modern medicine is that it is supplementing the care of the individual person with a collective approach applicable to the community as a whole. This change is rendered the more imperative and urgent because social and technological innovations are rapidly creating new medical problems that will be encountered by all industrialized societies. Needless to say, the less developed countries must deal first with the problems posed by malnutrition, poor sanitation, and infection, but they cannot ignore the new problems posed by industrialization.

The disease patterns of industrialized societies are directly relevant to those of the less developed countries, because at the rate at which social and technologic changes occur, the future begins today. Wherever a new industry is established, there will be changes in nutritional habits and in other social practices, pollution of the water and the air, occupational dangers for the workers and their families, emotional upsets arising from unfamiliar working conditions and from disruption of ancestral customs. These disturbances will add their deleterious effects to those of malaria, tuberculosis, schistosomiasis, etc. by creating physiological misery in the areas undergoing industrialization.

The health problems posed by social and technological changes have determinants that are peculiar to each area, indeed to each community. Smogs differ in composition and physical structure according to the amount of insolation, the topography of the place, the local type of technology, and the kind of fuel used. Likewise, each industrial process and each system of sewage disposal engenders its own kind of water pollution. Nutritional and psychic disturbances also are conditioned by local factors. Clearly then, public health problems cannot be solved by applying slavishly formulae developed under other conditions. Programs of control and research must take into account local peculiarities. In certain cases, of course, it will be sufficient to modify known procedures in order to adapt them to local needs and facilities, but complete departures from orthodox present-day practices will often be needed.

Since environmental pollution is inevitable wherever large factories and urban centers are created, industrial sites and automobile traffic should be planned in such a manner as not to duplicate in the cities now developing the horrible conditions that exist in areas of the world already urbanized. As we have seen, the disposal of sewage also demands entirely new thinking. Nutrition is another field where there are many opportunities for new departures (Chapters III and VI).

The need for adaptation and innovation rather than mere imitation is great not only in preventive medicine but also with regard to the treatment of disease, and more generally to the delivery of medical services. One specific example will serve to illustrate how the increase in knowledge renders obsolete many of the traditional practices.

Ever since the beginning of medical history, the treatment of tuberculosis has been one of the most complex and specialized aspects of the physician's art. Until recently, the tuberculous patient was best cared for in designated institutions, special wards, or preferably sanatoria, where he was nursed and watched for many months or years. Even after drug treatment first became possible the potential toxicity of streptomycin and the complex medical knowledge required for its administration made it imperative that the patient be constantly under physician's care. In contrast, the discovery of the curative and prophylactic properties of isoniazid 20 years ago completely changed the management of tuberculosis. Isoniazid is inexpensive, easy to administer, and sufficiently low in toxicity as to require only a minimum of surveillance for untoward effects. Recent studies have shown, furthermore, that the tuberculous patient need not be managed as an invalid while taking isoniazid and can indeed lead a fairly active life. As a result, tuberculosis can be treated at home instead of in sanatoria; except for diagnosis, general management, and occasional check-ups, most of the care can be delegated to auxiliary medical personnel. Thus, whereas each tuberculous patient had to be handled individually in the past, tuberculosis can now be managed on a population scale.

Admittedly, several practical problems still remain to be solved before tuberculosis can be effectively controlled in the population at large. For example, tubercle bacilli do develop resistance to isoniazid and there is great danger that the resistant strains will become widespread. It is also most probable that no drug, however potent, can control disease completely in populations with a low standard of living and poor nutrition. While it would be out of place to discuss here these still controversial problems, it is useful to mention them because they illustrate so well that some very crucial questions of medical management cannot be answered by the traditional studies focused on the care of the individual patient, but rather must be approached as population problems. They illustrate also that the application of existing medical techniques to the situation prevailing in the emerging countries demands a supplementary kind of knowledge that can be acquired only by studying the performance of these techniques under the very conditions in which they are to be used.

The general implication of the above remarks with regard to tuberculosis is that present concepts concerning the delivery of medical services, including diagnosis, therapeutic procedures, and after-care, must be recast to adapt them not only to present knowledge but especially to local conditions. The design and administration of hospitals and medical services must be formulated in terms of the needs and resources of each particular community. For example, it will prove more practical under certain local conditions to build a multiplicity of small health units distributed throughout the country, and under others to create a few large medical centers linked to the rest of the country by good roads. It would be tragic indeed if the countries now in the course of development were to copy a type of medical organization that was developed under another set of conditions and therefore is not suitable for them, precisely at the time when there is growing evidence that the delivery of medical services as practiced in the prosperous countries of Western civilization no longer meets their own requirements.

In the past, any progress in the control of disease was spread out over a period of centuries, and occurred in several different stages, each corresponding to a phase in the growth of medical sciences. The less developed countries have the opportunity to bypass several of these stages and to condense medical evolution within a lifetime. By taking advantage of past experiences, they can escape the frightful toll of disease that had to be paid during the nineteenth century before biologic and social adjustments could be made to the sanitary, nutritional, and infectious problems posed by the first industrial revolution. In his *History of the Russian Revolution,* Trotsky stated that an "underdeveloped" civilization has certain evolutionary potentials that an advanced one usually lacks. Trotsky wrote:

> Although compelled to follow after the advanced countries, a backward country does not take things in the same order. The privilege of historic backwardness—and such a privilege exists—permits, or rather compels, the adoption of whatever is ready in advance of any specified date, skipping a whole series of intermediate stages.

No community will find it possible to deal simultaneously with all the problems that threaten its health—those inherited from the past, and those brought about by modernization. Hence, there will have to be priorities, and these will involve difficult choices.

Humanitarian ideals might dictate that attention be first given to medical services for the sick and aged. However, concern for the future and for economic development may make it advisable to focus medical effort in the school system and even preferably on the very young, because experience has shown that the early years of life are the most critical for the creation of a healthy adult. These choices will naturally involve medical criteria, but they will also pose difficult ethical and social dilemmas.

In any case, medical programs for the emerging countries should adapt rather than copy the practices of Western civilization. More often than not, furthermore, entirely new knowledge and new techniques will be needed to cope with the local problems. Medicine and public health do not develop or function in a social void. They provide the social adaptive mechanisms that complement the biological adaptive responses to the conditions of life at a given time. They can fulfill their purpose, to improve the people's health, only if they are fitted to the needs and resources of the community as well as to the special conditions created by the total environment.

XVII. Curing, Helping, Consoling

When *Man Adapting* was first published in 1965, its main theme was that the states of health and disease are to a large extent the expressions of the success or failure of the adaptive responses made by the person to the challenges of the total environment. I developed this theme by discussing at length certain biological and social problems of human adaptation, but I mentioned only in passing the purely medical aspects of the problem. At that time, it was taken for granted that the medical professions contribute to health by making it easier for human beings to adapt to the various situations that produce biological and psychological stresses. For example, while I emphasized that improvements in health during the past century had been brought about less by therapeutic measures than by better nutrition and sanitation, I regarded these changes in the ways of life as part of medical policies, even if they had not been initiated by physicians. Skepticism concerning the usefulness of modern medicine has become so widespread during the past two decades, however, that it now seems useful to review the role of medicine in the adaptive processes that are essential for the maintenance of health, incorporating the more humanistic aspects of medical care.

It is a paradox of recent social history that skepticism concerning the effectiveness of modern medicine, and even hostility to it, is spreading through the general public precisely at the time when medical science can legitimately boast of its most spectacular achievements in the prevention and therapy of disease.

Chapter XVII is drawn from the first Caldwell B. Esselstyn Lecture at Yale University Medical School, delivered by René Dubos in April 1978. The theme for the lectureship, "Health and the Spirit of Man," was selected to provide "an opportunity for dialogue on relationships between social and ethical issues and how they affect health and the quality of life."

During the past few years, several books written by nonphysicians have popularized the view that scientific medicine is far less effective than is commonly assumed and, furthermore, that it is creating new pathological conditions—the so-called iatrogenic diseases. Modern medicine is also accused of weakening or destroying the sense of responsibility for one's own health, and even of having economic, racist, and sexual biases (Carlson, 1975; Ehrenreich, 1978; Illich, 1976; Lander, 1978).

While such books written by lay people might be dismissed as having been based on inadequate knowledge of clinical problems, similar criticisms have been presented in books and articles by eminent members of the medical profession, in Europe and in this country. In fact, some of the most radical suggestions for reform in medical science, education, and practice have come from the medical establishment itself, especially from academic medicine (*Daedalus, 106,* no. 1, 1977; Fuchs, 1968; Ingelfinger, 1978; McAuliffe, 1978; McKeown, 1976; White, 1977; Wolf and Berle, 1976; Yankauer, 1978).

There are many obvious reasons for dissatisfaction with the present status of medicine—for example, the slow rate of progress in the understanding and control of vascular disorders, cancers, rheumatoid arthritis, chronic nephritis, multiple sclerosis, and other chronic diseases; the increase in iatrogenic diseases; the cost of medical care; the poor physical conditions in many hospitals; the difficulty in obtaining necessary medical attention; the lack of medical facilities for underprivileged people; and the excessive use of drugs, diagnostic tests, and expensive equipment. But in my opinion much of the public disenchantment with scientific medicine has a deeper psychological basis, rooted in the history and traditions of medical practice. It comes from the belief that things were better in the past, when the physician served a "priestly" function, providing care in a paternalistic relationship with the patient, based on understanding and trust. There are, indeed, good historical and scientific reasons for the change in doctor-patient relationships.

Until the middle of this century, there were few specific therapies for diseases. The practice of medicine consisted chiefly in supportive care and encouragement, a form of service to the patient that usually required the physician's prolonged and repeated presence by the bedside. Sir William Osler, the most famous physician of the English-speaking world at the turn of the century, repeatedly stated that his effectiveness as a "healer" was due not to his scientific knowledge of medicine but to his personal influence on patients. In 1910, after he had become Regius Professor of Medicine in England at Oxford University, he organized a symposium, "The Faith That Heals," where he explained the therapeutic successes achieved at the Johns Hopkins Hospital in Baltimore during his chairmanship of the Department of Medicine: "Faith in *St. Johns Hopkins,* as we used to call him, an atmosphere of optimism, and cheerful nurses, worked just the same sort of cures as did Aesculapius at Epidaurus" (Osler's own italics; Cushing, 1925, 2:223). Dr. William Henry Welch, the architect of scientific medicine in the United States, also acknowledged that the very presence of the physician could have a healing effect independent of any intervention based on objective medical knowledge. Describing his own father, who had practiced medicine in Connecticut, Dr. Welch wrote, "The instant he entered the sickroom, the patient felt better. The art of healing seemed to surround his physical body like an aura; it was often not his treatment but his presence that cured" (Flexner and Flexner, 1941, p. 32). Francis Peabody's famous remark, "The secret of the care of the patient is in caring for the patient," is another way of stating that there is a miraculous moment when the very presence of the doctor is the most effective part of the treatment.

Belief in the healing power of certain persons—whether shamans, medicine men, curanderos, or scientific physicians—has been expressed not only in myths and in literature but also in the fine arts. The dignity of the ideal physician is admirably conveyed by Rembrandt in an etching dated 1651, now in the Philadelphia Museum. The etching depicts an Amsterdam physician standing at the foot of a stairway, where he has paused for a moment, presumably after seeing a patient. The physician's face expresses compassion for the plight of his patient and, even more, an immense effort to apprehend the complexities of disease. He seems to be pondering the problems of the body and the mind in their relation to the total undefinable experience of human life. The best-known portrait of a physician, however, is probably the sentimental one painted by Sir Luke Fildes in 1891. It shows a kindly, rotund family doctor, sitting passively but sympathetically at the bedside of a sick child. His pensive mood expresses more puzzlement than deep thought, and his passive attitude betrays his awareness that there would not be much he could do even if he understood the real nature of the child's disease.

A modern physician almost certainly would be able to diagnose the diseases that proved puzzling to the physicians portrayed by Rembrandt and Fildes. It is probable, furthermore, that the child shown in Fildes' painting would not be seriously ill nowadays because he would have been immunized against the most common childhood diseases or could be cured by antimicrobial therapy. But the traditional image of the "good ol' doc" leaning thoughtfully over his patient persists in the public mind, despite the fact that a modern physician's usefulness does not require his presence by the bedside for more than short periods of time. Many diagnostic tests and scientific measurements are carried out in laboratories. Immunization and other public health procedures prescribed by physicians are not necessarily performed by them. Sophisticated medical and surgical technologies involve difficult decisions, based on knowledge of complex medical sciences, but are likely to be executed largely by paramedical personnel. Whereas the *presence* of the physician was his chief contribution to therapy in the past, his *knowledge* is now the most important aspect.

Patients feel deprived and neglected, however, because modern medicine is no longer entirely based, as was prescientific medicine, on the traditional doctor-patient relationship. They have the nostalgic illusion that things were better in the horse-and-buggy days, that physicians were more compassionate and attentive—even though they were as perplexed and ineffective as the physician in the Fildes painting. In contrast, some contemporary physicians believe that the importance of the doctor-patient relationship decreases to the extent that medicine is based on better scientific knowledge of disease.

In his recent book *The Post-Physician Era: Medicine in the 21st Century,* Dr. J. S. Maxmen, of the Albert Einstein Medical School in New York City, discusses the remarkable advances that have been made in storing medical knowledge in computers and retrieving it in a form usable by clinicians. Computer programs have been designed for diagnosis, for taking medical histories, for prescribing therapy, and even for conducting psychotherapy. In these various functions, computers are not yet as good as the most experienced clinicians, but according to Dr. Maxmen they are in certain cases already better than ordinary physicians (Maxmen, 1976).

If medical knowledge and its application can thus be programmed in computers, Dr. Maxmen believes it likely that most of the duties now regarded as the prerogatives of physicians will eventually be carried out by what he calls "medics," namely, persons who have received only a limited amount of medical training. Highly trained physicians would then be needed only in a few specialties, such as neurosurgery, and in situations where the patient will desire personal contact with a doctor having the right kind of diploma.

The evolution of medicine is seen in a different light by Dr. Lewis Thomas, president of the Sloan–Kettering Cancer Center in New York City, who speaks from a long experience of medical teaching and of brilliant contributions to various fields of experimental pathology. Until 1930, Dr. Thomas states, medicine was almost useless and likely indeed to do more harm than good. Therapeutic effectiveness began with the use of insulin and other hormones, of sulfonamides and other antibacterial drugs, of vitamin B_{12} for the treatment of pernicious anemia, and of the sophisticated physiological knowledge that permitted new kinds of surgical intervention.

All these therapeutic advances have been derived from laboratory experimentation. They correspond to what Dr. Thomas calls the "high" technologies of medicine, based on precise scientific knowledge of etiology and pathogenesis. Not only are these high medical technologies very effective, but their use is relatively simple and inexpensive. For example, the treatment of tuberculosis with isoniazid is vastly more effective and much less expensive than prolonged periods of cure in a sanatorium; the same is true of polio vaccination compared with use of the iron-lung machine. Typhoid fever, which required prolonged hospitalization and exacting management by physicians and nurses, can now readily be cured by chloramphenicol.

In contrast, procedures such as surgery or radiation therapy for cancers, intensive care for coronary heart disease, renal dialysis, organ transplants, and psychotherapy can be regarded as "halfway" technologies, largely empirical because they are not based on a sufficient understanding of etiology and pathogenesis. These halfway technologies are at best questionably effective; they deal with the result of the disease rather than with its underlying mechanisms and are responsible for the high cost of medical care because they require the use of expensive equipment, elaborate hospital facilities, and the services of highly trained personnel for long periods of time. The only hope for the replacement of halfway technologies is research into the causation and mechanisms of disease conducted by the sophisticated methods of biological and medical sciences (Thomas, 1972, 1976, 1977).

Dr. Thomas' plea for more emphasis on theoretical medical sciences does not imply on his part a lack of appreciation for the humanistic aspects of medicine. In fact, he has recently urged that priority should be given to training in the humanities for college students who aspire to become physicians (Thomas, 1978).

While it is true that specific therapies were practically nonexistent before the 1930s, it is nevertheless certain that medicine has long been able to contribute to the management of disease and to the improvement of health without sophisticated scientific knowledge. I shall try to define the beneficial role of medicine in the past by referring to Edward Livingston Trudeau (1848–1915), an American physician, born in Louisiana, who was a contemporary and friend of William Osler.

Shortly after beginning medical practice in New York City, Trudeau was compelled by advanced tuberculosis to abandon his practice. As there seemed to be little hope for his recovery, he moved to the Adirondacks, intending to engage in his favorite sport—hunting—from the comfort of a canoe, operated by a guide. To his surprise, however, his health progressively improved, and he was able to resume the practice of medicine, remaining in the Adirondacks. He eventually settled in Saranac Lake, where he created the first tuberculosis sanatorium in the United States.

As a result of home management of tuberculosis by chemotherapy, the Trudeau Sanatorium closed its doors a few years ago; it has now been replaced by a biomedical research institute. In the library of the institute, one can still read a French motto often quoted by Trudeau during his days of medical practice:

Guérir quelquefois,
Soulager souvent,
Consoler toujours.

Freely translated, it means "to cure sometimes, to help often, to comfort and console always." This motto is probably the most exact expression of the best that medicine could do at the turn of the century.

As will be noted, there is no mention of prevention in the motto. The reason, of course, is that smallpox was then the only disease for which there was an effective method of prevention. The development since Trudeau's time of methods of prevention against many infectious diseases, nutritional deficiencies, occupational and environmental threats, and even a few genetic disorders provides a measure of scientific medicine's achievements in our century. Several dramatic achievements have also occurred in the cure of disease—by chemotherapy, replacement therapy, and surgical procedures. Such preventive and truly curative procedures correspond to what Dr. Thomas calls the high technologies of medicine. The third line of Trudeau's motto, "consoler toujours," represents the purely humanistic ideal of medicine, which is not likely to change with time or knowledge.

I shall now turn to Trudeau's "soulager souvent," because in my opinion it still corresponds to one of the most important roles of medicine in our times. The modern physician can often relieve the patient of the burden of disease even when the disease cannot be cured. The importance of this aspect of medicine can best be evaluated by considering first the meanings of the words health and disease.

In theory, health implies the absence of organic and mental disease. This scientific definition, however, is not really meaningful for the many persons who are more concerned with doing what they want to do and becoming what they want to become than they are with the condition of their bodies. For them, disease is any condition that deprives them of the freedom to act as they want; health is the possession of this freedom, even if that means the presence of disease. Although I have translated Trudeau's verb *soulager* by the expression "to help," the French word has a much richer meaning. It implies the removal of a burden, a handicap—in other words, making it possible or easier for the patient to function as he or she elects to do. From this point of view, medicine can do a great deal, even when it cannot prevent or cure the disease: it can relieve anxiety, for example, when a woman who has observed a lump in her breast is told by her physician that the lump is not malignant; it can decrease the severity of symptoms, for example, when patients suffering from arthritis or hypertension are treated with the proper drugs and advised as to the proper ways of life; it can help the patient to mobilize the natural defense mechanisms of the body and the mind—*naturae vis medicatrix*—which can go far toward controlling the disease or, at least, alleviating its manifestations; it can facilitate reeducation by taking advantage of the enormous resiliency of the human organism and by helping the patient to compensate for one or another handicap. Rehabilitation implies active participation of the mind and the body in a creative process of adaptive change, which depends on volition but usually also needs medical guidance (McDermott, 1977, 1978).

Thus, in many different ways, the physician can help a patient suffering from a particular disease to function more or less effectively, even though the cause of the disease is not known and its pathogenesis poorly understood. However, what Trudeau meant by soulager demands of the physician an awareness of the patient's human peculiarities—a concern that transcends the precise information incorporated in a computer program. This was the quality that Hawthorne had in mind when he wrote of Dr. Chillingsworth in *The Scarlet Letter,* "He deemed it essential to know the man before attempting to do him good." Knowing a person implies awareness of fears and aspirations as well as knowledge of biological characteristics.

Most societies have been rather ambiguous in their concept of the scope of medicine. In ancient China, there were special remedies for particular diseases, but there was also the Book of the Yellow Emperor, which formulated a broad medical code of comportment according to the seasons. The Greek god of medicine, Asclepius, had two daughters who symbolized the two complementary aspects of the medical art: Panakeia symbolized the knowledge of drugs derived from the earth and from plants; Hygeia, the doctrine that the way to health is to avoid excesses and to live according to the laws of reason.

In our own times, as mentioned earlier, some physicians believe that the only worthwhile and legitimate role of medicine is the prevention and treatment of disease by methods of scientifically proven value. Since truly effective methods exist for only a very small percentage of the complaints that bring people to physicians, the purely scientific view of medical practice would limit its role to precise diagnosis and to the use of a few specific remedies. Practice of such kind might decrease the importance of the human quality in the doctor-patient relationship. In contrast, the physicians who formulated the philosophy of the World Health Organization asserted in the preamble to its Charter that health is "a state of complete physical, mental, and social well-being and not merely the absence of disease or infirmity." Such a sweeping definition makes health a virtual synonym for happiness. If taken literally, it would demand of physicians that they become involved as guarantors in nearly every phase of human development.

There are, of course, many intermediate positions between these two extreme social views of medicine—the one based entirely on hard scientific evidence, the other accepting responsibility for all aspects of human life. Some of these intermediate positions involve, for example, the containment of organic diseases that cannot be cured, the management of chronic noncontainable diseases, and advice about the general problems of living, from marital difficulties to the choice of a career. The role of medicine in the social system must thus be compatible with questions of values that differ from society to society and from time to time—values that cannot be determined by the medical profession itself because they are the prerogative of each particular society as a whole. Nevertheless, there are reasons to believe that medicine cannot be limited to the prevention and treatment of disease and must inevitably incorporate the spiritual problems of patients.

One reason is indicated by the fact that, in Western societies at least, the physician is usually referred to as "doctor," a word which etymologically means teacher. This usage seems proper because under the conditions of stress caused by disease most patients need guidance and encouragement along with purely medical care. Few are the individuals in our society to whom a patient can turn for counsel on personal matters, and it is probable that physicians, whether they want to or not, will continue to be expected to act as "doctors"—as teachers of individual patients and teachers of communities in many aspects of behavior.

Another reason for extending the role of medicine beyond the scope of precise scientific knowledge is that whatever affects the mind also affects the body, and vice versa. This interplay has always been known from simple experience and is now at last becoming the subject of scientific exploration. Who could have imagined, only a decade ago, that behavior and the perception of pain are affected by endorphins and other peptide hormones secreted in the brain itself and that acupuncture does influence the secretion of some of these hormones! The body-mind relationship is likely indeed to become one of the most active fields of medicine in the near future.

The chief trouble with today's scientific medicine is that it is too one-sided and, therefore, not scientific enough. Modern medicine will become truly scientific only when it has learned to manage the biological and psychological forces that operate as *naturae vis medicatrix* and when it has seriously committed itself to the doctrine that, in human life, the health of the body is linked to the health of the mind.

Rembrandt's portrait of a physician, mentioned earlier, admirably conveys this spiritual aspect of human medicine. Ministering to the sick does not only mean dealing with the living organism as a machine and with the environment in which this machine functions. It implies also compassionate sympathy for the patient, who is considered a sensitive, ethical being. Medicine will retain its unique position among the sciences only if it accepts some responsibility for the various aspects of life that determine our humanness.

Bibliography

Chapter I.1

Allison, A. C. 1954. Protection afforded by sickle cell trait against subtertian malarial infection. *Brit. Med. J.*, *1*:290–92.

——— 1959. Metabolic polymorphisms in mammals and their bearing on problems of biochemical genetics. *Amer. Natur.*, *93*:5–16.

Benedict, R. 1934. *Patterns of culture.* Boston, Houghton Mifflin.

Bergounioux, F. M. 1961. Notes on the mentality of primitive man, p. 106–18. *In* S. L. Washburn, *Social life of early man.* Chicago, Aldine.

Brooks, C., and P. Cranefield. 1959. *The historical development of physiological thought.* New York, Hafner.

Clark, J. Desmond. 1959. *The prehistory of Southern Africa.* Harmondsworth, Middlesex, Penguin Books.

Coon, C. S. 1959. Race and ecology of man. *Cold Spring Harbor Symp. Quant. Biol.*, *24*:153–59.

——— 1962. *The story of man.* New York, Knopf.

DeSonneville-Bordes, D. 1963. Upper Paleolithic cultures in Western Europe. *Science*, *142*:347–55.

Dice, L. R. 1955. *Man's nature and nature's man.* Ann Arbor, Univ. of Michigan Press.

Dobzhansky, Th. 1956. *The biological basis of human freedom.* New York, Columbia Univ. Press.

——— 1962(a). Genetics and equality. *Science*, *137*:112–15.

——— 1962(b). Genetics, society and evolution. *Bull. N. Y. Acad. Med.*, *38*:451–59.

——— 1962(c). *Mankind evolving.* New Haven, Yale Univ. Press.

——— 1963. Genetics of race equality. *Eugen. Quart.*, *10*:151–60.

——— and O. Pavlovsky. 1960. How stable is balanced polymorphism? *Proc. Nat. Acad. Sci.*, *46*:41–7.

Eiseley, L. C. 1957. *The immense journey.* New York, Random House.

Fry, R. 1956. *Vision and design.* New York, Meridian Books.

Giedion, S. 1962. *The eternal present.* New York, Pantheon Books.

Graziosi, P. 1960. *Paleolithic art.* New York, McGraw-Hill.

Haldane, J. B. S. 1932. *The causes of evolution.* London, Longmans Green.

——— 1959. Natural selection p. 101–49. *In* P. R. Bell, *Darwin's biological work: Some aspects reconsidered.* Cambridge Univ. Press.

Hoagland, H., and R. W. Burhoe. 1962. *Evolution and man's progress.* New York, Columbia Univ. Press.

Hutchinson, G. E. 1962. A speculative consideration of certain possible forms of sexual selection in man; *and* Fifty years of man in the zoo. In *The enchanted voyage.* New Haven, Yale Univ. Press.

Huxley, J. S. 1941. *The uniqueness of man.* London, Chatto & Windus.

——— 1960. The evolutionary vision, p. 249–61. *In* S. Tax, [ed.], *Evolution after Darwin.* Univ. of Chicago Press.

——— 1961. *The humanist frame.* New York, Harper.

——— 1964. Psychometabolism: General and Lorenzian. *Perspect. Biol. Med.,* *7*:399–432.

James, E. O. 1957. *Prehistoric religion.* New York, Praeger.

Jennings, H. S. 1930. *The biological basis of human nature.* New York, W. W. Norton.

LaBarre, W. 1954. *The human animal.* Univ. of Chicago Press.

Metchnikoff, E. 1903. *The nature of man.* Translated by P. Chalmers Mitchell. London, G. P. Putnam's Sons.

Montagu, M. F. A. 1962. *Culture and the evolution of man.* New York, Oxford Univ. Press.

Ortega y Gasset, J. 1961. *History as a system.* New York, W. W. Norton.

Polanyi, M. 1959. *The study of man.* Univ. of Chicago Press.

Rensch, B. 1959. *Homo sapiens; vom Tier zum Halbgott.* Göttingen, Vandenhoeck and Ruprecht.

Sahlins, M. D., and E. R. Service. 1960. *Evolution and culture.* Ann Arbor, Univ. of Michigan Press.

Searles, H. 1960. *The nonhuman environment.* New York, International Univ. Press.

Simpson, G. G. 1964. *This view of life.* New York, Harcourt, Brace & World.

Smith, H. W. 1959. The biology of consciousness, p. 109–36. *In* C. Brooks and P. Cranefield, [eds.], *The historical development of physiological thought.* New York, Hafner.

Vallois, H. V. 1961. The social life of early man: The evidence of skeletons, p. 214–35. *In* S. L. Washburn, [ed.], *Social life of early man.* Chicago, Aldine.

Waddington, C. H. 1960. *The ethical animal.* London, Allen & Unwin.

Washburn, S. L. 1961. *Social life of early man.* Chicago, Aldine.

Chapter 1.2

Ader, R., and P. Conklin. 1963. Handling of pregnant rats: Effects on emotionality of their offspring. *Science,* *142*:411–12.

Ainsworth, M. D. 1962. Deprivation of maternal care. Geneva, World Health Organization Public Health Papers, No. 14.

Allee, W. C. 1951. *Cooperation among animals.* New York, Henry Schuman.

Barnett, S. A. 1961. The behavior and needs of infant mammals. *Lancet, 1:*1067–71.

——— 1963. Instinct. *Daedalus, 92:*564–80.

——— 1964. The biology of aggression. *Lancet, 2:*803–7.

Bennett, E. L., M. C. Diamond, D. Krech, and M. R. Rosenzweig. 1964. Chemical and anatomical plasticity of brain. *Science, 146:*610–19.

Bliss, E. L. 1962. *Roots of behavior.* New York, Harper.

Bovard, E. W. 1958. The effects of early handling on viability of the albino rat. *Psychol. Rev., 65:*257–71.

Bowlby, J. 1958. The nature of the child's tie to his mother. *Int. J. Psychoanal., 39:*350–73.

——— 1961. Tavistock seminar on mother-infant interaction, p. 301–3. *In* B. M. Foss, [ed.], *Determinants of infant behavior.* London, Methuen.

Denenberg, V. H. 1962. The effects of early experience, p. 109–38. *In* E. Hafez, [ed.], *The behavior of domestic animals.* Baltimore, Williams and Wilkins.

——— 1963. Experience and emotional development. *Sci. Amer., 208:*138–46.

——— and G. G. Karas. 1959. Effects of differential infantile handling upon weight gain and mortality in the rat and mouse. *Science, 130:*629–30.

——— and A. E. Whimbey. 1963. Behavior of adult rats is modified by the experiences their mothers had as infants. *Science, 142:*1192–93.

Fiske, D. W., and S. R. Maddi. 1961. *Functions of varied experience.* Homewood, Ill., Dorsey Press.

Guppy, N. 1958. *Wai-Wai: Through the forests north of the Amazon.* London, John Murray.

Harlow, H. F. 1959. Basic social capacity of primates, p. 40–58. *In* J. N. Spuhler, [ed.], *The evolution of man's capacity for culture.* Detroit, Wayne State Univ. Press.

——— and R. R. Zimmerman. 1958. The development of affectional responses in infant monkeys. *Proc. Amer. Philosoph. Soc., 102:*501.

——— and ——— 1959. Affectional responses in the infant monkey. *Science, 130:*421–32.

Hebb, D. O. 1961. *In* P. Solomon et al., *Sensory deprivation,* p. 6–7. Cambridge, Harvard Univ. Press.

Hediger, H. P. 1961. The evolution of territorial behavior, p. 34–57. *In* S. L. Washburn, [ed.], *Social life of early man.* Chicago, Aldine.

Hunt, H. F., and L. Otis. 1962. Early experience and its effects on later behavioral processes in rats: I. Initial experiments. *Trans. N. Y. Acad. Sci., 25*:858–70.

Hutchings, D. E. 1962. Early experience and its effects on later behavioral processes in rats: III. Effects of infantile handling and body temperature reduction on later emotionality. *Trans. N. Y. Acad. Sci., 25*:890–901.

Huxley, J. S. 1964. See Chapter I.1.

Kalter, H., and J. Warkany. 1959. Experimental production of congenital malformations in mammals by metabolic procedures. *Physiol. Rev., 39*:691–715.

Keeley, K. 1962. Prenatal influence on behavior of offspring of crowded mice. *Science, 135*:44–45.

Levine, S. 1956. A further study of infantile handling and adult avoidance learning. *J. Personality, 25*:70–80.

——— 1960. Stimulation in infancy. *Sci. Amer., 202*:81–86.

——— 1962. Psychophysiological effects of infantile stimulation, p. 246–53. *In* E. L. Bliss, [ed.], *Roots of behavior.* New York, Harper.

Lieberman, M. W. 1963. Early developmental stress and later behavior. *Science, 141*:824–25.

Lorenz, K. 1952. *King Solomon's ring.* New York, Crowell.

Miller, N. E. 1964. Physiological and cultural determinants of behavior. *Proc. Nat. Acad. Sci., 51*:94–154.

Pasamanick, B., M. E. Rogers, and A. M. Lilienfeld. 1956. Pregnancy experience and the development of behavior disorder in children. *Amer. J. Psychol., 112*:613–17.

Pollin, W., P. V. Cardon, and S. S. Kety. 1961. Effects of amino acid feedings in schizophrenic patients treated with iproniazid. *Science, 133*:104–5.

Schaefer, T., Jr. 1962. Early experience and its effects on later behavioral processes in rats: II. A critical factor in the early handling phenomenon. *Trans. N. Y. Acad. Sci., 25*:871–89.

Scott, J. P. 1962. Critical periods in behavioral development. *Science, 138*:949–58.

Solomon, P., P. E. Kubzansky, P. H. Leiderman, J. H. Mendelson, R. Trumbull, and D. Wexler. 1961. *Sensory deprivation.* Cambridge, Harvard Univ. Press.

Stechler, G. 1964. Newborn attention as affected by medication during labor. *Science, 144*:315–17.

Thompson, W. R. 1957. Influence of prenatal maternal anxiety on emotionality in young rats. *Science, 125*:698–99.

——— 1960. Early environmental influences on behavioral development. *Amer. J. Orthopsychiat., 30*:306–14.

——— and L. Goldenberg. 1962. Some physiological effects of maternal adrenalin injection during pregnancy in rat offspring. *Psychol. Rep., 10*:759–74.

Washburn, S. L. 1961. See Chapter I.1.
Young, R. D. 1964. Drug administration to neonatal rats: Effects on later emotionality and learning. *Science, 143*:1055–57.

Chapter I.3

Boyer, S. H. 1963. Cultural determinants of biochemical evolution. *Hum. Biol., 35*:292–98.
Bovard, E. W. 1962. The balance between negative and positive brain system activity. *Perspect. Biol. Med., 6*:116–27.
Fiske, D. W., and S. R. Maddi. 1961. See Chapter I.2.
Hatch, A., T. Balazs, G. S. Wiberg, and H. C. Grice. 1963. Long-term isolation stress in rats. *Science, 142*:507.
Hebb, D. O. 1949. *The organization of behavior.* New York, John Wiley & Sons.
——— 1961. See Chapter I.2.
Riesen, A. H. 1961. Excessive arousal effects of stimulation after early sensory deprivation, p. 34–40. *In* P. Solomon et al., *Sensory deprivation.* Cambridge, Harvard Univ. Press.
Solomon, P. P. E. Kubzansky, P. H. Leiderman, J. H. Mendelson, R. Trumbull, and D. Wexler. 1961. See Chapter I.2.
Thompson, W. R., and T. Schaefer. 1961. Early environmental stimulation, p. 81–105. *In* D. W. Fiske and S. R. Maddi, [eds.], *Functions of varied experience.* Homewood, Ill., Dorsey Press.
Zubek, J. P. 1963. Counteracting effects of physical exercises performed during prolonged perceptual deprivation. *Science, 142*:504–6.
——— G. Welch, and M. G. Saunders. 1963. Electroencephalographic changes during and after 14 days of perceptual deprivation. *Science, 139*:490–92.
——— J. Flye, and M. Aftanas. 1964. Cutaneous sensitivity after prolonged visual deprivation. *Science, 144*:1591–93.

Chapter I.4

Bogdonoff, M. D., E. H. Estes, Jr., and D. Trout. 1959. Acute effect of psychologic stimuli upon plasma non-esterified fatty acid level. *Proc. Soc. Exp. Biol. Med., 100*:503–4.
Brod, J. 1963. Haemodynamic basis of acute pressor reactions and hypertension. *Brit. Heart J., 25*:227–45.
Cannon, W. B. 1915. *Bodily changes in pain, hunger, fear and rage.* New York, D. Appleton.
Dodds, E. R. 1951. *The Greeks and the irrational.* Berkeley, Univ. of California Press.
Friedman, M., and R. H. Rosenman. 1959. Association of specific overt

behavior pattern with blood and cardiovascular findings. *J.A.M.A., 169*:1286–96.

Gage, S. H., and P. A. Fish. 1924. Fat digestion, absorption, and assimilation in man and animals as determined by the dark-field microscope, and a fat-soluble dye. *Amer. J. Anat., 34*:1–77.

Hamburg, D. A. 1961. The relevance of recent evolutionary changes to human stress biology, p. 278–86. *In* S. L. Washburn, [ed.], *Social life of early man*. Chicago, Aldine.

——— 1962. Plasma and urinary corticosteroid levels in naturally occurring psychologic stresses, p. 406–13. In *Ultrastructure and metabolism of the nervous system,* Vol. XL, Research Publications.

Hediger, H. 1955. *Studies of the psychology and behavior of captive animals in zoos and circuses.* Translated by G. Sircom. London, Butterworths.

Klein, P. D., and R. A. Martin. 1962. Environmental stress and cholesterol esterification in plasma and liver. *Nutr. Rev., 20*:88–90.

Malm, O. J. 1958. Calcium requirement and adaptation in adult men. *Scand. J. Clin. Lab. Invest.,* Suppl. 36, *10*:1–290.

Miller, N. E. 1964. See Chapter I.2.

Morris, D. 1962. *The biology of art*. London, Methuen & Co.

Ogston, D., G. A. McDonald, and H. W. Fullerton. 1962. The influence of anxiety in tests of blood coagulability and fibrinolytic activity. *Lancet, 2*:521–23.

Ratcliffe, H. L., and M. T. I. Cronin. 1958. Changing frequency of arteriosclerosis in mammals and birds at the Philadelphia zoological garden. *Circulation, 18*:41–52.

——— T. G. Yerasimides, and G. A. Elliott. 1960. Changes in the character and location of arterial lesions in mammals and birds in the Philadelphia zoological garden. *Circulation, 21*:730–38.

——— and R. L. Snyder. 1962. Patterns of disease, controlled populations and experimental design. *Circulation, 26*:1352–57.

Sarnoff, S. 1963. The circulation: An evolutionary dilemma. *Man under stress symposium,* San Francisco, November 15–17.

Selye, H. 1956. *The stress of life*. New York, McGraw-Hill.

Tanner, J. M. 1960. *Stress and psychiatric disorder*. Oxford, Blackwell.

Timberline Lodge (Oregon) Symposium: Psychophysiologic aspects of cardiovascular disease. 1963. *Science, 142*:601–2.

Venning, E. H., I. Dyrenfurth, and J. C. Beck. 1957. Effect of anxiety upon aldosterone excretion in man. *J. Clin. Endocr. 17*:1005–08.

Washburn, S. L. 1961. See Chapter I.1.

Wolff, H. G. 1953. *Stress and disease*. Springfield, Ill. Charles C Thomas.

——— 1960. Stressors as a cause of disease in man, p. 17–33. *In* J. M. Tanner, *Stress and psychiatric disorder*. Oxford, Blackwell.

Chapter II.1

Allen-Price, E. D. 1960. Uneven distribution of cancer in West Devon. *Lancet, 1*:1235–38.

Edelstein, L. 1939. The genuine works of Hippocrates. *Bull. Hist. Med.* 7:236–48.

Huntington, E. 1924. *The character of races as influenced by physical environment, natural selection and historical development.* New York, C. Scribner.

Madsen, T. 1937. *Lectures on the epidemiology and control of syphilis, tuberculosis and whooping cough, and other aspects of infectious disease.* Baltimore, Williams and Wilkins.

Miller, G. 1962. "Airs, waters, and places" in history. *J. Hist. Med., 17*:129–40.

Chapter II.2

Barnwell, F. H. 1960. A solar daily variation in oxygen consumption of the embryonated egg. *Proc. Soc. Exp. Biol. Med., 105*:312–20.

Bartter, F. C., and C. S. Delea. 1963. Circadian aspects of human adrenal function. *Proc. Roy. Soc. Med. 56*:257–59.

Benoit, J., I. Assenmacher, and E. Brard. 1955. Evolution testiculaire du canard domestique maintenu à l'obscurité totale pendant une longue durée. *C. R. Acad. Sci. 241*:251–53.

———, ———, and ——— 1956. Etude de l'évolution testiculaire du canard domestique soumis très jeune à un éclairement artificiel permanent pendant deux ans. *C. R. Acad. Sci. 242*:3113–15.

Berliner, M. D. 1961. Diurnal periodicity of luminescence in three basidiomycetes. *Science, 134*:740.

Bloch, M. 1964. Rhythmic diurnal variation in limb blood flow in man. *Nature, 202*:398–99.

Brown, F. A., Jr. 1959(a). Living clocks. *Science, 130*:1535–44.

——— 1959(b). The rhythmic nature of animals and plants. *Amer Sci., 47*:147–68.

Bruce, V. G., and C. S. Pittendrigh. 1956. Temperature independence in a unicellular "clock." *Proc. Nat. Acad. Sci. 42*:676–82.

Buckell, M., and F. A. Elliott. 1959. Diurnal fluctuation of plasma-fibrinolytic activity in normal males. *Lancet, 1*:660–62.

Bugard, P., and M. Henry. 1961. Quelques aspects de la fatigue dans l'aviation de transport. *La Presse Medicale, 69*:1093–96.

Bullough, W. S., and E. B. Lawrence. 1961. Stress and adrenaline in relation to the diurnal cycle of epidermal mitotic activity in adult male mice. *Proc. Roy. Soc. (Biol.) 154*:540–46.

Bünning, E. 1964. *The physiological clock, endogenous diurnal rhythms and biological chronometry.* New York, Academic Press.

Burn, J. H., and H. W. Ling. 1928. The effect of insulin on acetonuria. *J. Physiol. 65*:191–203.

Campbell, H. L. 1945. Seasonal changes in food consumption and rate of growth of the albino rat. *Amer. J. Physiol. 143*:428–33.

Chitty, H. 1961. Variations in the weight of the adrenal glands of the field vole, *Microtus agrestis. J. Endocr., 22*:387–93.

Cloudsley-Thompson, J. L. 1961. *Rhythmic activity in animal physiology and behavior.* New York, Academic Press.

Coburn, A. F., P. F. Frank, and J. Nolan. 1957. Studies on the pathogenicity of *Streptococcus pyogenes:* IV. The relation between the capacity to induce fatal respiratory infections in mice and epidemic respiratory diseases in man. *Brit. J. Exp. Path., 38*:256–67.

Cori, C. F. 1921. Untersuchungen über die Ursachen der Unterschiede in der Herznervenerregbarkeit bei Fröschen zu verschiedenen Jahreszeiten. *Arch. Exper. Path. Pharm., 91*:130–55.

Cori, G. T., and C. F. Cori. 1927. The fate of sugar in the animal body; seasonal occurrence of ketonuria in fasting rats accompanied by changes in carbohydrate metabolism. *J. Biol. Chem.,* 72:615–25.

Cowgill, U. M., A. Bishop, R. J. Andrew, and G. E. Hutchinson. 1962. An apparent lunar periodicity in the sexual cycle of certain prosimians. *Proc. Nat. Acad. Sci., 48*:238–41.

Duke-Elder, S. 1934. Diurnal variation in intraocular pressure, p. 493–94. In *Textbook of opthalmology,* Vol. 1. London, Kimpton.

Ehret, C. F. 1959. Photobiology and biochemistry of circadian rhythms in non-photosynthesizing cells. *Fed. Proc. 18*:1232–40.

Eleftheriou, B. E., and M. X. Zarrow. 1962. Seasonal variation in thyroid gland activity in deermice. *Proc. Soc. Exp. Biol. Med. 110*:128–31.

Fiske, V. M. 1964. Serotonin rhythm in the pineal organ. *Science, 146*:253–54.

Flink, E. B., and R. P. Doe. 1959. Effect of sudden time displacement by air travel on synchronization of adrenal function. *Proc. Soc. Exp. Biol. Med. 100*:498–501.

Gaskell, J. F. 1927. Seasonal variation in pneumococcal virulence. *J. Path. Bact. 30*:568–69.

Glick, J. L., and W. D. Cohen. 1964. Nocturnal changes in oxidative activities of rat liver mitochondria. *Science, 143*:1184–85.

Halberg, F. 1953. Some physiological and clinical aspects of 24 hour periodicity. *Lancet, 73*:20–32.

——— 1960(a). Temporal coordination of physiologic function. *Cold Spring Harbor Symp. Quant. Biol., 25*:289–310.

——— 1960(b). The 24-hour scale: A time dimension of adaptive functional organization. *Perspect. Biol. Med., 3*:491–525.

——— 1963. Circadian (about 24-hour) rhythms in experimental medicine. *Proc. Roy. Soc. Med., 56*:253–57.

——— C. Barnum, R. Silber, and J. Bittner. 1958. 24-hour rhythms of several levels of integration in mice on different lighting regimens. *Proc. Soc. Exp. Biol. Med., 97*:897–900.

——— E. Halberg, C. Barnum, and J. Bittner. 1959. Physiologic 24-hour periodicity in human beings and mice, the lighting regimen. *In* R. B. Withrow, [ed.], *Photoperiodism and related phenomena in plants and animals.* Washington, D. C., A.A.A.S.

——— E. Johnson, B. Brown, and J. Bittner. 1960. Susceptibility rhythm to *E. coli* endotoxin and bioassay. *Proc. Soc. Exp. Biol. Med., 103*:142–44.

Hale, H., J. Ellis, and R. McNee. 1961. Summer-fall variation in human urinary excretion of electrolytes, nitrogen fractions and 17-hydroxycorticosteroids. *Fed. Proc.,* Part I, *20*:208.

Hamilton, L., C. Gubler, G. Cartwright, and M. Wintrobe. 1950. Diurnal variation in the plasma iron level of man. *Proc. Soc. Exp. Biol. Med., 75*:65–68.

Harker, J. E. 1961. Biological clocks. *Discovery, 22*:138–42.

Hastings, J. W. 1959. Unicellular clocks. *Ann. Rev. Microbiol., 13*:297–312.

——— 1960. Biochemical aspects of rhythms: Phase shifting by chemicals. *Cold Spring Harbor Symp. Quant. Biol., 25*:131–43.

Hawking, F. 1963. Ciradian rhythms in filariasis. *Proc. Roy. Soc. Med., 56*:260.

Heller, H., G. Herdan, and S. M. Zaidi. 1957. Seasonal variations in the response of rats to antidiuretic hormone. *Brit. J. Pharmacol., 12*:100–3.

Hughes, E. 1931. *Seasonal variations in man.* London, H. K. Lewis.

Karakashian, M., and J. W. Hastings. 1962. The inhibition of a biological clock by actinomycin D. *Proc. Nat. Acad. Sci., 48*:2130–37.

Kleitman, N. 1939. *Sleep and wakefulness as alternating phases in the cycle of existence.* Univ. of Chicago Press.

Kowarzyk, H., J. Kaniak, and M. Kotschy. 1960. Diurnal fluctuations of plasma fibrinolytic activity. *Lancet, 1*:176–77.

Lewis, P. R. 1959. Adaptation of man to a change in day length. *Proc. Roy. Soc. Med., 52*:676–80.

Llanos, J. M. E. 1964. *Z. Zellforsch, 61*:824. *(In* Regenerating liver keeps its rhythm, *New Scientist, 22*:304.)

Lobban, M. C. 1960. The entrainment of circadian rhythms in man. *Cold Spring Harbor Symp. Quant. Biol., 25*:325–32.

Maqsood, M. 1951. Effects of the thyroid, castration and season on adrenals in the male rabbit. *Nature, 167*:323.

Marte, E., and F. Halberg. 1961. Circadian susceptibility rhythm of mice to librium. *Fed. Proc.,* Part I, *20*:305.

McCarthy, J. L., R. C. Corley, and M. X. Zarrow. 1960. Diurnal rhythm in plasma corticosterone and lack of diurnal rhythm in plasma compound F-like material in the rat. *Proc. Soc. Exp. Biol. Med. 104*:787–89.

Menaker, W., and A. Menaker. 1959. Lunar periodicity in human reproduction: A likely unit of biological time. *Amer. J. Obstet. Gynec., 77*:905–14.

Michel, M. C. 1961. Activité métabolique de la flore totale isolée de l'intestin de porc. *Ann. Biol. Anim. Bioch. Biophys., 1*:16–28.

Mills, C. 1949. Climate in health and disease, p. 453–500. *In* H. A. Christian, [ed.], *The Oxford medicine,* Vol. 1. New York, Oxford Univ. Press.

Mills, J. N. 1963. The part played by the adrenals in human circadian renal rhythms. *Proc. Roy. Soc. Med., 56*:259–60.

Nowell, N. W., and D. C. White. 1963. Seasonal variation of magnesium and calcium in serum of the hypothermic rat. *J. Appl. Physiol., 18*:967–69.

Palmai, G. 1962. Diurnal and seasonal variation in deep body temperature. *Med. J. Aust., 2*:989–91.

Petersen, W. F., and M. E. Milliken. 1937. *The patient and the weather.* Ann Arbor, Edwards Brothers.

Pittendrigh, C. S. 1961. On temporal organization in living systems. *Harvey Lectures,* Ser. 56, 93–125.

Pizzarello, D. J., D. Isaak, and K. E. Chua. 1964. Circadian rhythmicity in the sensitivity of two strains of mice to whole-body radiation. *Science, 145*:286–91.

Richter, C. P. 1960. Biological clocks in medicine and psychiatry: Shock-phase hypothesis. *Proc. Nat. Acad. Sci., 46*:1506–30.

Rose, G. 1961. Seasonal variation in blood pressure in man. *Nature, 189*:235.

Sharp, G. W. G. 1961. Reversal of diurnal temperature rhythms in man. *Nature, 190*:146–48.

Snyder, S. H., M. Zweig, and J. Axelrod. 1964. Control of the circadian rhythm in serotonin content of the rat pineal gland. *Life Sci., 3*:1175–79.

Sweeney, B. M., and F. T. Haxo. 1961. Persistence of a photosynthetic rhythm in enucleated *Acetabularia. Science, 134*:1361–63.

——— 1963. Biological clocks in plants. *Ann. Rev. Plant Physiol., 14*:411–37.

Tromp, S. W. 1964. Weather, climate, and man, p. 283–93. In *Handbook of physiology, Section 4, Adaptation to the environment.* Washington, D.C., Amer. Physiol. Soc.

Weiss, H. S. 1961. Seasonal variation in avian blood pressure. *Fed. Proc.,* Part I, *20*:115.

CHAPTER II.3

Barnett, S. A. 1961. Some effects of breeding mice for many generations in a cold environment. *Proc. Roy. Soc., 155*:115–35.
Benzinger, T. H. 1963. Peripheral cold and central warm-reception, main origins of human thermal discomfort. *Proc. Nat. Acad. Sci. 49*:832–39.
Boe, A. A., and D. K. Salunkhe. 1963. Effects of magnetic fields on tomato ripening. *Nature, 199*:91–92.
Bortels, H. 1942. Meteorobiologische reaktionen einiger mikroorganismen. *Zbl. Bakt., 105*:305–8.
——— 1950. Mikrobiologie and witterungsablauf. *Zbl. Bakt., 155*:160–70.
Brown, F. 1963. How animals respond to magnetism. *Discovery, 24*:18–22.
Brown, G. M., R. Semple, C. Lennox, G. Bird, and C. Baugh. 1963. Response to cold of Eskimos of the eastern Canadian Arctic. *J. Appl. Physiol., 18*:970–74.
Buettner, K. J. K. 1957. Present knowledge on correlations between weather changes, sferics and air electric space charges, and human health and behavior. *Fed. Proc., 16*:631–37.
Burch, G. E. 1956. Influence of a hot and humid environment on the patient with coronary heart disease. *J. Chronic Dis., 4*:350–63.
Carlson, L. D. 1962. Temperature. *Ann. Rev. Physiol., 24*:85–101.
Clark, H. F., and C. C. Shepard. 1963. Effect of environmental temperatures on infection with *Mycobacterium marinum (balnei)* of mice and a number of poikilothermic species. *J. Bact., 86*:1057–69.
Coon, C. S. 1961. Man against the cold. *Natural History, 70*:56–69.
Cowles, R. B. 1959. Some ecological factors bearing on the origin and evolution of pigment in the human skin. *Amer. Natur., 63*:283–93.
Crisp, D. J., and P. S. Meadows. 1964. Adsorbed layers: The stimulus to settlement in barnacles. *Proc. Roy. Soc., 158*:364–87.
Dordick, I. 1958. The influence of variations in atmospheric pressure upon human beings. *Weather, 13*:359–64.
Duell, G., and B. Duell. 1954. *Organic responses to cosmic rays and their secondaries.* Meteorological Monographs, Amer. Meteorological Soc., 2:61–67.
Dunican, L. K., and H. W. Seeley, Jr. 1963. Temperature-sensitive dextransucrase synthesis by a lactobacillus. *J. Bact., 86*:1079–83.
Edholm, O. G., and H. E. Lewis. 1964. Terrestrial animals in cold: Man in polar regions, p. 435–46. In *Handbook of physiology, Section 4, Adaptation to the environment.* Washington, D.C., Amer. Physiol. Soc.
Edstrom, G. 1948. Investigations into the effects of hot, dry micro-

climate on peripheral circulation, etc. in arthritic patients. *Ann. Rheum. Dis. 7*:76–82.

Friedman, H., and R. O. Becker. 1963. Geomagnetic parameters and psychiatric hospital admissions. *Nature, 200*:626–28.

Glaser, E. M. 1960. Influence of climate on life. *Nature, 188*:1080–81.

Gross, L., and L. W. Smith. 1961. Effect of magnet fields on wound healing in mice. *Fed. Proc. 20*:164.

Hammel, H. T. 1964. Terrestrial animals in cold: Recent studies of primitive man, p. 413–34. In *Handbook of physiology, Section 4, Adaptation to the environment.* Washington, D.C., Amer. Physiol. Soc.

——— R. Elsner, D. LeMessurier, H. Andersen, and F. Milan. 1959. Thermal and metabolic responses of the Australian aborigine exposed to moderate cold in summer. *J. Appl. Physiol., 14*:605–15.

Hannon, J., and A. Larson. 1961. The site and mechanism of norepinephrine-calorigenesis in the cold acclimatized rat. *Fed. Proc., 20*:209.

Hicks, C. S. 1963. Climatic adaptation and drug habituation of the Central Australian aborigine. *Perspect. Biol. Med., 7*:39–57.

Hollander, J. P., and S. Yeostros. 1963. The effect of simultaneous variations of humidity and barometric pressure on arthritis. *AIBS Bull., 13*:24–28.

Kingdon, K. H. 1961. Possible biological effects of electrically charged particles in tobacco smoke. *Nature, 189*:180–82.

Krueger, A. P., and R. F. Smith. 1959. An enzymatic basis for the acceleration of ciliary activity by negative air ions. *Nature, 183*:1332–33.

Ladell, W. S. S. 1964. Terrestrial animals in humid heat: Man, p. 625–59. In *Handbook of physiology, Section 4, Adaptation to the environment.* Washington, D.C., Amer. Physiol. Soc.

Lee, D. J. K. 1963. Biometeorology in occupational health. *AIBS Bull. 13*:29–31.

Lwoff, A., 1962. The thermosensitive critical event of the viral cycle. *Cold Spring Harbor Symp. Quant. Biol., 27*:159–74.

——— and M. Lwoff. 1958. L'inhibition du développement du poliovirus à 39° C. *C. R. Acad. Sci. (Paris), 246*:190–92.

Macpherson, R. K. 1958. Acclimatization status of temperate-zone man. *Nature, 182*:1240–41.

Marshall, I. D. 1959. The influence of ambient temperature on the course of myxomatosis in rabbits. *J. Hyg., 57*:484–97.

Mills, C. A. 1939. *Medical climatology.* Springfield, Ill. Charles C Thomas.

——— 1949. Climate in health and disease, p. 453–500. *In* H. A.

Christian, [ed.], *The Oxford medicine,* Vol. 1. New York, Oxford Univ. Press.

Miraglia, G. and L. J. Berry. 1963. Possible source of secondary invading staphylococci in mice exposed to acute cold. *J. Bact., 85*:345–48.

Moriyama, I. M., and L. P. Herrington. 1938. The relation of diseases of the cardiovascular and renal systems to climatic and socioeconomic factors. *Amer. J. Hyg., 28*:423–36.

Petersen, W. F., and M. E. Milliken. 1937. See Chapter II.2.

Pollard, M. 1964. Germfree animals and biological research. *Science, 145*:247–51.

Regli, J., and R. Stämpfli. 1947. Die Kapillarresistenz als objektives Mass für die Wettereinflüsse auf den Menschen. *Helv. Physiol. Pharma. Acta, 5*:40–63.

Rodahl, K. 1963. *The last of the few.* New York, Harper and Row.

Roots, B. I. 1961. Temperature acclimation and the nervous system in fish. *Fed. Proc.,* Part I, *20*:209.

Sabin, A. B. 1961. Reproductive capacity of poliovirus of diverse origins at various temperatures, p. 90–108. *In* M. Pollard, [ed.], *Perspectives in virology,* Vol. II. Minneapolis, Burgess.

Sarasin, V. G. 1959. Zum Organotropismus der Spirochaete B. duttoni gegenüber der übertragenden Zecke. *Acta Trop. (Basel), 16*:218–43.

Sargent, F., and R. G. Stone. 1954. *Recent studies in bioclimatology.* Meteorological Monographs, 2:121 p.

Schaefer, H. J. 1964. Man and radiant energy. Ionizing radiation, p. 989–97. In *Handbook of physiology, Section 4, Adaptation to the environment.* Washington, D.C., Amer. Physiol. Soc.

Scholander, P. F. 1958. Studies on man exposed to cold. *Fed. Proc., 17*:1054–57.

Smith, R. E. 1963. International symposium on temperature acclimation. *Fed. Proc., 22*:687–960.

Thompson, E. M., M. G. Staley, M. A. Kight, and M. E. Mayfield. 1959. The effect of high environmental temperature on basal metabolism and serum ascorbic acid concentration of women. *J. Nutr., 68*:35–47.

Tromp, S. W. 1963. *Medical biometeorology: Weather, climate, and the living organism.* Amsterdam, Elsevier.

VonMuralt, A. 1957. 25 Jahre Hocalpine Forschungsstation Jungfraujoch. *Experientia Supplementum VI,* 86 p.

——— 1960. Influence of climate on life. *Nature, 188*:1080–81.

Wellington, W. G. 1946. The effects of variations in atmospheric pressure upon insects. *Canad. J. Res., 24*:51–70.

Worden, J. L., and J. R. Thompson. 1956. Air-ion concentration and the growth of cells in vitro. *Anat. Rec., 124*:500.

Chapter III.1

Birnbaum, S., M. Greenstein, M. Winitz, and J. Greenstein. 1958. Quantitative nutritional studies with water-soluble chemically defined diets: VI. Growth studies on mice. *Arch. Biochem., 78*:245–47.

Boyd-Orr, J. B. 1958. *Feast and famine*. London, Rathbone.

——— and J. L. Gilks. 1931. *Studies of nutrition: The physique and health of two African tribes*. London, Research Council, No. 155.

Brock, J. F. 1963. Sophisticated diets and man's health, p. 36–56. *In* G. Wolstenholme, [ed.], *Man and his future*. Boston, Little, Brown.

Cassel, J. 1955. A comprehensive health program among South African Zulus, p. 15–41. *In* B. D. Paul, [ed.], *Health, culture, and community: Case studies of public reactions to health programs*. New York, Russell Sage Foundation.

Cuthbertson, D. P. 1963. Nutritional problems in infancy and childhood. *Proc. Nutr. Soc., 22*:119–21.

Davis, C. M. 1928. Self selection of diet by newly weaned infants; experimental study. *Amer. J. Dis. Child., 36*:651–79.

Day, P. L. 1962. *Nutrient requirements of domestic animals: X. Nutrient requirements of laboratory animals*. Washington, D.C., Nat. Acad. Sci., Nat. Research Council Publication No. 990.

Drummond, J. C., and A. Wilbraham. 1957. *The Englishman's food: A history of five centuries of English diet*. London, Jonathan Cape.

Fenton, P. F., M. Dowling, and J. Mershon. 1954. Relation of dietary fat level to fatty livers in several strains of mice. *J. Nat. Cancer Inst., 15*:429–32.

Gordon, H., and A. Ganzon. 1959. On the protein allowances for young infants. *J. Pediat., 54*:503–28.

György, P. 1959(a). Nutrition in infancy. *Fed. Proc. 18*:9–16.

——— 1959(b). Symposium on protein requirement and its assessment in man. *Fed. Proc. 18*:1125–1231.

Harper, A. E. 1959. Amino acid balance and imbalance: I. Dietary level of protein and amino acid imbalance. *J. Nutr., 68*:405–18.

Holt, L. E. 1959. The protein requirement of infants. *J. Pediat., 54*:496.

Konishi, F., J. Hawkins, F. Berger, G. Isaac, and T. Friedemann. 1957. The dietary composition and adequacy of the food consumed by young men on an *ad libitum* regimen. *J. Nutr., 63*:41–56.

Liener, I. E. 1962. Toxic factors in edible legumes and their elimination. *Amer. J. Clin. Nutr., 11*:281–98.

McCance, R. A., and E. M. Widdowson. 1962. Nutrition and growth. *Proc. Roy. Soc., 156*:326–37.

Mellander, O., B. Vahlquist, and T. Melbin. 1962. Diet of the newborn human being. *Nature, 196*:120–21.

Mertz, E. T. 1959. Recent research on human protein requirements and the amino acid supplementation of foods. *Proc. 11th Res. Conf.*, Amer. Meat Inst. Foundation, Chicago.

Neel, J. V., F. M. Salzano, P. C. Jungqueria, F. Keiter, and D. Maybury-Lewis. 1964. Studies on the Xavante Indians of the Brazilian Mato Grosso. *Human Genetics, 16*:52–140.

Reichenow, E. 1920. Den wiederkauer-Infusorien verwandte Formen aus Gorilla und Schimpanse. *Arch. Protistenk, 41*:1–33.

Richards, A. I. 1932. *Hunger and work in a savage tribe.* London, George Routledge & Sons.

Scrimshaw, N. 1963. Factors influencing protein requirements, p. 181–216. In *The Harvey Lectures.* New York, Academic Press.

Sunderman, S. E., E. Roitman, A. Boyer, and L. E. Holt, Jr. 1961. Essential amino acid requirements of infants. *Amer. J. Dis. Child., 102*:157–62.

Waterlow, J. C., and J. M. L. Stephen. 1955. *Human protein requirements and their fulfillment in practice.* Bristol, England, Wright.

Wilson, J. L. 1959. Nutrition in childhood and adolescence. *Fed. Proc., 18*:17–21.

Wolstenholme, G. E. W., and M. O'Connor. 1964. *Diet and bodily constitution.* Ciba Foundation Study Group No. 17. Boston, Little, Brown.

Wostmann, B. S. 1959. Nutrition of the germfree mammal. *Ann. N. Y. Acad. Sci., 78*:175–82.

Chapter III.2

Bean, W. B. 1959. The endeavor of internal medicine, 1859–1959. *Arch. Intern. Med., 104*:851.

Berg, B. N., and H. S. Simms. 1960. Nutrition and longevity in the rat: II. Longevity and onset of disease with different levels of food intake. *J. Nutr., 71*:255–63.

Brock, J. F. 1959. Nutrition and the clinician. *Lancet, 2*:923–27.

Bruch, H. 1957. *The importance of overweight.* New York, W. W. Norton.

Caius, J. 1912. A boke or counseill against the disease commonly called the sweate or sweatyng sicknesse (1552), p. 1–36. In *The works of John Caius, M.D.* London, Cambridge Univ. Press.

Cassel, J. 1955. See Chapter III.1.

Champagnat, A., C. Vernet, B. Lainé, and J. Filosa. 1963. Biosynthesis of protein-vitamin concentrates from petroleum. *Nature, 197*:13–14.

Cook, B., E. Lau, and B. Bailey. 1963. The protein quality of waste-grown green algae: I. Quality of protein in mixtures of algae, nonfat powdered milk, and cereals. *J. Nutr., 81*:23–29.

Drogat, N. 1962. *The challenge of hunger.* Translated by J. R. Kirwan. Westminster, Md., Newman Press.

Forbes, G. 1957. Overnutrition for the child: Blessing or curse? *Nutr. Rev., 15*:193–96.

Gordon, J. E., I. Chitkara, and J. Wyon. 1963. Weanling diarrhea. *Amer. J. Med. Sci., 245*:345–77.

Gunther, M. 1963. The comparative merits of breast and bottle feeding. *Proc. Nutr. Soc., 22*:134–39.

György, P. 1959. See Chapter III.1.

Hundley, J. M. 1959. Malnutrition—a global problem. *Fed. Proc., 18*:76–81.

Jelliffe, D. B. 1959. Protein-calorie malnutrition in tropical preschool children. *J. Pediat., 54*:227–56.

Keys, A., and M. Keys. 1963. *Eat well and stay well.* New York, Doubleday.

Kumta, U. S., and A. E. Harper. 1962. Amino acid balance and imbalance: IX. Effect of amino acid imbalance on blood amino acid pattern. *Proc. Soc. Exper. Biol. Med., 110*:512–17.

Lane, P., and M. Dickie. 1958. The effect of restricted food intake on the life span of genetically obese mice. *J. Nutr., 64*:549–54.

Liener, I. E. 1962. See Chapter III.1.

MacKeith, R. C. 1963. Is a big baby healthy? *Proc. Nutr. Soc., 22*:128–34.

McCance, R. A. 1953. Overnutrition and undernutrition. *Lancet, 2*:685–90.

——— 1962. Food, growth and time. *Lancet, 2*:671–75.

McCay, C. M., G. Sperling, and L. Barnes. 1943. Growth, ageing, chronic diseases and life span in rats. *Arch. Biochem., 2*:469–79.

Pirie, N. W. 1961. Progress in biochemical engineering broadens our choice of crop plants. *Econ. Botany, 15*:302–10.

——— 1963. The selection and use of leafy crops as a source of protein for man, p. 53–60. In *Proc. 5th Internat. Cong. of Biochem.,* Vol. VIII. Oxford, Pergamon.

Rinzler, S. H. 1962. Lessons from the anticoronary club study. *Fed. Proc., 21*:33–35.

Ross, M. H. 1959. Proteins, calories and life expectancy. *Fed. Proc., 18*:1190–1207.

Scrimshaw, N. S. 1963. World-wide importance of protein malnutrition and progress toward its prevention. *Amer. J. Public Health, 53*:1781–94.

———, R. Bressani, M. Béhar, and F. Viteri. 1958. Supplementation of cereal proteins with amino acids: I. Effect of amino acid supplementation of corn-masa at high levels of protein intake on the nitrogen retention of young children. II. Effect of amino acid supplementation of corn-masa at intermediate levels of protein intake on the nitrogen retention of young children. *J. Nutr.*, *66*:485–99; 501–13.

——— and M. Béhar. 1959. World-wide occurrence of protein malnutrition. *Fed. Proc.*, *18*:82–88.

Sebrell, W. H., Jr., and D. B. Hand. 1957. Protein malnutrition as a world problem, p. 47. *In* W. H. Cole, [ed.], *Amino acid malnutrition*. New Brunswick, Rutgers Univ. Press.

Sidransky, H. 1960. Chemical pathology of nutritional deficiency induced by certain plant proteins. *J. Nutr.*, *71*:387–95.

Silberberg, M., and R. Silberberg. 1955. Diet and life span. *Physiol. Rev.*, *35*:347–62.

Sinclair, H. M. 1955. Too rapid maturation of children as a cause of ageing. *Ciba Foundation Colloquia on Ageing*, *1*:194–201.

Stare, F. J. 1963. Overnutrition. *Amer. J. Public Health*, *53*:1795–1802.

Stout, C., J. Morrow, E. Brandt, and S. Wolf. 1964. Unusually low incidence of death from myocardial infarction. *J.A.M.A.*, *188*:845–49.

Tepperman, J. 1958. Etiologic factors in obesity and leanness. *Persp. Biol. Med.*, *1*:293–306.

Waterlow, J. C. 1953. *United Nations, Food and Agriculture Organization, Protein Malnutrition Conference*. Cambridge, England.

——— 1962. Protein metabolism in human protein malnutrition. *Proc. Roy. Soc.*, *156*:345–51.

——— and J. M. L. Stephen. 1955. See Chapter III.1.

Yudkin, J. 1964. Dietary fat and dietary sugar in relation to ischaemic heart-disease and diabetes. *Lancet*, *2*:4–5.

——— and J. Roddy. 1964. Levels of dietary sucrose in patients with occlusive atherosclerotic disease. *Lancet*, *2*:6–8.

Chapter III.3

Boyne, A. W. 1960. Secular changes in the stature of adults and the growth of children, with special reference to changes in intelligence of 11-year-olds, p. 97–120. *In* J. M. Tanner, [ed.], *Human growth*. New York, Pergamon Press.

Brewer, T. H. 1962. Role of malnutrition, hepatic dysfunction and gastrointestinal bacteria in the pathogenesis of acute toxemia of pregnancy. *Amer. J. Obstet. Gynec.*, *84*:1253.

Brit. Med. J. 1961. Early maturing and larger children. 2:502–3.

Brock, J. F. 1959. See Chapter III.2.

Brozek, J. 1959. Experimental studies on the impact of deficient diet on behavior. *Borden Rev. Nutr. Res.*, *20*:75–88.

Cone, T. E., Jr. 1961. De pondere infantum recens natorum. The history of weighing the newborn infant. *Pediatrics*, *28*:490–98.

Cuthbertson, D. P. 1963. Nutritional problems in infancy and childhood. *Proc. Nutr. Soc.*, 22:119–21.

Ershoff, B. H. 1955. Nutrition and stress. *Nutr. Rev.*, *13*:33–36.

Formal, S., H. Noyes, and H. Schneider. 1960. Experimental shigella infections: III. Sensitivity of normal, starved and carbon tetrachloride treated guinea pigs to endotoxin. *Proc. Soc. Exp. Biol. Med.*, *103*:415–18.

Lancet. 1960. Continuing secular trend in growth. *1*:1336.

Lát, J., E. M. Widdowson, and R. A. McCance. 1961. Some effects of accelerating growth: III. Behaviour and nervous activity. *Proc. Roy. Soc.*, *153*:347–56.

Loo, Y. H., E. Diller, and J. E. Owen. 1962. Effect of phenylalanine diet on learning in the rat. *Nature*, *194*:1286.

McCance, R. A. 1962. See Chapter III.2.

——— and E. M. Widdowson. 1962. See Chapter III.1.

Mitchell, H. H., and M. Edman. 1951. *Nutrition and climatic stress, with particular reference to man.* Springfield, Ill., Charles C Thomas.

National Animal Diseases Laboratory, Ames, Iowa: USDA Agricultural Research, *12*, 11. (*As reviewed in* Taming ferrets with food, *New Scientist*, 1964, 22:293.)

Olewine, D. A., and C. H. Barrows, Jr. 1961. Random and voluntary activity following prolonged dietary restriction. *Fed. Proc.*, 20:364.

Oppers, V. M. 1963. *Analyse van de Acceleratie van de Menselijke Lengtegroei door Bepaling van het Tijdstip van de Groeifasen.* Univ. of Amsterdam.

Pollin, W., P. V. Cardon, and S. S. Kety. 1961. See Chapter I.2.

Ross, M. H. 1959. See Chapter III.2.

Sinclair, H. M. 1955. See Chapter III.2.

Sutter, J., and M. Goux. 1962. Evolution de la consanguinité en France de 1926 à 1958, avec des données récentes détaillées. *Population*, *17*:683–802.

Tanner, J. M. 1960. *Human growth.* New York, Pergamon Press.

Vaughan, D. A. 1959. Arctic survival rations: VI. The physiological effects of restricted diets during successive winter field trials. Arctic aeromedical laboratory. (As reviewed in *Nutr. Rev.*, 1961, *19*:73–75.)

Widdowson, E. M., and R. A. McCance. 1963. The effect of finite periods of undernutrition at different ages on the composition and subsequent development of the rat. *Proc. Roy. Soc. (Biol.)*, *158*: 329–42.

Wolstenholme, G. E. W., and M. O'Connor. 1964. See Chapter III.1.

Chapter III.4

Baur, L. D., and L. J. Filer, Jr. 1959. Influence of body composition of weanling pigs on survival under stress. *J. Nutr.*, *69*:128–34.

Bruch, H. 1957. See Chapter III.2.
Grande, F. 1964. Man under caloric deficiency, p. 911–37. In *Handbook of physiology, Section 4, Adaptation to the environment*. Washington, D. C., Amer. Physiol. Soc.
György, P. 1959. See Chapter III.1 (a and b).
Huggett, A. 1949. Nutrition and viable young. *Brit. J. Nutr., 3*:96–107.
Kaunitz, H. 1958. Non-energetic powers of nutrition. *Perspect. Biol. Med., 2*:75–83.
Lancet. 1963. Protein stores. *1*:811–12.
Malm, O. J. 1958. See Chapter I.4.
McCance, R. A. 1962. See Chapter III.2.
Mellander, O., B. Vahlquist, and T. Melbin. 1962. See Chapter III.1.
Pathak, C. L. 1958. Nutritional adaptation to low dietary intakes of calories, proteins, vitamins, and minerals in the tropics. *Amer. J. Clin. Nutr., 6*:151–58.
Spolter, P. D., and A. E. Harper. 1962. Adaptation of rats to diets containing ethionine or excess leucine. *Fed. Proc., 21*:8.
Vaughan, D. A. 1959. See Chapter III.3.
Waterlow, J. C. 1962. See Chapter III.2.

Chapter III.5

Lee, D. 1962. Food and human existence. *Nutr. News, 25*:9–10.
Pope, G. S., M. J. McNaughton, and H. E. H. Jones. 1960. Estrogens in pasture forages. *Nutr. Rev., 18*:14–15.

Chapter IV.1

Allee, W. C. 1951. See Chapter I.2.
Allen, O. N., and E. K. Allen. 1954. Morphogenesis of the leguminous root nodule, p. 209–32. In *Abnormal and pathological plant growth: Report of symposium, 1953, Washington, D. C.* U. S. Atomic Energy Commission, Unclassified Reports BNL-258.
Bates, M. 1960. *The forest and the sea*. New York, Random House.
——— 1961. *Man in nature*. Englewood Cliffs, N. J., Prentice-Hall.
Braun, A. C. 1954. The physiology of plant tumors. *Ann. Rev. Plant Physiol., 5*:133–62.
——— 1959. Demonstration of the recovery of the crown-gall tumor cell with the use of complex tumors of single-cell origin. *Proc. Nat. Acad. Sci., 45*:932–38.
Brooks, M. B. 1963. Symbiosis and aposymbiosis in arthropods, p. 200–31. *In* P. S. Nutman and B. Mosse, [eds.], *Symbiotic associations*. London, Cambridge Univ. Press.
Cleveland, L. R., and A. V. Grimstone. 1964. The fine structure of the

flagellate *Mixotrichia paradoxa* and its associated micro-organisms. *Proc. Roy. Soc. (Biol.), 159*:668–86.

DeHarven, E. 1964. Virus particles in the thymus of conventional and germfree mice. *J. Exp. Med., 120*:857–68.

Dubos, R. 1961. Integrative and creative aspects of infection, p. 200–05. *In* M. Pollard, [ed.], *Perspectives in virology, II*. Minneapolis, Burgess.

Fernandes, M. V., T. Wiktor, and H. Koprowski. 1964. Endosymbiotic relationship between animal viruses and host cells: a study of rabies virus in tissue culture. *J. Exp. Med., 120*:1099–1116.

Freeman, V. J., and I. U. Morse. 1952. Further observations on the change to virulence of bacteriophage-infected avirulent strains of *Corynebacterium diphtheriae*. *J. Bact., 63*:407–14.

Gibor, A., and S. Granick. 1964. Plastids and mitochondria: Inheritable systems. *Science, 145*:890–97.

Humm, H. J. 1944. Bacterial leaf nodules. *J. N. Y. Bot. Gard., 45*: 193–99.

Jacob, F., and E. L. Wollman. 1961. *Sexuality and the genetics of bacteria*. New York, Academic Press.

Koprowski, H. 1963. The role of SV_{40} (simian virus 40) in the transformation of human cells. *Proc. Roy. Soc. Med., 56*:252.

Kropotkin, P. 1902 (1955). *Mutual aid: A factor of evolution*. Boston, Extending Horizons.

Lederberg, J. 1952. Cell genetics and hereditary symbiosis. *Physiol. Rev., 32*:403–26.

Limbaugh, C. 1961. Cleaning symbiosis. *Sci. Amer., 205*:42–50.

Lwoff, A. 1953. Lysogeny. *Bact. Rev., 17*:269–337.

Milne, L. J., and M. Milne. 1960. *The balance of nature*. New York, Knopf.

Nutman, P. S., and B. Mosse. 1963. *Symbiotic associations*. The 13th Symposium of the Society for General Microbiology. London, Cambridge Univ. Press.

Paul, J. 1963. Spontaneous transformation in cultured cells. *Proc. Roy. Soc. Med., 56*:251.

Pollard, M. 1964. See Chapter II.3.

Salser, W. 1961. Non-genetic biological information mechanisms. *Perspect. Biol. Med., 4*:177–98.

Seronde, J. 1954. Resistance of rats to inoculation with corynebacterium pathogenic in pantothenate deficiency. *Proc. Soc. Exp. Biol. Med., 85*:521–24.

Shein, H. M., J. F. Enders, J. D. Levinthal, and A. E. Burket. 1963. Transformation induced by simian virus 40 in newborn Syrian hamster renal cell cultures. *Proc. Nat. Acad. Sci., 49*:28–34.

Siegal, R. W. 1960. Hereditary endosymbiosis in *Paramecium bursaria*. *Exp. Cell Res., 19*:239–52.

Sonneborn, T. M. 1961. Kappa particles and their bearing on host-parasite relations, p. 5–17. *In* M. Pollard, [ed.], *Perspectives in virology, II.* Minneapolis, Burgess.

Stoker, M. 1963. Neoplastic transformation induced *in vitro* by Polyoma virus. *Proc. Roy. Soc. Med., 56*:251–52.

Weber, N. A. 1957. Fungus-growing ants and their fungi: *Cyphomyrmex costatus. Ecology, 38*:480–94.

Zabriskie, J. B. 1964. The role of temperate bacteriophage in the production of erythrogenic toxin by Group A streptococci. *J. Exp. Med., 119*:761–80.

Chapter IV.2

Allee, W. C. 1951. *Cooperation among animals.* New York, Henry Schuman.

Barnett, S. A. 1960. Social behaviour among tame rats and among wild-white hybrids. *Proc. Zool. Soc. Lond., 134*:611–21.

——— 1963. *The rat: A study in behaviour.* Chicago, Aldine.

——— 1964. Social stress, p. 170–218. In *Viewpoints in biology.* London, Butterworths.

Barrow, J. H., Jr. 1955. Social behavior in fresh-water fish and its effect on resistance to trypanosomes. *Proc. Nat. Acad. Sci., 41*:676–79.

Benoit, J., I. Assenmacher, and E. Brard. 1955, 1956. See Chapter II.2.

Bernardis, L., and F. Skelton. 1963. Effect of crowding on hypertension and growth in rats bearing regenerating adrenals; *and* Effect of gentling on development of adrenal regeneration hypertension in immature female rats. *Proc. Soc. Exp. Biol. Med., 113*:952–57.

Bronson, F. H., and B. E. Eleftheriou. 1965(a). Adrenal response to fighting in mice: Separation of physical and psychological causes. *Science, 147*:627–28.

——— and ——— 1965(b). Relative effects of fighting on bound and unbound corticosterone in mice. *Proc. Soc. Exp. Biol. Med., 118*:146–49.

Calhoun, J. B. 1949. A method for self-control of population growth among mammals living in the wild. *Science, 109*:333–35.

——— 1962. Population density and social pathology. *Sci. Amer., 206*:139–48.

Carpenter, C. R. 1958. Territoriality: A review of concepts and problems, p. 224–50. *In* A. Roe and G. G. Simpson, [eds.], *Behavior and evolution.* New Haven, Yale Univ. Press.

Chitty, D. 1958. Self-regulation of numbers through changes in viability. *Cold Spring Harbor Symp. Quant. Biol., 22*:277–80.

Christian, J. J., and D. E. Davis. 1956. The relationship between adrenal weight and population status of urban Norway rats. *J. Mammal., 37*:475–86.

——— and H. O. Williamson. 1958. Effect of crowding on experimental granuloma formation in mice. *Proc. Soc. Exp. Biol. Med., 99*:385–87.

———, V. Flyger, and D. E. Davis. 1960. Factors in mass mortality of a herd of Sika deer (*Cervus nippon*). *Chesapeake Science, 1*:79–95.

Curry-Lindahl, K. 1963. New theory on a fabled exodus. *Natural Hist., 122*:46–53.

Davis, D. E., and C. P. Read. 1958. Effect of behavior on development of resistance in trichinosis. *Proc. Soc. Exp. Biol. Med., 99*:269–72.

Deevey, E. S. 1960. The hare and the haruspex: A cautionary tale. *Amer. Sci., 48*:415–29.

Ellis, P. E., and J. B. Free. 1964. Social organization of animal communities. *Nature, 201*:861–63.

Elton, C. S. 1958. *The ecology of invasions by animals and plants.* New York, John Wiley & Sons.

Etkin, W. 1964. *Social behavior and organization among vertebrates.* Univ. Chicago Press.

Flickinger, G., and H. Ratcliffe. 1961. The effect of grouping on the adrenals and gonads of chickens. *Fed. Proc., 20*:176.

Greenwood, M. 1935. *Epidemics and crowd-diseases.* London, Williams and Norgate.

Hall, E. T. 1959. *The silent language.* New York, Doubleday.

——— 1964. Silent assumptions in social communication, p. 41–55. *In* D. Rioch and E. A. Weinstein, [eds.], *Disorders of communication.* Baltimore, Williams & Wilkins.

Hediger, H. 1950. *Wild animals in captivity.* London, Butterworths.

Hinde, R. A. 1960. An ethological approach, p. 49–58. *In* J. M. Tanner, [ed.], *Stress and psychiatric disorder.* Oxford, Blackwell.

Keeley, K. 1962. Prenatal influence on behavior of offspring of crowded mice. *Science, 135*:44–45.

Koford, C. B. 1963. Rank of mothers and sons in bands of rhesus monkeys. *Science, 141*:356–57.

Lasagna, L. 1962. Some explored and unexplored psychological variables in therapeutics. *Proc. Roy. Soc. Med., 55*:773–76.

Mackintosh, J. H. 1962. Effect of strain and group size on the response of mice to "seasonal" anaesthesia. *Nature, 194*:1304.

Mason, J. W. 1959. Psychological influences on the pituitary-adrenal cortical system. *Recent Progr. Hormone Res., 15*:345–89.

McDonald, D. G., J. Stern, and W. Hahn. 1963. Effects of differential housing and stress on diet selection, water intake, and body weight in the rat. *J. Appl. Physiol., 18*:937–42.

McKissick, G. E., G. L. Flickinger, Jr., and H. L. Ratcliffe. 1961. Coronary arteriosclerosis in isolated, paired, and grouped chickens. *Fed. Proc., 20*:91.

Siegal, H. S. 1959. The relation between crowding and weight of adrenal glands in chickens. *Ecology, 40*:495–98.

Thiessen, D. D. 1963. Varying sensitivity of C57BL/Crgl mice to grouping. *Science, 141*:827–28.

Tinbergen, N. 1953. *Social behaviour in animals*. London, Methuen.

Washburn, S. L., and L. Devore. 1961. The social life of baboons. *Sci. Amer., 204*:62–71.

Welty, C. 1957. The geography of birds. *Sci. Amer., 197*:118–28.

Zeuner, F. E. 1963. *A history of domesticated animals*. London, Hutchinson.

Chapter V.1

Bishop, R. F., and E. Allcock. 1960. Bacterial flora of the small intestine in acute intestinal obstruction. *Brit. Med. J., 1*:766–70.

Clapper, W. E., and G. H. Meade. 1963. Normal flora of the nose, throat, and lower intestine of dogs. *J. Bact., 85*:643–48.

Cushing, H., and L. E. Livingood. 1900. Experimental and surgical notes upon the bacteriology of the upper portions of the alimentary canal, with observations on the establishment there of an amicrobic state. *Johns Hopkins Hospital Reports, 9*:543–91.

Dixon, J. M. S. 1960. The fate of bacteria in the small intestine. *J. Path. Bact., 79*:131–40.

Dubos, R., and R. W. Schaedler. 1964. The digestive tract as an ecosystem. *Amer. J. Med. Sci., 248*:267–71.

———, ——— and R. Costello. 1963. Composition, alteration, and effects of the intestinal flora. *Fed. Proc., 22*:1322–29.

———, ——— and M. Stephens. 1963. The effect of antibacterial drugs on the fecal flora of mice. *J. Exp. Med., 117*:231–43.

Evenson, A., E. McCoy, B. R. Geyer, and C. A. Elvehjem. 1946. The cecal flora of white rats on a purified diet and its modification by succinylsulfathiazole. *J. Bact., 51*:513–21.

Gustafsson, B. E., and R. J. Fitzgerald. 1960. Alteration in intestinal microbial flora of rats with tail cups to prevent coprophagy. *Proc. Soc. Exp. Biol. Med., 104*:319–22.

Gyllenberg, H., and P. Roine. 1957. The value of colony counts in evaluating the abundance of "Lactobacillus" bifidus in infant faeces. *Acta. Path. Microbiol. Scand., 41*:144–50.

Haenel, H. 1960. Aspekte der mikroökologischen Beziehungen des makroorganismus. *Zbl. Bakt., 176*:1–121.

Horsfall, F. L., Jr. 1959. Viral infections of the respiratory tract. *Amer. Rev. Resp. Dis., 80*:315–25.

Kenworthy, R., and W. E. Crabb. 1963. The intestinal flora of young pigs, with reference to early weaning, *E. coli* and scours. *J. Comp. Path. Ther., 73*:215–28.

Lecce, J. G., and B. R. Reep. 1962. *Escherichia coli* associated with colostrum-free neonatal pigs raised in isolation. *J. Exp. Med., 115*:491.

Levesque, J., and M. Gautier. 1959. Rôle du facteur Bifidus 2 dans l'installation et le maintien de la flore bifidus du nourrisson. *Sem. Hop. Paris, 35*:30–36.

Masson, P., and Cl. Regaud. 1918. Sur l'existence de nombreux microbes vivant à l'état normal dans le tissu des follicules lymphoides de l'intestin, chez le lapin. *C. R. Soc. Biol., 81*:1256–60.

——— and ——— 1919(a). Apparition et pullulation des microbes dans le tissu lymphoide de l'appendice caecal du lapin au cours du développement. *C. R. Soc. Biol., 82*:30–34.

——— and ——— 1919(b). Sur la manière dont pénètrent des microbes de la cavité intestinale dans l'épithélium de revètement des follicules lymphoides, chez le lapin. *C. R. Soc. Biol., 82*:144–46.

———and ——— 1919(c). Sur les microbes du tissu lymphoide de l'intestin du lapin normal: Rectification à propos de leur découverte. *C. R. Soc. Biol., 82*:304.

Metchnikoff, E. 1908. Études sur la flore intestinale. Putréfaction intestinale. *Ann. Inst. Pasteur, 22,* No. 2.

Nelson, J. B., and G. R. Collins. 1961. The establishment and maintenance of a specific pathogen-free colony of Swiss mice. *Proc. Anim. Care Panel, 11*:65–72.

Newton, W. L., H. Steinman, and M. Brandriss. 1964. Absence of lethal effect of penicillin in germfree guinea pigs. *J. Bact., 88*:537–40.

Petuely, F. 1962. Der heutige Stand unseres Wissens über das Bifidusproblem in bakteriologischer, chemischer und klinischer Hinsicht. *Arch. Kinderheilk., 165*:209–18.

——— and G. Lindner. 1962. Ernahrung und darmflora. *Deutsche. Med. J., 13*:116–19.

Porter, J. R., and L. F. Rettger. 1940. Influence of diet on the distribution of bacteria in the stomach, small intestine and cecum of the white rat. *J. Infect. Dis., 66*:105–10.

Regaud, Cl. 1919. Mitochondries et symbiotes. *C. R. Soc. Biol., 82*:244–51.

Rettger, L. F., and G. D. Horton. 1914. A comparative study of the intestinal flora of white rats kept on experimental and ordinary mixed diets. *Zbl. Bakt. Parasit., 73*:362–72.

———, M. N. Levy, L. Weinstein, and J. E. Weiss. 1935. *Lactobacillus acidophilus and its therapeutic application.* New Haven, Yale Univ. Press.

Rose, C. S., and P. György. 1963. Bifidus factor 2 for growth of lactobacillus bifidus. *Proc. Soc. Exp. Biol. Med., 112*:923–26.

Rosebury, T. 1962. *Microorganisms indigenous to man.* New York, McGraw-Hill.

Schaedler, R. W., and R. Dubos. 1962. The fecal flora of various strains of mice: Its bearing on their susceptibility to endotoxin. *J. Exp. Med., 115*:1149–59.

Schneider, H. A. 1946. Nutrition and resistance to infection—the strategic situation. *Vitamins Hormones, 4*:35–70.

Seeliger, H., and H. Werner. 1963. Recherches qualitatives et quantitatives sur la flore intestinale de l'homme. *Ann. Inst. Pasteur., 105*:911–36.

Sieburth, J. 1959. Gastrointestinal microflora of antarctic birds. *J. Bact., 77*:521–23.

——— 1961. Antibiotic properties of acrylic acid, a factor in the gastrointestinal antibiosis of polar marine animals. *J. Bact., 82*:72–79.

Slanetz, L. W., and C. H. Bartley. 1957. Numbers of enterococci in water, sewage, and feces determined by the membrane filter technique with an improved medium. *J. Bact., 74*:591–95.

Smith, H. W., and W. E. Crabb. 1956. The typing of *Escherichia coli* by bacteriophage. *J. Gen. Microbiol., 15*:556–74.

——— and ——— 1961. The faecal bacterial flora of animals and man: Its development in the young. *J. Path. Bact., 82*:53–66.

——— and J. E. T. Jones. 1963. Observations on the alimentary tract and its bacterial flora in healthy and diseased pigs. *J. Path. Bact., 86*:387–412.

Tissier, H. 1899. Recherches sur la flore intestinale des nourrissons. Thèse de Paris.

——— 1905. Répartition des microbes dans l'infection intestinale chez le nourrisson. *Ann. Inst. Pasteur, 14*:109–23.

Tomarelli, R. M., R. Hartz, and F. W. Bernhart. 1960. The effect of lactose feeding on the body fat of the rat. *J. Nutr., 71*:221–28.

Wilbur, R. D., D. Catron, L. Quinn, V. Speer, and V. Hays. 1960. Intestinal flora of the pig as influenced by diet and age. *J. Nutr., 71*:168–75.

Zubrzycki, L., and E. H. Spaulding. 1962. Studies on the stability of the normal human fecal flora. *J. Bact., 83*:968–74.

Chapter V.2

Anderson, G. W., M. Hauser, M. Wright, and J. Couch. 1956. The effect of dietary enterococci and chlortetracycline hydrochloride on the intestinal flora and growth of chicks. *Canad. J. Microbiol., 2*:733–39.

Barber, R. S., R. Braude, and K. Mitchell. 1955. Antibiotic and copper supplements for fattening pigs. *Brit. J. Nutr., 9*:378–81.

Bare, L. N., R. Wiseman, and O. Abbott. 1964. Effect of dietary antibiotics upon coliform bacteria and lactobacilli in the intestinal tract of uric acid-fed chicks. *J. Bact., 87*:329–31.

Barnes, R. H., G. Fiala, and E. Kwong. 1963. Decreased growth rate resulting from prevention of coprophagy. *Fed. Proc.*, *22*:125–28.

Coates, M. E., M. Davies, and S. Kon. 1955. The effect of antibiotics on the intestine of the chick. *Brit. J. Nutr.*, *9*:110–19.

Daft, F. S., E. McDaniel, L. Herman, M. Romine, and J. Hegner. 1963. Role of coprophagy in utilization of B vitamins synthesized by intestinal bacteria. *Fed. Proc.*, *22*:12–33.

Donaldson, R. M. 1964. Normal bacterial populations of the intestine and their relation to intestinal function. *New Eng. J. Med.*, *270*:938–45; 994–1001; 1050–56.

Draper, H. H. 1958. The absorption of radiolysine by the chick as affected by penicillin administration. *J. Nutr.*, *64*:33–42.

Dubos, R., R. W. Schaedler, and R. L. Costello. 1963. The effect of antibacterial drugs on the weight of mice. *J. Exp. Med.*, *117*:245–57.

Eyssen, H., and P. deSomer. 1963. The mode of action of antibiotics in stimulating growth of chicks. *J. Exp. Med.*, *117*:127–38.

Fauve, R. M. 1963. L'emploi, en pathologie expérimentale, de souris bactériologiquement propres. La souche NCS. *Med. et Hyg.*, *21*:178–79.

Forbes, M., J. Park, and M. Lev. 1959. Role of the intestinal flora in the growth response of chicks to dietary penicillin. *Ann. N. Y. Acad. Sci.*, *78*:321–27.

Foy, H., A. Kondi, and P. Manson-Bahr. 1955. Penicillin in megaloblastic anaemias of Africans. *Lancet*, *2*:693–99.

Francois, A. C. 1962. Mode of action of antibiotics on growth. *World Rev. Nutr. Diet.*, *3*:25–64.

——— and M. Michel. 1955. Relations entre l'influence des antibiotiques sur la croissance du porc et l'inhibition des désaminases de la flore intestinale. *C. R. Soc. Biol. (Paris)*, *240*:808–10.

——— and ——— 1958. Les antibiotiques dans l'alimentation animale. *Ann. Nutr. (Paris)*, *10*:96–99.

Gustafsson, B. E., and R. J. Fitzgerald. 1960. See Chapter V.1.

——— and L. S. Lanke. 1960. Bilirubin and urobilins in germfree, exgermfree, and conventional rats. *J. Exp. Med.*, *112*:975–81.

Haight, T. H., and W. E. Pierce. 1955. Effect of prolonged antibiotic administration on weight of healthy young males. *J. Nutr.*, *56*:151–61.

Henry, S. M. 1962. The significance of microorganisms in the nutrition of insects. *Trans. N. Y. Acad. Sci.*, *24*:676–83.

Hill, C. H., A. Keeling, and J. Kelly. 1957. Studies on the effect of antibiotics on the intestinal weights of chicks. *J. Nutr.*, *62*:255–67.

Lev, M., and M. Forbes. 1959. Growth response to dietary penicillin of germ-free chicks and of chicks with a defined intestinal flora. *Brit. J. Nutr.*, *13*:78–84.

Levenson, S. M., and B. Tennant. 1963. Some metabolic and nutritional studies with germfree animals. *Fed. Proc., 22*:109–19.

Libby, D. A., R. Evans, S. Bandemer, and P. Schaible. 1956. Arsenicals as growth promoters. *Nutr. Rev., 14*:206–9.

——— and P. Schaible. 1955. Observations on growth responses to antibiotics and arsonic acids in poultry feeds. *Science, 121*:733–35.

Luckey, T. D. 1959. Antibiotics in nutrition, p. 174–321. *In* H. S. Goldberg, [ed.], *Antibiotics; Their chemistry and nonmedical uses.* New York, Van Nostrand.

——— 1963. *Germfree life and gnotobiology.* New York, Academic Press.

Metchnikoff, E. 1908. *The prolongation of life.* New York, G. P. Putnam's Sons.

Michel, M. 1961. See Chapter II.2.

——— and A. C. Francois. 1956. Influence de la chlortétracycline sur les décarboxylases de la flore intestinale du porc. *C. R. Soc. Biol. (Paris), 242*:1770–72.

Mickelsen, O. 1956. Intestinal synthesis of vitamins in the nonruminant. *Vitamins Hormones, 14*:1–95.

——— 1962. Nutrition—germfree animal research. *Ann. Rev. Biochem., 31*:515–48.

Miyakawa, M. 1959. The lymphatic system of germfree guinea pigs. *Ann. N. Y. Acad. Sci., 78*:221–36.

New Eng. J. Med., 1962. The colon: Cesspool or nutrient factory? *267*:891–92.

Pollard, M. 1964. Germfree animals and biological research. *Science, 145*:247–51.

Rettger, L. F., M. N. Levy, L. Weinstein, and J. E. Weiss. 1935. See Chapter V.1.

Tomarelli, R. M., R. Hartz, and F. W. Bernhart. 1960. See Chapter V.1.

Walshe, J. M. 1958. Biochemical studies in hepatic coma. *Lectures on scientific basis of medicine, 8*:407–28.

Wostmann, B. S. 1959. Nutrition of the germfree mammal. *Ann. N. Y. Acad. Sci., 78*:175–82.

——— and N. L. Wiech. 1961. Total serum and liver cholesterol in germfree and conventional male rats. *Amer. J. Physiol., 201*:1027–29.

———, P. Knight, L. Keeley, and D. Kan. 1963. Metabolism and function of thiamine and naphthoquinones in germfree and conventional rats. *Fed. Proc., 22*:120–24.

Chapter V.3

Bauer, H., R. Horowitz, and S. Levenson. 1962. Quantitative morphology of the lymphatic system in germfree animals. *Fed. Proc., 21*:280.

Dubos, R., and R. W. Schaedler. 1964. The digestive tract as an ecosystem. *Amer. J. Med. Sci., 248*:267–71.

Gordon, H. A. 1959. Morphological and physiological characterization of germfree life. *Ann. N. Y. Acad. Sci., 78*:208–20.

——— 1965. Demonstration of a bioactive substance in caecal contents of germ-free animals. *Nature, 205*:571–72.

——— and B. S. Wostmann. 1959. Responses of the animal host to changes in the bacterial environment, p. 336–39. In *Recent progress in microbiology*. Springfield, Ill., Charles C Thomas.

Hudson, J. A., and T. D. Luckey. 1964. Bacteria induced morphologic changes. *Proc. Soc. Exp. Bio. Med., 116*:628–31.

Lesher, S., H. Walburg, and G. Sacher. 1964. Generation cycle in the duodenal crypt cells of germ free and conventional mice. *Nature, 202*:884–86.

Luckey, T. D. 1963. See Chapter V.2.

Paget, G. E. 1962. The pathological state of specific pathogen-free animals. *Proc. Roy. Soc. Med., 55*:262–63.

Skelly, B. J. 1963. Effect of intestinal bacteria upon cecal size of gnotobiotic mice. *Bact. Proc.*, p. 67.

———, P. Trexler, and J. Tanami. 1962. Effect of a clostridium species upon cecal size of gnotobiotic mice. *Proc. Soc. Exp. Biol. Med., 110*:455–58.

Sprinz, H. 1962. The morphological response of intestinal mucosa to entire bacteria and its implication for sprue and Asiatic cholera. *Fed. Proc., 21*:57–64.

———, D. Kundel, G. Dammin, R. Horowitz, H. Schneider, and S. Formal. 1961. The response of the germ-free guinea pig to oral bacterial challenge with *Escherichia coli* and *Shigella flexneri*. *Amer. J. Path., 39*:681–95.

———, R. Sribhibhadh, E. Gangarosa, C. Benyajati, D. Kundel, and S. Halstead. 1962. Biopsy of small bowel of Thai people. *Amer. J. Clin. Path., 38*:43–51.

Thorbecke, G. J., and B. Benacerraf. 1959. Some histological and functional aspects of lymphoid tissue in germfree animals. *Ann. N. Y. Acad. Sci., 78*:247–53.

Wiseman, R. F., and H. A. Gordon. 1965. Reduced levels of a bioactive substance in the caecal content of gnotobiotic rats monoassociated with *Salmonella typhimurium*. *Nature, 205*:572–73.

Wostmann, B. S. 1959. Serum proteins in germfree vertebrates. *Ann. N. Y. Acad. Sci., 78*:254–60.

Chapter V.4

Ashburner, F. M., and R. Mushin. 1962. Experimental intestinal coliform infections in mice. *J. Hyg. (Camb.), 60*:175–81.

Berntsen, C. A., and W. McDermott. 1960. Increased transmissibility of staphylococci to patients receiving an antimicrobial drug. *New Eng. J. Med., 262*:637–42.

Black, P. H., L. Kunz, and M. Swartz. 1960. Salmonellosis—a review of some unusual aspects. *New Eng. J. Med., 262*:864–69.

Bohnhoff, M., C. P. Miller, and W. R. Martin. 1964. Resistance of the mouse's intestinal tract to experimental *salmonella* infection: I. Factors which interfere with the initiation of infection by oral inoculation. II. Factors responsible for its loss following streptomycin treatment. *J. Exp. Med., 120*:805–16; 817–28.

Boyden, S. V., and M. E. Andersen. 1955. Diet in experimental tuberculosis in the guinea-pig. *Acta Path. Microbiol. Scand., 37*:201–4.

——— and ——— 1956. Diet in experimental tuberculosis in the guinea-pig. *Acta Path. Microbiol. Scand., 39*:107–16.

Dineen, P. 1960. Effect of reduction of bowel flora on experimental staphylococcal infection in mice. *Proc. Soc. Exp. Biol. Med., 104*:760–62.

——— 1961. The effect of alterations in intestinal flora and host resistance to systematic bacterial infection. *J. Infect. Dis., 109*:280–86.

Donaldson, R. M. 1964. See Chapter V.2.

Dubos, R., R. W. Schaedler, and R. Costello. 1963. See Chapter V.1.

Fauve, R. M., C. H. Pierce-Chase, and R. Dubos. 1964. Corynebacterial pseudotuberculosis in mice: II. Activation of natural and experimental latent infections. *J. Exp. Med., 120*:283–304.

Formal, S. B., H. Noyes, and H. Schneider. 1960. See Chapter III.3.

———, G. Abrams, H. Schneider, and H. Sprinz. 1963. Experimental *Shigella* infection: VI. The role of the small intestine in an experimental infection in guinea pigs. *J. Bact., 85*:119–25.

———, ———, ———, and R. Laundy. 1963. Penicillin in germ-free guinea pigs. *Nature, 198*:712.

Freter, R. 1955. The fatal enteric infection in the guinea pig achieved by inhibition of normal enteric flora. *J. Infect. Dis., 97*:57–65.

——— 1956. Experimental enteric *Shigella* and *Vibrio* infections in mice and guinea pigs. *J. Exp. Med., 104*:411–18.

Gordon, H. W. 1964. Pharmacologically active substances in cecal contents of germfree animals. *Fed. Proc., 23*:200.

Gustaffsson, B. E. 1960. Properdin titers in sera from germ-free rats. *Proc. Soc. Exp. Biol. Med., 105*:598–600.

Jensen, S. B., S. Mergenhagen, R. Fitzgerald, and H. Jordan. 1963. Susceptibility of conventional and germfree mice to lethal effects of endotoxin. *Proc. Soc. Exp. Biol. Med., 113*:710–14.

Kilbourne, E., and J. Schulman. Personal communication.

Lancet. 1963. Coma and the colon. *1*:310–11.

Larson, N. L., and E. Hill. 1960. Amino formation and metabolic activity of microorganisms in the ileum of young swine fed chlortetracycline. *J. Bact.*, *80*:188–92.

Levenson, S. M., and B. Tennant. 1963. See Chapter V.2.

Lindstedt, G., S. Lindstedt, and B. E. Gustaffsson. 1965. Mucus in intestinal contents of germfree rats. *J. Exp. Med.*, *121*:201–13.

Melnykowycz, J., and K. R. Johansson. 1955. Formation of amines by intestinal microorganisms and the influence of chlortetracycline. *J. Exp. Med.*, *101*:507–17.

Meynell, G. G., and T. V. Subbaiah. 1963. Antibacterial mechanisms of the mouse gut: I. Kinetics of infection by *Salmonella typhi-murium* in normal and streptomycin-treated mice studied with abortive transductants. *Brit. J. Exp. Path.*, *64*:197–219.

Miller, C. P., and M. Bohnhoff. 1962. A study of experimental salmonella infection in the mouse. *J. Infect. Dis.*, *111*:107–16.

Mushin, R., and F. Ashburner. 1964. Ecology and epidemiology of coliform infections: I. The incidence of enteropathogenic and other specific serotypes of *Escherichia coli*. II. The biochemical reactions and drug sensitivity of coliform organisms. *Med. J. Aust.*, *1*:257–62; 303–8.

Pierce-Chase, C. H., R. M. Fauve, and R. Dubos. 1964. Corynebacterial pseudotuberculosis in mice: I. Comparative susceptibility of mouse strains to experimental infection with *Corynebacterium kutscheri*. *J. Exp. Med.*, *120*:267–81.

Rauss, K., and I. Kétyi. 1960. Beiträge zur antagonistischen Wirkung der Coli-Flora des Darmkanals. *Zbl. Bakt.*, *177*:161–75.

Rogers, W. F., M. Burdick, and G. Burnett. 1955. The effect of antibiotics on the excretion of phenolic compounds. *J. Lab. Clin. Med.*, *45*:87–96.

Rosoff, C. B. 1963. The role of intestinal bacteria in the recovery from whole body radiation. *J. Exp. Med.*, *118*:935–43.

Schaffer, J., P. Beamer, P. Trexler, G. Breidenbach, and D. Walcher. 1963. Response of germ free animals to experimental virus monocontamination: I. Observation of Coxsackie B virus. *Proc. Soc. Exp. Biol. Med.*, *112*:561–64.

Sieburth, J. 1959. See Chapter V.1.

Sprinz, H. D., D. Kundel, G. Dammin, R. Horowitz, H. Schneider, and S. Formal. 1961. See Chapter V.3.

Stewart, G. T., R. Holt, H. Coles, and K. Bhat. 1964. Replacement coliform flora in carriers of intestinal pathogens. *J. Hyg.* (*Camb.*), *62*:39–44.

Young, G., N. Underdahl, and R. Hinz. 1955. Procurement of baby pigs by hysterectomy. *Amer. J. Vet. Res.*, *16*:123–31.

Warren, K., and W. L. Newton. 1959. Portal and peripheral blood ammonia concentrations in germ-free and conventional guinea pigs. *Amer. J. Physiol.*, *197*:717–20.

Wiseman, R. F., and H. A. Gordon. 1964. Effect of *Salmonella typhimurium* on a pharmacologically active substance in gnotobiotic rat cecal content. *Bact. Proc.*, p. 46.

CHAPTER VI.1

Actor, P. 1960. Protein and vitamin intake and visceral leishmaniasis in the mouse. *Exp. Parasit.*, *10*:1–20.

Boyd-Orr, J. B., and J. L. Gilks. 1931. See Chapter III.1.

Boyden, S. V., and M. E. Andersen. 1955, 1956. See Chapter V.4.

Dubos, R. 1964. Acquired immunity to tuberculosis. *Amer. Rev. Resp. Dis.*, *90*:505–15.

——— and J. Dubos. 1952. *The white plague*. Boston, Little, Brown.

——— and R. W. Schaedler. 1958. Effect of dietary proteins and amino acids on the susceptibility of mice to bacterial infections. *J. Exp. Med.*, *108*:69–81.

——— and ——— 1959. Nutrition and infection. *J. Pediat.*, *55*:1–14.

Elberg, S. S. 1960. Cellular immunity. *Bact. Rev.*, *24*:67–95.

Formal, S., H. Noyes, and H. Schneider. 1960. See Chapter III.3.

Gordon, J. E. 1964. Acute diarrheal disease. *Amer. J. Med. Sci.*, *248*:345–65.

———, I. Chitkara, and J. Wyon. 1963. See Chapter III.2.

Gray, I. 1963. Lysine deficiency and host resistance to anthrax. *J. Exp. Med.*, *117*:497–508.

——— 1964. Effect of protein nutrition on leukocyte mobilization. *Proc. Soc. Exp. Biol. Med.*, *116*:414–16.

Hardwick, J. L., and R. L. Hartles. 1959. Dental decay. *Discovery*, *20*:417.

Hedgecock, L. W. 1958. The effect of diet on the inducement of acquired resistance by viable and nonviable vaccines in experimental tuberculosis. *Amer. Rev. Tub. Pul. Dis.*, *77*:93–105.

Hegner, R. W. 1924. The relations between a carnivorous diet and mammalian infections with intestinal protozoa. *Amer. J. Hyg.*, *4*:393–400.

Mackaness, G. B. 1964. The immunological basis of acquired cellular resistance. *J. Exp. Med.*, *120*:105–20.

O'Dell, B., D. Nabb, G. Garner, and W. Regan. 1961. A salmonellosis resistance factor for the guinea pig. *Proc. Soc. Exp. Biol. Med.*, *108*:512–14.

Rhoads, J. E., and C. Alexander. 1955. Nutritional problems of surgical patients. *Ann. N. Y. Acad. Sci.* *63*:268–75.

Schaedler, R. W., and R. Dubos. 1956. Reversible changes in the susceptibility of mice to bacterial infections: II. Changes brought about by nutritional disturbances. *J. Exp. Med.*, *104*:67–84.

Schneider, H. A. 1956. Nutritional and genetic factors in the natural resistance of mice to *Salmonella* infections. *Ann. N. Y. Acad. Sci., 66*:337–47.

———, J. Lee, and P. Olitsky. 1957. Effect of nutrition on the production of acute disseminated encephalomyelitis in mice. *J. Exp. Med., 105*:319–34.

Scrimshaw, N. S., C. E. Taylor, and J. E. Gordon. 1959. Interactions of nutrition and infection. *Amer. J. Med. Sci., 237*:367–403.

———, ——— and ——— 1965. Interactions of nutrition and infection. *World Health Organization, Monograph Series.* In press.

Sidransky, H. 1960. Chemical pathology of nutritional deficiency induced by certain plant proteins. *J. Nutr., 71*:387–99.

Smith, D. A. 1955. Parasitic infections and nutrition. *Vitamins Hormones. 13*:239–59.

Wolstenholme, G. E. W., and M. O'Connor. 1964. See Chapter III.1.

Yaeger, R., and O. N. Miller. 1961. The effect of threonine deficiency on susceptibility of rats to infection with *Trypanosoma cruzi. Fed. Proc., 20*:370.

——— and ——— 1962. Effect of biotin deficiency on parasitemia by *T. cruzi. Fed. Proc., 21*:390.

Chapter VI.2

Angel, J. L. 1961. Civilization and dental disease. *In* Notes and events, *J. Hist. Med., 16*:79.

Behrman, S. J., S. Mandel, and J. Schaffer. 1961. A means of control of dental caries: A hypothesis. *N. Y. J. Dent., 31*:157–63.

Drummond, J. C., and A. Wilbraham. 1957. See Chapter III.1.

Finn, S. B. 1952. Prevalence of dental caries. p. 119–73. In *A survey of the literature on dental caries.* Nat. Acad. Sci., Nat. Res. Coun. Publ. 225.

Fitzgerald, R. J. 1963. Gnotobiotic contribution to oral microbiology. *J. Dent. Res., 42*:549–52.

——— and E. G. McDaniel. 1960. Dental calculus in the germ-free rat. *Arch. Oral Biol., 2*:239–40.

Healy, W. B., T. G. Ludwig, and R. S. Malthus. 1962. An association between soil conditions and dental caries in rats. *Nature, 194*:456–58.

Lancet. 1956. Dental caries in Africa. 2:181–82.

Lewis, H. E. 1963. The Tristan islanders: A medical study of isolation. *New Scientist, 20*:720–22.

Neel, J. V., F. M. Salzano, P. C. Jungqueria, F. Keiter, and D. Maybury-Lewis. 1964. See Chapter III.1.

New Scientist. 1961. Tristan eruption ends striking dental study. *12*:411.

Price, W. A. 1939. *Nutrition and physical degeneration.* New York, Paul B. Hoeber.

Ratcliffe, H. L. 1951. Protein intake and pulmonary tuberculosis in rats and hamsters. *Fed. Proc., 10*:368–69.
——— 1954. Influence of protein intake on tuberculosis in guinea pigs and rats. *Fed. Proc., 13*:441.
——— and J. V. Merrick. 1957. Tuberculosis induced by droplet nuclei infection. *Amer. J. Path., 33*:107–29.
Young, H. B. 1962. A study of dental health in American children of Italian origin and comparable groups in Italy. *New Eng. J. Med., 267*:843–49.

Chapter VII

Allison, A. C. 1954, 1959. See Chapter I.1.
Berger, H., and W. H. Linkenheimer. 1962. Activation of *Bartonella muris* infection in X-irradiated rats. *Proc. Soc. Exp. Biol. Med., 109*:271–73.
Bourke, P. M. A. 1964. Emergence of potato blight, 1843–46. *Nature, 203*:805–08.
Burnet, F. M. 1962. *Natural history of infectious disease.* 3rd ed. London, Cambridge Univ. Press.
Caius, J. 1912. See Chapter III.2.
Collart, P., L. Borel, and P. Durel. 1962. Etude de l'action de la pénicilline dans la syphilis tardive. *Ann. Inst. Pasteur, 103*:953–59.
———, ——— and ——— 1964. Reviewed in *Lancet,* p. 453.
Dubos, R. 1963. Staphylococci and infection immunity. *Amer. J. Dis. Child., 105*:643–45.
——— 1964. See Chapter VI.1.
——— and J. Dubos. 1952. See Chapter VI.1.
——— and R. W. Schaedler. 1959. See Chapter VI.1.
Fauve, R. M., C. H. Pierce-Chase, and R. Dubos. 1964. See Chapter V.4.
Fenner, F. 1959. Myxomatosis in Australian wild rabbits—evolutionary changes in an infectious disease. p. 25–55. *The Harvey Lectures.* New York, Academic Press.
——— and F. N. Ratcliffe. 1965. *Myxomatosis.* Cambridge Univ. Press, in press.
Ferguson, R. G. 1955. *Studies in tuberculosis.* Univ. of Toronto Press.
Fracastorii, H. *De Contagione et Contagiosis Morbis et Eorum Curatione, Libri III.* Translated by W. C. Wright. 1930. New York, G. P. Putnam's Sons.
Greenwood, M. 1935. See Chapter IV.2.
Grigg, E. R. N. 1958. The arcana of tuberculosis. *Amer. Rev. Tub. Pul. Dis., 78*:151–72.
Guze, L. B., and G. M. Kalmanson. 1964. Persistence of bacteria in "protoplast" form after apparent cure of pyelonephritis in rats. *Science, 143*:1340–41.

Hinkle, L. E. 1961. Ecological observations of the relation of physical illness, mental illness and the social environment. *Psychosom. Med., 23*:289–97.

Horsfall, F. L., Jr. 1959. See Chapter V.1.

Huebner, R. 1961. Cancer as an infectious disease. p. 45–62. *The Harvey Lectures.* New York, Academic Press.

Langer, W. L. 1964. The Black Death. *Sci. Amer. 210*:114–21.

Large, E. C. 1940. *The advance of the fungi.* London, Jonathan Cape.

Larsen, P., H. Hoffmeyer, J. Kieler, E. Thaysen, J. Thaysen, P. Thygesen, and M. Wulff. 1952. *Famine disease in German concentration camps.* Copenhagen, Munksgaard.

Lurie, M. 1941. Heredity, constitution and tuberculosis: An experimental study. *Amer. Rev. Tub., 44*:1–125.

Lynch, C., C. Pierce-Chase, and R. Dubos. 1965. A genetic study of susceptibility to experimental tuberculosis in mice infected with mammalian tubercle bacilli. *J. Exp. Med., 121,* 1051–70.

McDermott, W. 1958. Microbial persistence. *Yale J. Biol. Med., 30*:257–91.

——— 1959. Inapparent infection. The R. E. Dyer Lecture. *Public Health Reports, 74*:485–99.

Metropolitan Life Insurance Co. Statistical Bulletin. Health progress in Puerto Rico. 1963, *44*:8–10.

Miraglia, G., and L. J. Berry. 1963. See Chapter II.3.

Moffet, H. L., and H. G. Cramblett. 1962. Viral isolations and illnesses in young infants attending a well-baby clinic. *New Eng. J. Med., 267*:1213–18.

Puffer, R. 1944. *Familial susceptibility to tuberculosis.* Harvard Univ. Monographs in Medicine and Public Health, 5. Cambridge, Harvard Univ. Press.

Rogers, D. E. 1959. The changing pattern of life-threatening microbial disease. *New Eng. J. Med., 261*:677–83.

Scrimshaw, N. S., C. E. Taylor, and J. E. Gordon. 1959, 1965. See Chapter VI.1.

Shinefield, H., J. Ribble, M. Boris, and H. Eichenwald. 1963. Bacterial interference: Its effect on nursery-acquired infection with *Staphylococcus aureus. Amer. J. Dis. Child., 105*:646–54.

Simon, H. J. 1960. *Attenuated infection.* Philadelphia, J. B. Lippincott.

Smadel, J. E. 1960. Some aspects of intracellular infections. *J. Immun., 84*:1–5.

——— 1963. Intracellular infections. *Bull. N. Y. Acad. Med., 39*:158–72.

Smillie, W. G. 1955. *Public health, its promise for the future.* New York, Macmillan.

Spink, W. W., and G. M. Bradley. 1960. Persistent parasitism in ex-

perimental brucellosis: Attempts to eliminate brucellae with long-term tetracycline therapy. *J. Lab. Clin. Med.*, *55*:535–47.
Smith, T. 1934. *Parasitism and disease.* Princeton Univ. Press.
Stearn, E. W., and A. E. Stearn. 1945. *The effect of smallpox on the destiny of the Amerindian.* Boston, Bruce Humphries.
Swellengrebel, N. H. 1940. *The efficient parasite.* Baltimore, Waverly.
Thomson, D. 1955. The ebb and flow of infection. *Monthly Bull. Ministry Health,* p. 106. In *Tubercle, 36*:244.
Webster, L. T. 1946. Experimental epidemiology. *Medicine, 25*:77–109.
World Health Organization Technical Report Series, No. 262. 1963. *In* The spread of gonorrhea, *Lancet,* 1964, p. 511.

Chapter VIII.1

Anderson, R. J. 1963. A forward stride in occupational health. *Public Health Reports, 78*:453–56.
Carson, R. 1962. *Silent spring.* Boston, Houghton Mifflin.
Herber, L. 1962. *Our synthetic environment.* New York, Knopf.
Ingalls, T. H., E. Tiboni, and M. Werrin, 1961. Lead poisoning in Philadelphia, 1955–1960. *Arch. Environ. Health, 3*:575–79.
Kehoe, R. A. 1961. Chemical aspects of environmental health. *Industr. Med. Surg., 30*:380–89.
MacLeod, C. M. 1963. *Use of pesticides.* President's Science Advisory Committee. Washington, D.C. U.S. Government Printing Office.
Simon, H. J. 1960. *Attenuated infection.* Philadelphia, J. B. Lippincott.
Stern, A. C. 1962. *Air pollution.* New York, Academic press.

Chapter VIII.2

Cannon, H. H., and J. M. Bowles. 1962. Contamination of vegetation by tetraethyl lead. *Science, 137*:765–66.
Cantarow, A., and M. Trumper. 1944. *Lead poisoning.* Baltimore, Williams & Wilkins.
Evelyn, J. *The inconvenience of the aer and smoake of London dissipated.* London, 1661. Reprinted in 1933 by the National Smoke Abatement Society, Manchester.
Haagen-Smit, A. J. 1958. Air conservation. *Science, 128*:869–78.
Hofreuter, D. H., E. Catcott, R. Keenan, and C. Xintaras. 1961. The public health significance of atmospheric lead. *Arch. Environ. Health, 3*:568–74.
Katz, M. 1961. Some aspects of the physical and chemical nature of air pollution, p. 97–158. In *Air pollution.* Geneva, World Health Organization Monograph Series, No. 46.
Kershaw, K. A. 1963. Lichens. *Endeavour. 22*:65–69.

Lawther, P. J., A. Martin, and E. Wilkins. 1962. *Epidemiology of air pollution*. World Health Organization Public Health Papers, No. 15.

McCabe, L. C. 1961. The identification of the air pollution problem, p. 39–48. In *Air pollution*. Geneva, World Health Organization Monograph Series, No. 46.

Mendenhall, R. N. 1958. Tolerance and cross-tolerance development to the atmospheric pollutants, ketene and ozone. *Ohio Valley Sect. Soc. for Exp. Biol. Med.*, Columbus, Ohio.

Pincus, S. 1961. *Sewerage and pollution control program of New York City*. Background Research on the Top Structure of the Government of the City of New York, Report No. 29.

Rupp, W. H. 1956. Air pollution sources and their control, p. 1. *In* P. L. Magill, F. R. Holden, and C. Ackley, [eds.], *Air pollution handbook*, New York.

Stern, A. C. 1964. Summary of existing air pollution standards. *A.P.C.A. Journal*, *14*:5–15.

U. S. Public Health Service. 1961. *Six years of research in air pollution, July 1, 1955—June 30, 1961*. Washington, D.C., Dept. Health, Education and Welfare.

——— 1963. *National conference on air pollution*. Washington, D.C., Dept. Health, Education and Welfare.

Wanta, R. C., and H. E. Heggestad. 1959. Occurrence of high ozone concentrations in the air near metropolitan Washington. *Science*, *130*:103–04.

World Health Organization Monograph Series, No. 46. 1961. *Air pollution*. Geneva.

Chapter VIII.3

Amdur, M. O. 1962. Air pollution and human health—chronic biologic effects. *New Eng. J. Med.*, *266*:555–56.

Arch. Environ. Health. 1965. Seventh annual air pollution medical research conference, Los Angeles, Feb. 10–11, 1964.

Beard, R. R., J. Horton, and R. McCaldin. 1964. Observations on Tokyo-Yokohama asthma and air pollution in Japan. *Public Health Reports*, *79*:439–44.

Bower, G. 1961. Deaths and illness from bronchitis, emphysema, and asthma. *Amer. Rev. Resp. Dis.*, *83*:894–98.

Catcott, E. J. 1961. Effects of air pollution on animals, p. 221–32. In *Air pollution*. Geneva, World Health Organization Monograph Series, No. 46.

Ciocco, A., and D. Thompson. 1961. A follow-up of Donora ten years after: Methodology and findings. *Amer. J. Public Health*, *51*:155–64.

Dunn, D. B. 1959. Some effects of air pollution on *Lupinus* in the Los Angeles area. *Ecology*, *40*:621–25.

Edwards, R. W. 1958. Aquatic and air pollution. *Nature, 182*:1208–10.
Fletcher, C. M. 1959. Chronic bronchitis: Its prevalence, nature, and pathogenesis. *Amer. Rev. Resp. Dis., 80*:483–94.
———, N. L. Jones, B. Burrows, and A. H. Niden. 1964. American emphysema and British bronchitis. *Amer. Rev. Resp. Dis., 90*:1–13.
Goldsmith, J. R. 1963. Air pollution and medical research. *Science, 141*:832–34.
Gorham, E. 1958. Bronchitis and the acidity of urban precipitation. *Lancet, 2*:691.
Grayson, R. R. 1956. Silage gas poisoning: nitrogen dioxide pneumonia, a new disease in agricultural workers. *Ann. Intern. Med., 45*:393–408.
Heimann, H. 1961. Effects of air pollution on human health, p. 159–220. In *Air pollution*. Geneva, World Health Organization Monograph Series, No. 46.
———, L. O. Emik, R. Prindle, and W. Fisher. 1958. *Progress in air pollution medical research*. Washington, D.C., Public Health Service.
Kobayashi, M., M. Stahmann, J. Rankin, and H. Dickie. 1963. Antigens in moldy hay as the cause of farmer's lung. *Proc. Soc. Exp. Biol. Med., 113*:472–76.
Lancet. 1962. The cigarette as co-carcinogen. *1*:85–86.
Lowry, T., and L. M. Schuman. 1956. "Silo-fillers' disease": A syndrome caused by nitrogen dioxide. *J.A.M.A., 162*:153–60.
Mauer, E. F. 1959. Harvey in London. *Bull. Hist. Med., 33*:21–36.
McCabe, L. C. 1961. See Chapter VIII.2.
McDermott, W. 1961. Air pollution and public health. *Sci. Amer., 205*:49–57.
McNamara, M. J., E. Thomas, A. Strobl, and E. Kilbourne. 1960. Minor disease in ten "healthy" adults: An intensive, day-by-day, clinical and virologic study. *Amer. Rev. Resp. Dis., 82*:469–81.
Miller, S., and R. Ehrlich. 1958. Susceptibility to respiratory infections of animals exposed to ozone: I. Susceptibility to *Klebsiella pneumoniae*. *J. Infect. Dis., 103*:145–49.
Morris, J. 1963. Ticket to Huddersfield. *Encounter, 20*:16–27.
Morrow, P. E. 1964. Animals in toxic environments: Mammals in polluted air, p. 795–808. In *Handbook of physiology, Section 4, Adaptation to the Environment*. Washington, D.C., Amer. Physiol. Soc.
Motley, H. L. 1962. From a physiologist with an epidemiologic bent. *Arch. Environ. Health, 4*:118–119.
Oshima, Y., T. Ishizaki, T. Miyamoto, S. Shimizu, T. Shida, and I. Kabe. 1964. Air pollution and respiratory diseases in the Tokyo-Yokohama area. *Amer. Rev. Resp. Dis., 90*:572–81.
Passey, R. D. 1962. Some problems of lung cancer. *Lancet, 2*:107–12.
Prindle, R. A., and E. Landau. 1962. Health effects from repeated

exposures to low concentrations of air pollutants. *Public Health Reports*, 77:901–09.

———, G. Wright, R. McCaldin, S. Marcus, T. Lloyd, and W. Bye. 1963. Comparison of pulmonary function and other parameters in two communities with widely different air pollution levels. *Amer. J. Public Health*, 53:200–17.

Purvis, M., S. Miller, and R. Ehrlich. 1961. Effect of atmospheric pollutants on susceptibility to respiratory infection: I. Effect of ozone. *J. Infect. Dis.*, 109:238–42.

Sandage, C., and K. Back. 1962. Effects on animals of 90-day continuous inhalation exposure to toxic compounds. *Fed. Proc.*, 21:451.

Stern, A. C. 1964. See Chapter VIII.2.

Thomas, M. D. 1961. Effects of air pollution on plants, p. 233–78. In *Air pollution*. Geneva, World Health Organization Monograph Series, No. 46.

Thomas, P. T., H. Evans, and D. Hughes. 1956. Chemically induced neoplasms in fungi. *Nature*, 178:949–51.

Watson, W. 1963. The changing face of the urban north. *Discovery*, 24:28–35.

Wilkins, E. T. 1954. Air pollution and the London fog of December, 1952. *J. Roy. San. Inst.*, 74:1–21.

World Health Organization Monograph Series, No. 46. 1961. *Air pollution*. Geneva.

World Health Organization Technical Report Series, No. 271. 1964. *WHO Expert Committee on Atmospheric Pollutants*. Geneva.

Chapter VIII.4

Asatoor, A. M., A. Levi, and M. Milne. 1963. Tranylcypromine and cheese. *Lancet*, 2:733–34.

Cohlan, S., G. Bevelander, and T. Tiamsic. 1964. *In* Tetracycline and bone growth. *Nutr. Rev.*, 22:11–12.

Dimond, A. E., H. Reynolds, and W. B. Ennis, Jr. 1964. Pesticide research. *Science*, 143:151–55.

Egler, F. E. 1964. Pesticides—in our ecosystem. *Amer. Sci.*, 52:110–36.

Frost, D. V. 1960. Arsenic and selenium in relation to the food additive law of 1958. *Nutr. Rev.*, 18:129–32.

Gerarde, H. W. 1964. Animals in toxic environments: Man and industrial chemicals, p. 829–34. In *Handbook of physiology, Section 4, Adaptation to the environment*. Washington, D.C., Amer. Physiol. Soc.

Henkind, P., and N. F. Rothfield. 1963. Ocular abnormalities in patients treated with synthetic antimalarial drugs. *New Eng. J. Med.*, 269:433–39.

Hueper, W. C. 1962. Environmental and occupational cancer hazards. *Clin. Pharmacol. Ther., 3*:776–813.

Lancet. 1963. Toxicity of tetracyclines. *2*:283–84.

Lasagna, L. 1964. The diseases drugs cause. *Persp. Biol. Med., 7*:457–70.

Litwin, M. S., and K. M. Earle. 1965. Proceedings of the first annual conference on biologic effects of laser radiation. *Fed. Proc.,* Suppl. 14.

Meyler, L. 1964. *Side effects of drugs.* New York, Excerpta Medica Foundation.

——— and H. M. Peck. 1962. *Drug induced diseases.* Springfield, Ill., Charles C Thomas.

Quilligan, J. J., R. Boche, L. Romig, P. Kotin, and H. Falk. 1959. Experimental synthetic air pollutants and respiratory infections. *Fed. Proc., 18*:2328.

Richter, C. P. 1959. Lasting after-effects produced in rats by several commonly used drugs and hormones. *Proc. Nat. Acad. Sci., 45*:1090–95.

Seevers, M., and G. Deneau. 1964. Animals in toxic environments: Mammals and narcotic analgesics, p. 809–27. In *Handbook of Physiology, Section 4, Adaptation to the environment.* Washington, D.C., Amer. Physiol. Soc.

Spain, D. M. 1963. *The complications of modern medical practices.* New York, Grune and Stratton.

Zbinden, G. 1963. Experimental and clinical aspects of drug toxicity. *Advances Pharmacol., 2*:1–112.

Chapter IX

Ackerknecht, E. H. 1947. The role of medical history in medical education. *Bull. Hist. Med., 21*:135–45.

Audy, J. R. 1961. Man-made maladies, p. 100–12. *In* S. M. Farber and R. H. L. Wilson, [eds.], *The air we breathe.* Springfield, Ill., Charles C Thomas.

Bates, M. 1955. *The prevalence of people.* New York, Scribner.

Bigelow, J. 1867. *Modern inquiries: classical, professional, and miscellaneous.* Boston, Little, Brown.

Bland, E. F. 1960. Declining severity of rheumatic fever: A comparative study of the past four decades. *New Eng. J. Med., 262*:597–99.

Bogdonoff, M. D., E. H. Estes, Jr., and D. Trout. 1959. See Chapter I.4.

Burkitt, D. 1962. Determining the climatic limitations of a children's cancer common in Africa. *Brit. Med. J., 2*:1019–23.

Burnet, F. M. 1962. See Chapter VII.

Christian, J. J., and D. E. Davis. 1956. See Chapter IV.2.

Coburn, A. F. 1963. The pathogenesis of rheumatic fever—a concept. *Persp. Biol. Med., 6*:493–511.

Cruz-Coke, R. 1963. Ecologia humana de la Isla de Pascua. *Revista Medica de Chile, 91*:773–79.

——— and E. Covarrubias. 1962. Factors influencing blood-pressure in a rural Chilean community. *Lancet, 2*:1138–40.

Dalldorf, G. 1962. Lymphomas of African children. *J.A.M.A., 181*:1026–28.

———, C. Linsell, F. Barnhart, and R. Martyn. 1964. An epidemiologic approach to the lymphomas of African children and Burkitt's sarcoma of the jaws. *Persp. Biol. Med., 7*:435–49.

Davies, C., H. Drysdale, and R. Passmore. 1963. Does exercise promote health? *Lancet, 2*:930–32.

Dean, G. 1963. *The porphyrias.* Philadelphia, J. B. Lippincott.

Dublin, L., and M. Spiegelman. 1952. Factors in the higher mortality of our older age groups. *Amer. J. Public Health, 42*:422–29.

Dungal, N. 1961. The special problem of stomach cancer in Iceland. *J.A.M.A., 178*:789–98.

Engel, C. 1956. Changing of disease patterns: Disappearance of diseases: New diseases. *Med. J. Aust., 1*:90–97.

Fanconi, G. 1962. Le changement des maladies depuis le début de notre siècle. *Revue Médicale de Tours, 3*:241–60.

Ferguson, T. 1960. Mortality in Shetland a hundred years ago. *Scot. Med. J., 5*:107–12.

Garn, S. M. 1963. Culture and the direction of human evolution. *Hum. Biol., 35*:221–36.

Gordon, J. E. 1963. Changing accents in community disease. *Amer. J. Public Health, 53*:141–47.

Groen, J. J., K. Tijong, M. Koster, A. Willebrands, G. Verdonck, and M. Pierloot. 1962. The influence of nutrition and ways of life on blood cholesterol and the prevalence of hypertension and coronary heart disease among Trappist and Benedictine monks. *Amer. J. Clin. Nutr., 10*:456–70.

Gsell, D., and J. Mayer. 1962. Low blood cholesterol associated with high caloric, high saturated fat intakes in a Swiss Alpine village population. *Amer. J. Clin. Nutr., 10*:471–79.

Haenszel, W., D. Lovelace, and M. Sirken. 1962. Lung-cancer mortality as related to residence and smoking histories: I. White males. *J. Nat. Cancer Inst., 28*:947–1001.

Halford, F. J. 1954. *Nine doctors and God.* Univ. of Hawaii Press.

Herxheimer, H. 1964. Asthma in American Indians. *New Eng. J. Med., 270*:1128–29.

Holmes, O. W. 1861. *Currents and countercurrents in medical science, with other addresses and essays.* Boston, Ticknor and Fields.

Hueper, W. C. 1962. See Chapter VIII.4.

Keys, A., and M. Keys. 1963. See Chapter III.2.

Klein, P. D., and R. M. Dahl. 1961. The sensitivity of cholesterol esterification to environmental stress. *J. Biol. Chem., 236*:1658–60.

Lancet. 1962. Central African lymphomas. *2*:1363–64.

Lehner, T. 1963. Involvement of the jaws in African lymphoma. *Lancet, 2*:39.

Leighton, A. H., T. Lambo, C. Hughes, D. Leighton, J. Murphy, and D. Macklin. 1963. *Psychiatric disorder among the Yoruba.* A report from the Cornell-Aro mental health research project in the Western Region, Nigeria. Ithaca, N. Y., Cornell Univ. Press.

Lewis, H. E. 1963. See Chapter VI.2.

Metchnikoff, E. 1908. See Chapter V.2.

Metropolitan Life Insurance Co. Statistical Bulletin. 1963. See Chapter VII.

Montagu, M. F. A. 1962. See Chapter I.1.

Morris, J. N. 1959. Occupation and coronary heart disease. *Arch. Intern. Med., 104*:903–07.

——— 1963. Some current trends in public health. *Proc. Roy. Soc., 159*:65–80.

Motulsky, A. 1960(a). Metabolic polymorphisms and the role of infectious diseases in human evolution. *Hum. Biol., 32*:28–62.

——— 1960(b). Population genetics of glucose-6-phosphate dehydrogenase deficiency of the red cell. *Proc. Conf. Genetic Polymorphisms,* U.S. Dept. Health, p. 258–92.

Neel, J. V., F. M. Salzano, P. C. Jungqueria, F. Keiter, and D. Maybury-Lewis. 1964. See Chapter III.1.

New Scientist. 1962. Trout show susceptibility to cancer-producing agents. *15*:662.

Nutrition Reviews. 1963. Exercise and heart disease. *21*:178.

Opler, M. K. 1959. *Culture and mental health.* New York, Macmillan.

Ratcliffe, H. L., and M. T. I. Cronin. 1958. See Chapter I.4.

Roberts, F. 1956. Morbidity yesterday and today. *Brit. Med. J., 2*:653:55.

Rosen, S., M. Bergman, D. Plester, A. El-Mofty, and M. Satti. 1962. Presbycusis study of a relatively noise-free population in the Sudan. *Ann. Otol., 71*:727–43.

Sellers, A. H. 1960. The lengthening life span. *Can. J. Public Health, 51*:171–86.

Shattuck, L. 1850. *Report of the sanitary commission of Massachusetts.* Cambridge, Harvard Univ. Press. (Reviewed in *Bull. Johns Hopk. Hosp., 84*:405–06.)

Smillie, W. G., and E. Kilbourne. 1963. *Preventive medicine and public health.* New York, Macmillan.

Southwood, A. R. 1963. Death rate from cardiac diseases. *Brit. Med. J., 2*:1339.

Srole, L., T. Langner, S. Michael, M. Opler, and T. Rennie. 1962. *Mental health in the metropolis: The midtown Manhattan study.* New York, Blakiston.

Taylor, H., E. Klepetar, A. Keys, W. Parlin, H. Blackburn, and T. Puchner. 1962. Death rates among physically active and sedentary employees of the railroad industry. *Amer. J. Public Health, 52*:1697–1707.

Vallois, H. V. 1961. See Chapter I.1.

Vertue, H. St. H. 1953. An enquiry into venereal disease in Greece and Rome. *Guy's Hospital Reports, 102*:277–302.

——— 1955. Chlorosis and stenosis. *Guy's Hospital Reports, 104*:329–48.

Walker, A. 1963. Extremes of coronary heart disease mortality in ethnic groups in Johannesburg, South Africa. *Amer. Heart J., 66*:293–95.

Weidenreich, F. 1939. The duration of life of fossil man in China and the pathological lesions found in his skeleton. *Chin. Med. J., 55*:34–44. (Reprinted in *Anthropological Papers of Franz Weidenreich,* New York, The Viking Fund, 1949.)

Wells, C. 1964. *Bones, bodies, and disease.* New York, Praeger.

White, P. D. 1959. Charles W. Thiery—1850 to 1958. *New Eng. J. Med., 260*:77–80.

World Health Organization Technical Report Series, No. 276. 1964. *Prevention of cancer.* Geneva.

Wynder, E. L., F. Lemon, and I. Bross. 1959. Cancer and coronary artery disease among Seventh-day Adventists. *Cancer, 12*:1016–28.

Chapter X

Barber, H. N. 1954. Genetic polymorphism in the rabbit in Tasmania. *Nature, 173*:1227–29.

Beecher, H. K. 1961. Surgery as placebo. *J.A.M.A., 176*:1102–07.

——— 1962. Nonspecific forces surrounding disease and the treatment of disease. *J.A.M.A., 179*:437–40.

Cannon, W. B. 1932. *The wisdom of the body.* New York, W. W. Norton.

Chapman, L., H. Goodell, and H. G. Wolff. 1959. Increased inflammatory reaction induced by central nervous system activity. *Trans. Assoc. Amer. Physicians, 62*:84–109.

Dill, D. B. [ed.]. 1964. *Handbook of physiology, Section 4, Adaptation to the environment.* Washington, D.C., Amer. Physiol. Soc.

Dobzhansky, Th., and O. Pavlovsky. 1960. See Chapter I.1.

Drabkin, D. L. 1959. Imperfection: Biochemical phobias and metabolic ambivalence. *Persp. Biol. Med., 2*:473–517.

Frenster, J. H. 1962. Load tolerance as a quantitative estimate of health. *Ann. Intern. Med.* 57:788–94.

——— 1963. Human throughput systems. *Proceedings of the 16th Annual Conference on Engineering in Medicine and Biology, Baltimore, Maryland, November 18, 1963.*

Hill, S. R., F. Goetz, H. Fox, B. Murawski, L. Krakauer, R. Reifenstein, S. Gray, W. Reddy, S. Hedberg, J. St. Marc, and G. Thorn. 1956. Studies on adrenocortical and psychological response to stress in man. *Arch. Intern. Med.*, 97:269–98.

Hinkle, L. E., and H. G. Wolff. 1958. Ecologic investigations of the relationship between illness, life experiences and the social environment. *Ann. Intern. Med.*, 49:1373–88.

Jane, C. (translator). 1960. *The journal of Christopher Columbus.* London, A. Blond.

Liddell, H. S. 1960. Experimental neuroses in animals, p. 59–66. *In* J. M. Tanner, *Stress and psychiatric disorder.* Oxford, Blackwell.

Meares, A. 1962. What makes a patient better? Atavistic regression as a basic factor. *Lancet*, 1:151–53.

Morris, J. 1963. See Chapter VIII.3.

Richards, D. W. 1960. Homeostasis: Its dislocations and perturbations. *Persp. Biol. Med.*, 3:238–51.

Rosen, S., M. Bergman, D. Plester, A. El-Mofty, and M. Satti. 1962. See Chapter IX.

Selye, H. 1956. See Chapter I.4.

Watson, W. 1963. See Chapter VIII.3.

Welch, W. H. 1897. *Adaptation in pathological processes.* Reprinted 1937. Baltimore, Johns Hopkins Univ. Press.

Wolf, S. 1961. Disease as a way of life: Neural integration in systemic pathology. *Persp. Biol. Med.*, 4:288–305.

Wolff, H. G. 1953. See Chapter I.4.

——— 1960(a). The mind-body relationship, p. 41–72. *In* L. Bryson, [ed.], *An outline of man's knowledge.* New York, Doubleday.

——— 1960(b). See Chapter I.4.

CHAPTER XI.1

Baade, F. 1962. *The race to the year 2000, our future.* New York, Doubleday.

Bates, M. 1955. See Chapter IX.

Clark, C. 1963. Agricultural productivity in relation to population, p. 23–35. *In* G. Wolstenholme, *Man and his future.* Boston, Little, Brown.

Frederiksen, H. 1960. Malaria control and population pressure in Ceylon. *Public Health Reports*, 75:865–68.

Hauser, P. M. 1960. Demographic dimensions of world politics. *Science, 131*:1641–47.

——— 1963. *The population dilemma.* Englewood Cliffs, N. J., Prentice-Hall.

Hanlon, J. J. 1956. The public health worker and the population question. Amer. J. Public Health, *46*:1397–1404.

Langer, W. L. 1963. Europe's initial population explosion. *Amer. Hist. Rev.* (Reprinted in *Harvard Today,* Spring 1964, p. 2–10.)

Malthus, T. R. 1926. *First essay on population, 1798.* London, Macmillan.

Mayer, J. 1964. Food and population: The wrong problem? *Daedalus, 93*:830–44.

Morris, J. 1963. See Chapter VIII.3.

Mudd, S. 1964. *The population crisis and the use of world resources.* The Hague, W. Junk.

Nat. Acad. Sci. Publication 1091. 1963. *The growth of world population.* Washington, D.C.

Sax, K. 1955. *Standing room only: The challenge of overpopluation.* Boston, Beacon.

Tanner, J. M. 1960. See Chapter I.4.

Toynbee, A. J. 1957. *A study of history.* New York, Oxford Univ. Press.

Zirkle, C. 1957. Benjamin Franklin, Thomas Malthus and the United States census. *Isis, 48*:58–62.

Chapter XI.2

Birch, L. C. 1957. The role of weather in determining the distribution and abundance of animals. *Cold Spring Harbor Symp. Quant. Biol., 22*:203–18.

Bronson, F. H., and B. E. Eleftheriou. 1965(a and b). See Chapter IV.2.

Bruce, H. 1964. The language of odour. *Discovery, 25*:19–22.

Calhoun, J. B. 1962. See Chapter IV.2.

Chitty, D. 1958. See Chapter IV.2.

Christian, J. J. 1950. The adreno-pituitary system and population cycles in mammals. *J. Mammalogy, 31*:247–59.

——— 1961. Phenomena associated with population density. *Proc. Nat. Acad. Sci., 47*:428–49.

———, V. Flyger, and D. E. Davis. 1960. See Chapter IV.2.

——— and C. D. Lemunyan. 1958. Adverse effects of crowding on lactation and reproduction of mice and two generations of their progeny. *Endocrinology, 63*:517–29.

Curry-Lindahl, K. 1962. The irruption of the Norway lemming in Sweden during 1960. *J. Mammalogy, 43*:171–84.

——— 1963. See Chapter IV.2.

Deevey, E. S. 1958. The equilibrium population, p. 64–86. *In* R. G. Francis, [ed.], *The population ahead.* Minneapolis, Univ. of Minnesota Press.

Eleftheriou, B. E., F. H. Bronson, and M. X. Zarrow. 1962. Interaction of olfactory and other environmental stimuli on implantation in the deer mouse. *Science, 137*:764.

Elton, C. 1942. *Voles, mice and lemmings, problems in population dynamics.* Oxford, Clarendon.

Errington, P. L. 1957. Of population cycles and unknowns. *Cold Spring Harbor Symp. Quant. Biol., 22*:287–300.

Huntington, E. 1924. See Chapter II.1.

Lack, D. 1954. *The natural regulation of animal numbers.* Oxford, Clarendon.

Milne, L. J., and M. Milne. 1960. See Chapter IV.1.

Parkes, A. S., and H. M. Bruce. 1961. Olfactory stimuli in mammalian reproduction. *Science, 134*:1049–54.

Pitelka, F. A. 1957. Some aspects of population structure in the short-term cycle of the brown lemming in northern Alaska. *Cold Spring Harbor Symp. Quant. Biol., 22*:237–51.

Snyder, R. L. 1961. Evolution and integration of mechanisms that regulate population growth. *Proc. Nat. Acad. Sci., 47*:449–55.

Strecker, R. L. 1954. Regulatory mechanics in house-mouse populations: The effect of limited food supply on a confined population. *Ecology, 35*:249–53.

——— and J. T. Emlen. 1953. Regulatory mechanisms in house-mouse populations: The effect of limited food supply on a confined population. *Ecology, 44*:375–85.

Wynne-Edwards, V. C. 1962. *Animal dispersion in relation to social behaviour.* New York, Hafner.

——— 1963. Intergroup selection in the evolution of social systems. *Nature, 200*:623–26.

——— 1964. Population control in animals. *Sci. Amer., 211*:68–74.

Chapter XI.3, 4.

Bradley, C. C. 1962. Human water needs and water use in America. *Science, 138*:489–91.

Coale, A. J. 1961. Population growth. *Science, 134*:827–29.

Cole, L. C. 1957. Sketches of general and comparative demography. *Cold Spring Harbor Symp. Quant. Biol., 22*:1–16.

Cragg, J. B., and N. W. Pirie. 1955. *The numbers of man and animals.* London, Oliver and Boyd.

Davis, K. 1963. Population. *Sci. Amer., 209*:62–71.

Ferguson, T., B. Benjamin, A. Daley, D. V. Glass, T. McKeown, G. Z.

Johnson, and J. Mackintosh. 1964. *Public health and urban growth.* London, Centre for Urban Studies.

Finch, B. E., and H. Green. 1963. *Conception through the ages.* London, Peter Owen.

Francis, R. G. 1958. *The population ahead.* Minneapolis, Univ. of Minnesota Press.

Glass, D. V. 1960. Population growth, fertility and population policy. *Nature, 187*:849–50.

Gordon, J. E. 1961. Population problems in a contracting world. *Yale J. Biol. Med., 34*:60–69.

Gupta, P. B. 1964. Fertility and economic growth. *Sci. Amer., 210*:56.

Hardin, G. 1963. The cybernetics of competition: A biologist's view of society. *Persp. Biol. Med., 7*:58–84.

Hauser, P. M. 1963. See Chapter XI.1.

Himes, N. E. 1963. *Medical history of contraception.* New York, Gamut.

Hogg, Q. 1964. Specific present and foreseeable impacts of science on political life. *Nature, 203*:119–23.

Langer, W. L. 1964. See Chapter VII.

McKeown, T., and R. G. Record. 1963. Population studies. *In* Living and dying in the nineteenth century, *Lancet, 2*:870.

Meier, R. L. 1959. *Modern science and the human fertility problem.* New York, John Wiley & Sons.

Montagu, M. F. A. 1962. See Chapter I.1.

Pearson, L. C. 1964. Population planning. *Science, 143*:1124.

Rostow, W. W. 1960. *The stages of economic growth.* London, Cambridge Univ. Press.

Spengler, J. J. 1960. Population and world economic development. *Science, 131*:1497–1502.

Stott, D. H. 1962. Cultural and natural checks on population-growth, p. 355–76. *In* M. F. A. Montagu, [ed.], *Culture and the evolution of man.* New York, Oxford Univ. Press.

Stycos, J. M. 1964. The outlook for world population. *Science, 146*:1435–40.

Tietze, C. 1962. Population studies. *New Scientist, 16*:188.

Turnbull, C. M. 1961. *The forest people.* New York, Simon and Schuster.

United Nations Demographic Yearbook. 1961. Mortality statistics. Issue 13.

United Nations Population Series, No. 17. 1953. *The determinants and consequences of population trends.* New York, U. N. Dept. Social Affairs.

Woytinsky, W. S., and E. S. Woytinsky. 1953. *World population and production.* New York, Twentieth Century Fund.

Yaukey, D. 1961. *Fertility differences in a modernizing country: A survey of Lebanese couples.* Princeton Univ. Press.

Chapter XII

Ackerknecht, E. H. 1942(a). Primitive medicine and culture pattern. *Bull. Hist. Med., 12*:545–74.

———. 1942(b). Problems of primitive medicine. *Bull. Hist. Med., 11*:503–21.

Bean, W. B. 1962. *Aphorisms from Latham. See under* Latham, P. M.

Cannon, W. B. 1957. "Voodoo" death. *Psychosom. Med., 19*:182–90.

Edelstein, L. 1956. The professional ethics of the Greek physician. *Bull. Hist. Med., 30*:391–419.

Edelstein, E. J., and L. Edelstein. 1945. *Asclepius.* Baltimore, Johns Hopkins Univ. Press.

Engel, G. L. 1962(a). The nature of disease and the care of the patient: The challenge of humanism and science in medicine. *Rhode Island Med. J., 45*:245–51.

Engel, G. L. 1962(b). *Psychological development in health and disease.* Philadelphia, W. B. Saunders.

Gage, S. H., and P. A. Fish. 1924. See Chapter I.4.

Harlow, H. F., and R. R. Zimmerman. 1958, 1959. See Chapter I.2.

Hinkle, L. E., and H. G. Wolff. 1958. See Chapter X.

Latham, P. M. In *Aphorisms from Latham.* (1962) Collected and edited by W. B. Bean. Iowa City, Prairie Press.

Lorenz, K. 1952. See Chapter I.2.

Reilly, J. 1942. *Le Rôle du système nerveux en pathologie rénale.* Paris, Masson.

Richter, C. P. 1943. Total self-regulatory functions in animals and human beings. p. 63–104. *The Harvey Lectures.* Lancaster, Pa., The Science Press.

——— 1957. On the phenomenon of sudden death in animals and man. *Psychosom. Med., 19*:191–98.

Saunders, J. B. 1963. *The transitions from ancient Egyptian to Greek medicine.* Lawrence, Univ. of Kansas Press.

Schmale, A. H., Jr. 1958. Relationship of separation and depression to disease. *Psychosom. Med., 20*:259–77.

Selye, H. 1956. See Chapter I.4.

Sudhoff, K. 1926. *Essays in the history of medicine.* New York, Medical Life Press.

Whitehorn, J. C., and B. Betz. 1960. Studies of the doctor as a crucial factor for the prognosis of schizophrenic patients. *Int. J. Soc. Psychiat., 6*:71–77.

Wolf, S. 1961. See Chapter X.

——— 1963. A new view of disease. *J.A.M.A., 184*:143–44.

Wolff, H. G. 1953, 1960 (a and b). See Chapter X.

Chapter XIII

Benedict, R. 1934. See Chapter I.1.

Cassedy, J. H. 1962. *Charles V. Chapin and the public health movement.* Harvard Univ. Press.

——— 1962. Hygeia: A mid-Victorian dream of a city of health. *J. Hist. Med., 17*:217–28.

Gordon, I. 1958. That damned word health. *Lancet, 2*:638–39.

Greenwald, I. 1958. Notes on the history of goiter in Spain and among Jews. *Bull. Hist. Med., 32*:121–36.

Hutchinson, G. E., and E. M. Low. 1951. The possible role of sexual selection in the etiology of endemic goiter. *Science, 114*:482.

Jane, C. 1960. See Chapter X.

Metchnikoff, E. 1908. See Chapter V.2.

Neel, J. V., F. M. Salzano, P. C. Jungqueria, F. Keiter, and D. Maybury-Lewis. 1964. See Chapter III.1.

Pettenkofer, M. 1941. The value of health to a city: Two popular lectures delivered on March 26 and 29, 1873, in the *Verein fur Volksbildung,* in Munich. *Bull. Hist. Med., 10*:487–503.

Richardson, B. W. 1876. *Hygeia: A city of health.* London, Macmillan.

Rosen, S., M. Bergman, D. Plester, A. El-Mofty, and M. Satti. 1962. See Chapter IX.

Veith, I. (translator). 1949. *Yellow Emperor's classic of internal medicine.* Baltimore, Williams & Wilkins.

Chapter XIV

Andrews, J. M., and A. D. Langmuir. 1963. The philosophy of disease eradication. *Amer. J. Public Health, 53*:1–6.

Arden House Conference on Tuberculosis. 1959. Public Health Service Publication No. 784. Washington, D. C., U. S. Government Printing Office.

Brieger, E. M. 1949. The host parasite relationship in tuberculosis infection. *Tubercle, 30*:227–36; 242–53.

———, H. B. Fell, and B. R. Smith. 1951. A comparative study of the reaction *in vivo* and *in vitro* of rabbit tissues to infection with bovine tubercle bacilli. *J. Hyg., 49*:189–200.

Cockburn, T. A. 1961. Eradication of infectious diseases. *Science, 133*:1050–58.

——— 1963. *The evolution and eradication of infectious diseases.* Baltimore, Johns Hopkins Univ. Press.

———, K. MacLeod, H. Nimitz, F. Kellogg, K. Curfman, G. Baum, J. Skavlem, P. Cohn, and C. Ludlow. 1964. Tuberculosis eradication: The Cincinnati project. *Amer. Rev. Resp. Dis., 90:*116–19.

Contacos, P. G., J. S. Lunn, G. R. Coatney, J. Kilpatrick, and F. E. Jones. 1963. Quartan-type malaria parasite of New World monkeys transmissible to man. *Science, 142:*676.

Gillies, M. T. 1964. Selection for host preference in *Anopheles gambiae. Nature, 203:*852–54.

Henkind, P., and N. F. Rothfield. 1963. See Chapter VIII.4.

James, G. 1964. Health challenges today. *Amer. Rev. Resp. Dis., 90:*349–58.

Kerr, J. A. 1963. Lessons to be learned from failures to eradicate. *Amer. J. Public Health, 53:*27–30.

Lancet. 1963. See Chapter VIII.4.

Loddo, B., M. Schivo, and W. Ferrari. 1963. Development of vaccinia virus resistant to 5-IODO-2[1]-Deoxyuridine. *Lancet, 2:*914–15.

MacLeod, C. M. 1963. Biological implications of eradication and control. *Amer. Rev. Resp. Dis., 88:*763–68.

McDermott, W. 1958. See Chapter VII.

——— 1959. See Chapter VII.

Nature. 1962. Malaria epidemic in Mauritius. *194:*438–39.

Pampana, E. 1963. *A textbook of malaria eradication.* New York, Oxford Univ. Press.

Szepsenwol, J. 1957. Presence of a carcinogenic substance in hens' eggs. *Proc. Soc. Exp. Biol. Med., 96:*332–35.

——— 1964. Carcinogenic effect of ether extract of whole egg, alcohol extract of egg yolk and powdered egg free of the ether extractable part in mice. *Proc. Soc. Exp. Biol. Med., 116:*1136–39.

Wharton, R., D. Eyles, M. Warren, and D. Moorhouse. 1962. *Anopheles leucosphyrus* identified as a vector of monkey malaria in Malaya. *Science, 137:*758.

Chapter XV

Ackerknecht, E. H. 1953. *Rudolf Virchow.* Madison, Univ. of Wisconsin Press.

Carson, R. 1962. See Chapter VIII.1.

Edelstein, L. 1944. Sydenham and Cervantes. *Bull. Hist. Med.,* Suppl. *3:*55–61.

——— 1956. See Chapter XII.

Ellard, J. 1963. Psychotropic drugs in general practice. *Med. J. Aust., 2:*773–77.

Engel, C. 1956. See Chapter IX.

Herber, L. 1962. See Chapter VIII.1.

Hunter, T. D. 1963. New view of the hospital: A centre of social health. *Lancet, 2*:933–35.
James, G. 1964. See Chapter XIV.
Jewkes, J., and S. Jewkes. 1963. *Value for money in medicine*. Oxford, Blackwell.
Lancet. 1959. Medical costs in perspective. 2:655.
Life. 1959. Treatment for modern medicine—some sympathy added to science. October 12:145.
McDermott, W., K. Deuschle, J. Adair, H. Fulmer, and B. Loughlin. 1960. Introducing modern medicine into a Navajo community. *Science, 131*:197–205; 280–87.
Osler, W. 1914. *Aequanimitas*. Philadelphia, Blakiston.
Porterfield, J. D. 1964. The merchandising of public health. *Bull. N. Y. Acad. Med., 49*:130–41.
Powell, E. 1961. *Lancet, 2*:969.
Roberts, F. 1952. *The cost of health*. London, Turnstile Press.
Romano, J. 1950. Basic orientation and education of the medical student. *J.A.M.A., 143*:409–12.
Rostow, W. W. 1960. See Chapter XI.3, 4.
Seale, J. R. 1959. A general theory of national expenditure on medical care. *Lancet, 2*:555–59.
Stewart, W. H., and P. E. Enterline. 1961. Effects of the national service on physician utilization and health in England and Wales. *New Eng. J. Med., 265*:1187–94.
Strauss, D. B. 1960. Can we afford to be healthy? p. 43–57. *In* M. Sanders, [ed.], *The crisis in American medicine*. New York, Harper.
Weiss, P. 1964. Interface between basic medical sciences and their university roots. *J.A.M.A., 189*:209–16.

Chapter XVI

Bean, W. B. 1962. See Chapter XII.
Biancifiori, C., and R. Ribacchi. 1962. Pulmonary tumours in mice induced by oral isoniazid and its metabolites. *Nature, 194*:488–89.
Blake, W. 1927. Annotations to Sir Joshua Reynolds's discourses, London MDCCXCVIII, p. 980. *In* G. Keynes, [ed.], *Poetry and prose of William Blake*. London, Nonesuch Press.
Bovard, E. W. 1962. See Chapter I.3.
Boyer, S. H. 1963. See Chapter I.3.
Dubos, R. 1964. See Chapter VI.1.
Gage, S. H., and P. A. Fish. 1924. See Chapter I.4.
Giedion, S. 1962. *Space, time, and architecture*, 4th edition. Cambridge, Harvard Univ. Press.
Grundy, F., and J. M. Mackintosh. 1957. *The teaching of hygiene and*

public health in Europe: A review of trends in undergraduate and postgraduate education in nineteen countries. Geneva, World Health Organization Monograph Series, No. 34.

Haldane, J. B. S. 1963. Biological possibilities for the human species in the next ten thousand years, p. 337–83. *In* G. Wolstenholme, [ed.], *Man and his future*. Boston, Little, Brown.

Hardin, G. 1963. See Chapter XI.3,4.

Hediger, H. 1955. See Chapter I.4.

Katz, S. L. 1964. Efficacy, potential and hazards of vaccines. *New Eng. J. Med., 270*:884–89.

Koprowski, H. 1963. Future of infectious and malignant diseases, p. 196–216. *In* G. Wolstenholme, [ed.], *Man and his future*. Boston, Little, Brown.

Lederberg, J. 1963. Biological future of man, p. 263–74. *In* G. Wolstenholme, [ed.], *Man and his future*. Boston, Little, Brown.

Lancet. 1964. The future of smallpox vaccination. *1*:649.

McDermott, W. 1963. Science for the individual—the university medical center. *J. Chronic Dis., 16*:105–10.

——— 1964. The role of biomedical research in international development. *J. Med. Educ., 39*:655–69.

———, K. Deuschle, J. Adair, H. Fulmer, and B. Loughlin. 1960. See Chapter XV.

McKusick, V. A. 1963. Natural selection and contemporary cardiovascular disease. *Circulation, 27*:161–63.

Medawar, P. B. 1960. *The future of man*. New York, Basic Books.

——— 1961. Letters to the editor. *Persp. Biol. Med., 4*:385–86.

——— 1963. *In* G. Wolstenholme, [ed.], *Man and his future*, p. 382. Boston, Little, Brown.

Millikan, M., and W. W. Rostow. 1957. *A proposal: Key to an effective foreign policy*. New York, Harper.

Morris, D. 1962. See Chapter I.4.

Muller, H. J. 1950. Our load of mutations. *Amer. J. Hum. Genet., 2*:111–76.

——— 1959. The guidance of human evolution. *Persp. Biol. Med., 3*:1–43.

——— 1961. Review of *The future of man*, P. B. Medawar, New York, Basic Books. *Persp. Biol. Med., 4*:377–80.

——— 1963. Genetic progress by voluntarily conducted germinal choice, p. 247–62. *In* G. Wolstenholme, *Man and his future*. Boston, Little, Brown.

Mumford, L. 1956. *From the ground up*. New York, Harcourt, Brace.

——— 1961. *The city in history*. New York, Harcourt, Brace.

Neel, J. V. 1961. A geneticist looks at modern medicine. *Harvey Lectures*, p. 127–50.

——— 1962. Diabetes mellitus: A "Thrifty" genotype rendered detrimental by "progress"? *Amer. J. Hum. Genet., 14*:353–62.

——— 1963. *Changing perspectives on the genetic effects of radiation.* Springfield, Ill., Charles C. Thomas.

Neutra, R. 1954. *Survival through design.* New York, Oxford Univ. Press.

Richter, C. P. 1959. Rats, man and the welfare state. *Amer. Psychol., 14*:18–28.

Rosen, G. 1957. The biological element in human history. *Med. Hist., 1*:150–59.

Wolstenholme, G., [ed.]. 1963. *Man and his future.* Boston, Little, Brown.

Chapter XVII

American Academy of Arts and Sciences. 1977. Doing better and feeling worse: health in the United States. *Daedalus, 106,* no. 1.

Carlson, R. 1975. *The End of Medicine.* New York, Wiley.

Cushing, H. 1925. *The Life of Sir William Osler.* 2 vols. Oxford, The Clarendon Press.

Ehrenreich, J., ed. 1978. *The Cultural Crisis in Modern Medicine.* New York, Monthly Review Press.

Flexner, S., and J. T. Flexner. 1941. *William Henry Welch and the Heroic Age of American Medicine.* New York, Viking Press.

Fuchs, V. 1968. The growing demand for medical care. *New Engl. J. Med., 279*:190–95.

Illich, I. 1976. *Medical Nemesis.* New York, Pantheon.

Ingelfinger, F. J. 1978. Medicine: meritorious or meretricious. *Science, 200*:942–46.

Lander, L. 1978. *Risk, Anger & the Malpractice Crisis.* New York, Farrar, Straus & Giroux.

McAuliffe, W. E. 1978. On the statistical validity of standards used in profile monitoring of health care. *Amer. J. Public Health, 68*:645–51.

McDermott, W. 1977. Evaluating the physician and his technology. *Daedalus, 106,* no. 1:135–57.

——— 1978. Medicine: the public good and one's own. *Persp. in Biol. and Med., 21*:167–87.

McKeown, T. 1976. *The Role of Medicine: Dream, Mirage or Nemesis.* London, Nuffield Provincial Hospital Trust.

Maxmen, J. S. 1976. *The Post-Physician Era: Medicine in the 21st Century.* New York, Wiley.

Thomas, L. 1972. Aspects of Biomedical Science Policy. An address to the Institute of Medicine, 9 Nov. 1972, Washington, D.C.

——— 1976. A meliorist view of disease and dying. *J. Med. and Philosophy, 1*:212–21.

——— 1977. On the science and technology of medicine. *Daedalus, 106*, no. 1:35–46.

——— 1978. Notes of a biology watcher: how to fix the premedical curriculum. *New Engl. J. Med., 298*:1180–81.

White, K. L. 1977. Health problems and priorities and the health professions. *Preventive Medicine, 6*:560–66.

Wolf, S. G. and B. B. Berle, eds. 1976. *Limits of Medicine: The Doctor's Job in the Coming Era*. New York, Plenum Press.

Yankauer, A. 1978. By their fruits ye shall know them. *Amer. J. Public Health, 68*:631–33.

Index

Silliman volumes in print

William Bateson, *Problems of Genetics*
Jacob Bronowski, *The Origins of Knowledge and Imagination*
S. Chandrasekhar, *Ellipsoidal Figures of Equilibrium*
Theodosius Dobzhansky, *Mankind Evolving*
Richard Goldschmidt, *The Material Basis of Evolution*
Anne McLaren, *Germ Cells and Soma: A New Look at an Old Problem*
John von Neumann, *The Computer and the Brain*
Lyman Spitzer, *Searching Between the Stars*
Karl K. Turekian, *Late Cenozoic Glacial Ages*